Dikes and Society in Rural China: the Jianghan Plain, 1788–2010s

The Social Sciences of Practice

Series Editor

Philip C.C. Huang
(*University of California, Los Angeles* and *Renmin University of China*)

Kathryn Bernhardt
(*University of California, Los Angeles*)

VOLUME 9

The titles published in this series are listed at *brill.com/ssop*

Dikes and Society in Rural China: the Jianghan Plain, 1788–2010s

By

Jiayan Zhang

BRILL

LEIDEN | BOSTON

Library of Congress Cataloging-in-Publication Data

Names: Zhang, Jiayan, 1964- author
Title: Dikes and society in rural China : the Jianghan Plain, 1788-2010s / by Jiayan Zhang.
Description: Leiden ; Boston : Brill, 2025. | Series: Social sciences of practice, 2214-952X ; volume 9 | Includes bibliographical references and index.
Identifiers: LCCN 2025020032 (print) | LCCN 2025020033 (ebook) | ISBN 9789004727236 hardback | ISBN 9789004739482 ebook
Subjects: LCSH: Reclamation of land—China—Jianghan Region—History | Dikes (Engineering)—China—Jianghan Region—History | Rural development—China—Jianghan Region | Agriculture—China—Jianghan Region | Central-local government relations—China—History | Jianghan Region (China)—Environmental conditions
Classification: LCC TC345.C6 Z428 2025 (print) | LCC TC345.C6 (ebook) | DDC 627/.5490951132—dc23/eng/20250724
LC record available at https://lccn.loc.gov/2025020032
LC ebook record available at https://lccn.loc.gov/2025020033

Typeface for the Latin, Greek, and Cyrillic scripts: "Brill". See and download: brill.com/brill-typeface.

ISSN 2214-952X
ISBN 978-90-04-72723-6 (hardback)
ISBN 978-90-04-73948-2 (e-book)
DOI 10.1163/9789004739482

This book is printed on acid-free paper and produced in a sustainable manner.

To those who have contributed to the safety of the Jianghan dikes

Contents

Illustrations

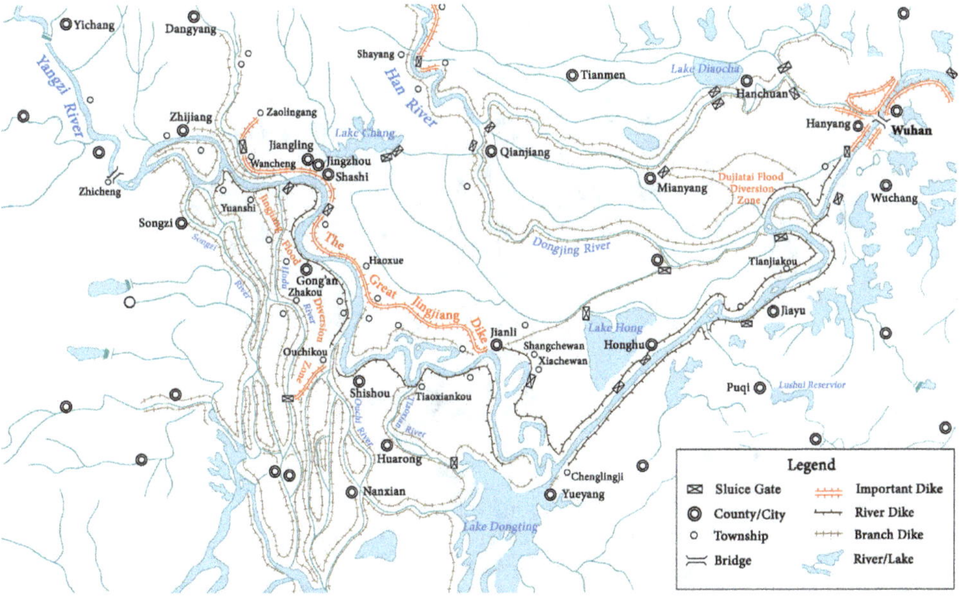

MAP 1 Situation map of the Great Jingjiang Dike, 1980s
JINGJIANG DADI ZHI, 1989, INSERT MAP

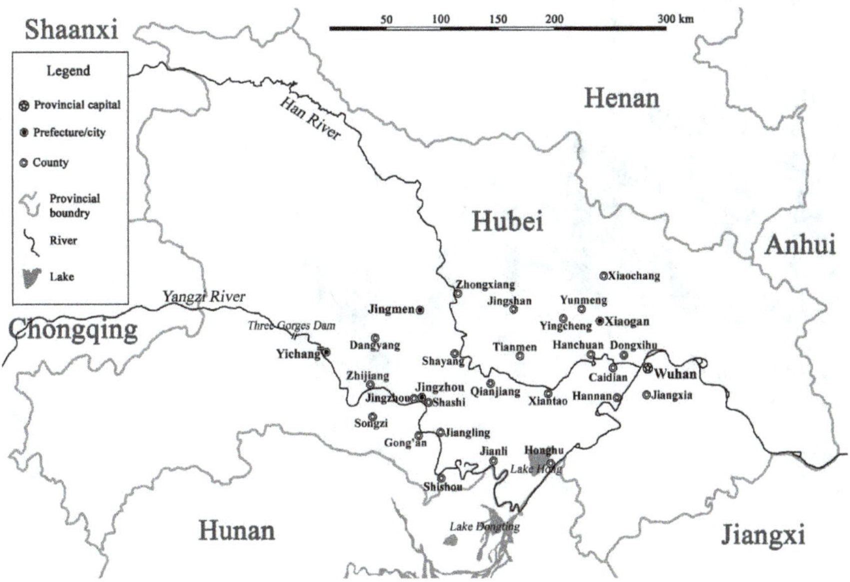

MAP 2 Administrative divisions of the Jianghan Plain, 1990s
HUBEI SHENG DITU YUAN, COMP. 1998: 2–3

PHOTO 1 Map of dikes and polders in Jianli County, 1920s
WANG ZHAOHU ET AL., COMP. 1925, VOL. 2, INSERT MAP

PHOTO 2 The Han River Dike
PHOTO BY THE AUTHOR, 2018

PHOTO 3
Aerial view of the Great Jingjiang Dike
PHOTO BY THE AUTHOR, 2017

PHOTO 4
The dike on the southern bank of the Yangzi River
PHOTO BY THE AUTHOR, 2024

Preface

This book explores the connections between dikes and society in rural China from an environmental perspective. In particular, it analyzes the dike systems of the Jianghan Plain in central Hubei and covers a three-century long swath of history stretching from the Qing dynasty, through the Republic, and into the People's Republic of China (PRC).

Dikes have played a very important role in Chinese history. This is especially the case in the Jianghan Plain. Thus it is surprising that only a few book-length writings have treated the Jianghan dikes in depth. Although in the late 1980s I started to study the agrarian history of the Jianghan Plain during the Ming-Qing era (1368–1911), I did not choose the Jianghan dikes as a research subject until the late 1990s, when I started my PhD program at UCLA. I decided to make these dikes the subject of my first graduate seminar paper. For that seminar, the instructor, Professor Kathryn Bernhardt, asked us students to provide her with an outline of our selected topic on the first day of class. I spent the entire summer going over materials and prepared a detailed outline, "Dikes and Society in Rural China," which I presented at the beginning of the fall semester. The outline included a discussion of both empirical data and theoretical issues. Professor Bernhardt liked the outline but thought it covered too much ground for a seminar paper, and in fact was more along the lines of a dissertation prospectus. At that time my main academic interest was connected with the notion of rural involutionary development proposed by Professor Philip Huang, and thus I did not expand that outline, but instead I only selected a small part of it, which I developed into a seminar paper. That paper later became a chapter of my dissertation.

In the 1990s, the notion of involution in China—namely, that rural development in modern China was trapped in a pattern of "involutionary growth" whereby per capita output stagnated and labor productivity declined—was influential but debatable. I went to UCLA with the intention of directly engaging in the debate by using the experience of the agrarian history of the Jianghan Plain during the Ming-Qing era to buttress or refute Professor Huang's analysis. However, I soon found that I could not meaningfully engage in the debate since income and expense data of Jianghan families was simply too sparse. On the other hand, there are ample materials on the constant changes in the environment in the Jianghan Plain and how the Jianghan peasants responded to those changes. Furthermore, it would relatively easy to compare the history of the Jianghan Plain in this regard with that of the North China Plain and the Yangzi delta. More importantly, almost no one had attempted to do this before.

Thus, I made that the subject of my dissertation. After many revisions, the dissertation was transformed into a book, *Coping with Calamity: Environmental Change and Peasant Response in Central China, 1736–1949*, published in 2014.

In that book I mentioned that building and maintaining the Jianghan dikes imposed a heavy burden on the local peasants, but dikes were not the book's main concern. Regarding the great importance of the dike systems to the local economy and society, there are obviously many issues worthy of further and more thorough exploration: crafting a more comprehensive critique of Karl Wittfogel's notion of "oriental despotism," reevaluating the imperial government system, assessing the impact of the enclosure of illegal polders (i.e., polders built without government approval or even in contravention of government rules) on local society and economy, investigating the ramifications of war for dike affairs, and systematically comparing the Qing-Republican era with the PRC in terms of dike management and water control and their effectiveness, and so on. Therefore, after I completed the manuscript of *Coping with Calamity*, I began to concentrate on studying the Jianghan dikes, which has led to this second book.

The primary materials for this book include gazetteers, treatises of water control, and government archives. Over the past two decades, I have visited the Hubei Provincial Archives many times and have collected a great deal of source materials, many of which were used in my first book as well as in the present book. In 2016, I thought I had completed most of the manuscript of the present book, but in the summer of that year, I visited the Songzi City Archives, where I found a trove of documents about the daily operations of the annual repairs and high-water control agencies of so-called people's polders (that is, polders paid for and built by local communities but registered with the government) in the 1940s. According to these materials, the directors of these agencies were mostly average people elected by their peers, not "local tyrants" who got their positions via bribery or by force in order to fatten themselves by bullying and exploiting others, as past scholarship and the PRC gazetteers had led us to believe. This discovery forced me to halt my writing. It was clear that it was essential to visit or revisit all the county archives in the Jianghan area and reread the source materials in order to avoid the possible biases in the new gazetteers (compiled in the PRC era, especially in the 1980s–1990s and after). In 2017 and 2018, supported by several grants, I traveled to China three times and in total visited twenty-one county-level archives and four prefectural-level archives in the Jianghan area in addition to revisiting the Hubei Provincial Archives many times. In these archives I read thousands of files and gathered an abundance of data. I also visited twenty-four libraries and read rare materials, such as township and village gazetteers. These materials—most of which have never been used by any scholar before—not only enriched my presentation but also largely reshaped my arguments and changed the structure of my manuscript. I virtually rewrote the entire manuscript, a process that was completed in 2021. All of this demonstrates what I presume most scholars eventually come to realize: writing a book is like venturing into uncharted waters, full of uncertainties.

This book makes several important contributions. It clarifies the role of the environment in the rural economy and the interrelation of humans and their environment in Hubei, and pays particular attention to the Jianghan dike systems. The book stands as the first and most thorough study from an environmental perspective of any of China's dike systems. Most importantly, it enriches our understanding of rural China over the past three centuries by reexamining some general understandings of, and assumptions about, rural China in imperial times, under the Nationalists (Guomindang) in the Republican period, in the Mao era, and during the post-Mao reform era. For example, the early PRC (or the Mao era) had features that are close to those described by the oriental despotism model. Second, with the development of the local economy since 1979, large-scale land reclamation from lakesides and riversides—a tradition that began in the Song dynasty (960–1279) and lasted for about a thousand years—seems to have ended. And third, with the increasing integrity of the dike systems since 1954, the Jianghan people have step by step changed their way of life: the cultivation of famine-relief crops (urgently planted to prevent starvation from floods or other disasters) has disappeared, there has been a reduction in disputes over dikes, and today no one takes the possibility of inundation into consideration when building their houses in low-lying places.

Beyond its academic contributions, this book also has some unique characteristics. First, its source materials are the most detailed by far in the field. Over the years I have collected an enormous amount of local gazetteers and archival materials. Many archival materials are handwritten and are difficult to decipher and even harder to translate but are of inestimable value. Second, my unique personal experiences (I personally participated in dike building and high-water control, and I also toured and surveyed many Jianghan dikes) and multidisciplinary background (I was trained in sciences in college) have helped me avoid or at least minimize some bizarre or absurd assertions. Third, the longue durée perspective and the in-depth and comprehensive analysis in this book help overcome many of the shortcomings of both Chinese and Western research. Chinese scholars oftentimes cite numerous sources without sufficient or in-depth analysis, while Western scholars tend to favor a theoretical approach but frequently lack in-depth knowledge of local realities.

In the long process of writing and rewriting, I have benefited from the help, guidance, and criticism of numerous people, too many to be named. Nonetheless, I want to express my gratitude to some of them. I wish to thank first Professors Bernhardt and Huang for their advice, inspiration, comments, and trust. This book, as I have mentioned, grew out of a paper prepared for a graduate seminar led by Professor Bernhardt. It has also greatly benefited from Professor Huang's comments and advice. As I wrote the book, these two mentors continued to comment on the manuscript—particularly on Chapter 4, which was presented at a conference on "The Social Sciences of Practice and China Research" in honor of Professor Huang's eightieth birthday (held at the School of Sociology of Renmin University in Beijing). After I completed the manuscript, they also provided insightful suggestions for revisions, and most importantly, they recommended that the monograph be included in the prestigious "Social Sciences of Practice" book series, published by Brill. I am grateful to Brill's referee, who wrote a very detailed report: I benefited tremendously from the referee's critique and revised/reorganized the manuscript accordingly. At Brill, I thank Iulia Ivana and Masja Horn, senior acquisitions editors, Stephanie Carta, Asian Studies editor, and Melissa Henson, production editor, for their enthusiasm, patience, guidance, and support. I also want to thank Brian DeMare, who read the manuscript and offered many excellent comments and suggestions. Richard Gunde deserves special thanks, not only for carefully copyediting the manuscript but also for providing many practical suggestions for revisions; without his editorial help, this book would not be what it is now.

Since this book heavily relies on archival materials, I particularly want to thank staff members in the archives I visited over the years. Many of them provided me with excellent professional services and made my work more effective and productive.

Chapter 5 was published in the *Journal of Social History* (2016); I thank its publisher—the Oxford University Press—for permitting me to include it in this book.

Financially, I have benefited from several travel funds from the Department of History and Philosophy and the College of Humanities and Social Sciences at Kennesaw State University, which supported my trip to China to collect research materials. In 2017 I won a faculty on-leave grant from Kennesaw State University and a research grant from the Chiang Ching-kuo Foundation. These made it possible for me to stay in China for several months, systematically reading and collecting archival materials in the Jianghan area. I sincerely thank them for their support.

Many of my siblings and other relatives in China not only hosted me but also helped me in multiple ways, especially by driving me to archives and dikes. My wife, Ningxia, provided moral support and made constructive comments on my research; she also drafted the maps used in this book. I sincerely thank them all.

Last but not least, I especially want to thank all the people who have contributed to the safety of the Jianghan dikes, including everyone from ancient times to the present, both local residents and people from afar, who were directly or indirectly involved in the maintenance and protection of the dikes and contributed to their integrity. This book is dedicated to them.

Illustrations and Tables

Maps

Photos

Tables

Measures

1 dan	100 *jin*, or about 50 kilograms
1 *dou*	10 *sheng*
1 *jin*	1.1 lb.
1 *li*	500 meters
1 *mu*	1/6 acre or 667 m^2
1 Qing *chi*	31.1 cm
1 Qing *cun*	3.11 cm
1 Qing *zhang*	3.11 meters
1 *sheng*	1 liter

Introduction

I was born in a small village on the Jianghan Plain in Hubei, central China. The Yangzi River dike, by any reckoning one of the most important in the country, is only a few hundred meters from my family's home. This dike and its protective forests are where my childhood friends and I played, picked up tree twigs and leaves for fuel, and collected cattle and horse droppings for fertilizer. Sometimes I went with adults rushing to harvest crops on the river beach outside the dike before the coming of the annual high water. The harvesting was rushed because this beachland would soon be submerged under muddy floodwater. During the slack farming season in winter, I occasionally went with my grandfather to fish in a small lake outside the dike, where there were flocks of wild ducks and other aquatic birds. The river beach was also an ideal breeding ground for the *Oncomelania* snail, a vector for schistosomiasis, or snail fever. To wipe out these snails, and also to respond to population pressure, in the mid-1970s this river beach was enclosed as a polder (*yuan*).[1] The annual high water thus no longer reached the dike, the lake outside the dike was transformed into a lake inside the polder, and the scene of flocks of wild ducks became a part of history.

But my childhood impression of the management of the Yangzi River dike came not from the high-water control but from the annual repairs of the dike. In some winters, villages that benefited from the protection of this dike were assigned a section for reinforcement, and some villagers from afar came to my village to work on repairing the dike section assigned to them. They had to walk from their homes and could not commute daily because the distance

1 In the local dialect in Hunan and Hubei, "yuan" usually refers to land encircled by dikes. The word can be translated as "protective embankments," "embanked fields," "diked lands," or "enclosures." I used the word "yuan" without translation in my first book (Jiayan Zhang 2014: 30–33, 230n3). But considering the ambiguity of the term and the confusion it may cause, in the present book, I translate it as "polder." Although both the yuan in the Jianghan Plain and the polders in the Netherlands are formed by surrounding land with dikes to prevent flooding, there are three fundamental physiographic differences between the two. First, the Dutch polders are usually very flat, whereas the yuan in the Jianghan Plain are mostly wok-shaped, with a high periphery and a low center. Second, the Dutch polders are mostly adjacent to the sea and their embankments mainly serve to block sea water, while the yuan in the Jianghan Plain are found throughout the region and their dikes are mostly intended to block river floods; and a few are used to block lake water. Third, the Dutch polders are basically below sea level, making continuous drainage crucial; the yuan in the Jianghan Plain, on the other hand, are above sea level (but lower than the river water level), and hence both drainage and irrigation are important.

 | DOI:10.1163/9789004739482_002

was too great. Instead, villagers who lived close to the dike put them up. Some of these laborers once stayed with my family, occupying all of the central room of the house. They laid rice straw on the ground, and then put their own cotton-wadded quilts on the straw. They worked throughout the day, and returned in the evening, exhausted. The heavy work gave them a huge appetite, which they satisfied by eating bowlful after bowlful of steamed rice, their main source of nutrition. As a reward, after they completed all their work and before they went home, there was a big meal (provided by their village) for them, which included pork braised in brown sauce, a fatty and greasy and very rare treat back then.

As a child, I also worked on dikes as did the rest of the village children. Our job mostly involved helping reinforce old dikes and building new ones. This child power was organized by our village school. It was probably a part of the curriculum of the so-called open-door schooling during the Cultural Revolution of 1966–1976. Although nowadays we think it is improper to have elementary school children do heavy physical labor, we did not do it very often and we did not work for long hours. As I remember it, school children seemed happier to go to work on dikes than to stay in the classroom.

My personal experience with high-water control of the Yangzi River can be traced back to when I was a high school student. After I graduated from high school and was admitted to college, I once went to the Yangzi River dike to participate in high-water control as a regular laborer. The Yangzi River has high water every year, and when it reached a certain level, there were people who watched over and maintained all the dikes along the river. Their responsibility was so great that no one could treat it lightly. The higher the high-water level, the greater the need for labor. Thus, I, a young man not yet seventeen years of age, answered one emergency call and went to the dike assigned to our production team to participate in that year's high-water control.[2] Though later on I went to high-water control sites on the Yangzi River dike many times, as a township (*xiang*) cadre when I served in the local government and as a bystander after I left the local government, my deepest impression is the first time I took part in high-water control. More than four decades later, some details still remain fresh in my memory.

When I arrived at the site, the water level of the Yangzi River had almost reached the top of the dike. The situation was so dangerous that laborers had to rush to build a small dike (called a *zidi*, or auxiliary dike) atop the existing

2 During the collective era (roughly 1956 to 1980), rural society below the county level was organized into (from the highest level to the lowest) people's communes, production brigades, and production teams.

dike to protect it from overflooding. My task and that of some other young villagers was to dig up and load dirt in a walking tractor. Some of this dirt was transported to the top of the dike and dumped there for the zidi. Standing atop the dike, all one could see was a vast body of muddy water—the other side of the Yangzi River was invisible. It took no imagination to recognize how devastating it would have been if the dike had been breached.

The segment of the dike I worked on was actually a polder dike, but it was located along the Yangzi River. Large-scale dike building and polder enclosure have a long history in the Jianghan area, stretching back to the Ming dynasty (1368–1644). The polder dike I went to work on for emergency high-water control was built in a new wave of dike building and polder enclosure in the early years of the People's Republic of China (PRC). Not surprisingly, it changed the life of the local people in multiple ways, not all of them good. Many villages that had owned river beachland before now lost part or all of this land, and many people who lived in different villages in the same commune were relocated to the newly reclaimed land to form new villages, which changed their lives forever. In a larger context, reclaiming land also exerted a powerful influence on the whole region as well as the entire Yangzi River valley by reshaping the ecology and society, as countless newly enclosed polders inevitably blocked waterways and increased the difficulty of discharging floodwater from the Yangzi River in its high-water season, particularly when the water level was unusually high. In this situation, the dikes surrounding many polders burst, or overflowed, or were intentionally opened to discharge floodwater. To reduce the pressure of the annual high water on the dikes along the Yangzi River and the Han River and protect the large cities along these rivers, since 1952 the government of Hubei has constantly reinforced these dikes and also built many flood diversion zones—when the river water reaches a certain level, the dikes in these zones will be opened to discharge floodwater. This, plus other state efforts, has made an important contribution to ensuring the stability of the Jianghan dikes, but has also brought many problems involving water control, environmental change, and rural development. All these are worthy of study.

1 The Jianghan Dikes

Water control has been of special importance throughout China's long history. The legendary sage-king Yu the Great "controlled the waters," securing the establishment of the Xia dynasty (ca. twenty-first to seventeenth century BCE), said to be the first dynasty in China's history. That Yu the Great employed a great number of laborers to control water demonstrates that society requires a well-organized system—in other words, a state—to mobilize people to participate in public affairs. The construction of the well-known Dujiang Weir (*Dujiangyan*), an irrigation system in the province of Sichuan built in the third century BCE, and other water-control projects have been widely regarded as greatly contributing to the strength of the Qin state and its eventual unification of China in 221 BCE. Yu the Great and Li Bin, the builder of the Dujiang Weir, thus became two of the few figures deified for controlling the waters and were worshiped by later generations in temples built in their honor. China's imperial governments had a tradition of paying great attention to water control. And when they fail to do so, the consequences could be cataclysmic. Some have even argued that unification or disunion in historical China depended on the control of the key economic zones—that is, areas that successfully developed public works for water control. Those who occupied these zones thus were able to unite China.[3] The rise and fall of water control therefore became a barometer of the rise and fall of dynasties.

Because of its crucial importance in Chinese history, there are numerous historical treatises pertaining to water control. In almost every official dynastic history of China there is a gazetteer of rivers (*hequ zhi*), and in many traditional local gazetteers (*difang zhi*) valuable space is also devoted to water control. Because China's history is long and its territory vast, the formation, function, and priority assigned to water-control projects varied tremendously across dynasties and regions. Yet, the dike system has always been one of the country's most important water-control measures. This is particularly true of the Yellow River dikes—their integrity was of especially great importance to the imperial state. In the Qing dynasty (1644–1911), the Yellow River Administration, the institution responsible for the management of the Yellow River and its dikes, was a crucial part of imperial administration, so crucial in fact that its director or river commissioner was usually a minister-level official (of the same level as, for example, the governors of China's provinces), or *erpin* (second rank) in Qing terminology. Some observers have even claimed that the fate of the Qing dynasty was directly related to the ups and downs of this institution—ten years after it was abolished, the Qing dynasty fell.[4] It is no wonder that past scholarship focused on the management and technology pertaining to the security of the Yellow River and its dikes. Even today the Yellow River is still a focus of Western scholarship. For example, four recent monographs on China's environmental history mainly deal with the Yellow River.[5]

3 Ch'ao-ting Chi 1963 (1936).
4 Ch'ang-tu Hu 1955.
5 Pietz 2015; Muscolino 2015; Ling Zhang 2016; Mostern 2021.

In the past several decades, however, the threat of breaches of the Yellow River dikes has been largely reduced—from the year 1972 to 1999, a section of the Yellow River in its lower reaches even frequently dried up. But the soundness of the Yangzi River dikes, particularly the Great Jingjiang Dike in the middle reaches of the river, was increasingly a national concern during the twentieth century. To some extent, the Yangzi River dikes have turned out to be more dangerous than the Yellow River dikes. It is well known that the Yellow River is a "hanging river": that is, its bed lies above the level of the surrounding land. For instance, at the city of Kaifeng, the Yellow River is 10 meters higher than the surrounding ground.[6] But few outsiders know that in the 1990s the water level of the Jing River (the section of the Yangzi River from Zhicheng, Hubei, to Chenglingji, Hunan) during its high-water season was 10–14 meters higher than the surrounding ground, and the Great Jingjiang Dike was 10–16 meters higher than the surrounding ground.[7] The Yangzi River dikes, including the Great Jingjiang Dike, which were far less important than the Yellow River dikes in imperial times, have become increasingly important since the twentieth century, particularly since the second half of the century.

Authors have frequently mentioned that there were "four rivers under heaven" in ancient China—the Yellow River and the Yangzi River being the two most prominent, but the other two rivers were variously identified. The Han River, a tributary of the Yangzi, was often mentioned as one of the "other two." In a draft proposal for realigning the Han River dike in 1938, the Han was listed as one of the famous four rivers in ancient China.[8] In a summary of the achievements in the management of the Han River dikes in the first ten years of the PRC, the Han was also listed as one of the "four rivers" of China.[9] A new (post-1949) local gazetteer pointed out that the Han River was one of the four famous rivers in ancient China and that it runs through the birthplace of Chinese civilization.[10] Both the Yangzi and the Han cross the Jianghan Plain, a fluvial plain in central and southern Hubei largely formed by the silt of these two rivers. The plain's name is a combination of the names of the two rivers: the Jiang (literally, "[The] River," an ancient name) and the Han.[11]

6 Pietz 2015: 19.

7 *Jingzhou diqu zhi* 1996: 2; *Jingjiang fenhong gongcheng zhi* 2000: 37.

8 LS31-4-221.

9 40-1-180.

10 *Jingzhou shuili zhi 1* 2016: 107.

11 Jiayan Zhang 2014: 20–23. Since the Six Dynasties (220–589 CE), the Jiang has been known as the Changjiang (lit., The Long River), although local names have been applied to various stretches. The word "Yangzi" (also "Yang-tse," "Yangtze," and others), used in the West, springs from local name of the river in the Jiangsu-Shanghai area.

But the official dynastic histories only occasionally mentioned the Jianghan dikes. It was not until the publication of the *Draft Qing History* in 1928 that these dikes were discussed in some depth. As far as the Qing state was concerned, the Jianghan dikes certainly were not nearly as important as the Yellow River dikes, the Grand Canal (which linked the Yangzi delta and the North China Plain), and the Sea Wall (located in Zhejiang and Jiangsu, protecting the Yangzi delta). Since the nineteenth century, however, officials had begun to pay increasing attention to issues related to the integrity of the dikes along the middle section of the Yangzi River, particularly on the Jianghan Plain. Ni Wenwei, the prefect of Jingzhou—a prefecture encompassing a key city and port on the north bank of the Yangzi River and the site of the Great Jingjiang Dike—from 1872 to 1880, even thought that, after the record flood of 1788, "the Jingzhou dike project is almost as important as the flood-prevention work [on dikes] on the Yellow River."[12] Although Ni was a very pragmatic official, his statement on the importance of Jingzhou's dikes is not entirely correct. It is true that more and more memorials in the nineteenth century had begun to address the soundness of Jingzhou's dikes—particularly the Wancheng Dike (renamed the Great Jingjiang Dike in 1918)—and some regulations in the Qing Code were specially drawn up for the maintenance of the Jianghan dikes, yet few Qing officials really thought that the Jianghan dikes were as important as the Yellow River dikes.

Only at the turn of the twentieth century did the Jianghan dikes become truly important at the national level. With the opening of the Beijing-Hankou railway in 1906 and the Beijing-Guangzhou railway in 1957, the enlargement of Wuhan (a tri-city metropolis at the confluence of the Han and Yangzi rivers) in the twentieth century, and the growth of the economy of the Jianghan Plain in the second half of the twentieth century, the security of the Jianghan dikes became increasingly important. Some observers in the early twentieth century had already begun to claim that the Great Jingjiang Dike was the nation's second most important dike, exceeded in importance only by the Yellow River dike.[13] In 1998, a record high flood hit the Yangzi River valley, and a great deal of resources and manpower were poured in to ensure the integrity of Jianghan's dikes, particularly the Great Jingjiang Dike. Two years later, the Communist Party secretary of Jingzhou Prefecture declared that the government of Jingzhou had accomplished a great deal and would do still more to improve the Great Jingjiang Dike to make sure it deserved the title of "China's

12 Ni Wenwei 2002 [1885]: 6.

13 *Jingjiang dizhi* 1937, vol. 4: 28.

number one dike."[14] In 2002, Wang Shengtie (a long-time leader in Jingzhou) in his preface to a new gazetteer of the Jianghan dikes also claimed that the Great Jingjiang Dike was "China's number one dike."[15] In a picture album published in 2003 dedicated to the epic 1998 flood-control victory, the caption of the photo of the Great Jingjiang Dike on the third insert reads: "China's Number One Dike—The Great Jingjiang Dike."[16] In the preface of a 2004 book recounting the history of Jiangling County, the iron ox (a statue and a talisman to ward off floods) that stands on the Great Jingjiang Dike in the town (*zhen*) of Haoxue was also described as "a symbol of China's number one dike."[17] All these examples demonstrate that the local people have widely accepted the idea that the Great Jingjiang Dike is the "number one dike" of China today. According to one source, the Great Jingjiang Dike, at 16 meters high, is the tallest dike in China.[18] From this point of view, it is really not an exaggeration to claim that the Great Jingjiang Dike is China's number one dike. According to another source, if one includes both river dikes and polder dikes, the total length of Hubei's dikes—most of which are on the Jianghan Plain—exceeds those of any other province.[19] But other than some gazetteers, there is no single book in any language devoted to the Jianghan dikes. The present book aims to fill this gap.

2 Dikes and Dike Management on the Jianghan Plain

Past scholarship has shown that although the Qing central government was never directly involved in the management of most of Jianghan's dikes, in the long run the Qing state—particularly local governments—in fact became increasingly involved in dike management. This included upgrading the ranks of officials responsible for dike security, appointing officials specially assigned to dike management, increasing the responsibilities of these officials, and diversifying the sources of funds for dike building and maintenance. Unfortunately, toward the end of the Qing, with a weakening of the central government and a deterioration of state finances, these efforts had little effect and failed to overcome the increasing number and magnitude of water calamities.

14 Liu Keyi 2000.
15 *Jingjiang difang zhi* 2012, Wang's preface.
16 Hubei sheng shuiliting 2003.
17 Luo Tinggui 2004: 2.
18 Li Jiansheng 1999: 227.
19 *Hubei shuili zhi* 2000: 543.

In the Republic (1912–1949), the central government began to directly manage some of the important river dikes on the Jianghan Plain, but the results were still meager. A major reason was the separation of departments responsible for technical services from the local administration.[20] Therefore, the important story was not the central government, but rather local government and local society.

Past scholarship on the relationship between the Chinese state and society, however, basically adopted a dualistic model, treating the state as a cohesive single entity and ignoring the complexity of multilateral interactions among different levels of government and other factors. My research, on the other hand, uses a triangular analytical framework consisting of the central state, the state at the local level, and local society, and considers, among other things, population pressure, peasant activity, and environmental change as well as the multilateral interactions among these various factors.

In my earlier book *Coping with Calamity: Environmental Change and Peasant Response in Central China, 1736–1949*, I also considered multilateral interactions among various factors, but with different key considerations. For instance, although *Coping with Calamity* discusses the Jianghan dikes, its main subject is environmental change and peasant responses, not the dike systems themselves. By concentrating on the Jianghan dikes and employing multiple perspectives, the present book explores the manifold relationships between dikes and society since 1788, including the relationships between society and state at different levels, between politics and economics, between environment and society, between high-water control and local culture, and between human activities and environmental change. During this period, China was transformed from an empire to a republic and then to a socialist state. This book examines what remains the same, what has changed, and the reasons for and results of changes in dike affairs. It particularly explores how the dike systems penetrated local society and changed the lifestyle of the Jianghan people.

We begin with the historic flood of 1788, a catastrophe that marked a watershed in the management of the Jianghan dikes. Before this monumental flood, there was no specific agency responsible for the management of the Wancheng Dike. After the flood, such an agency was assigned and was subject to new regulations such that the officials in charge of any dike project had to guarantee that it would remain sound for ten years. About two centuries later, in the 1970s, the annual repairs of this dike were included in the national plan for capital construction, that is, the funding was now to come

20 Jiayan Zhang 2014.

directly from the central government. Then, in 1998, following another huge flood, many more changes to dike management were introduced. One of the most important—"razing polder dikes so that floodwater can be discharged" (*pingyuan xinghong*)—can be viewed as a harbinger of the end of the reclamation of land from river beaches, a local tradition that had endured for nearly a millennium. The completion of the Three Gorges Dam on the Yangzi River just west of the city of Yichang, Hubei, in 2009 brought further changes to the management of the Jianghan dikes. One dramatic change was that the locals were usually no longer called upon to patrol the dikes during the high-water season, as they had done for centuries. These long-term changes can best be understood through the triangular framework model of the central state, local state, and local society, since local water control was, and still remains, unavoidably tied to the control of larger river systems, which require attention and coordination from the central level but also from the local level.

Since the eighteenth century, not only has China witnessed sweeping changes to its political system, but its socioeconomic situation has also changed dramatically: from peace and prosperity during what historians call the High Qing (roughly from 1683 to 1799), to decline for nearly a century thereafter (1840–1949), accompanied by disasters and chaos caused by unremitting foreign invasion and internal strife, to the era of socialist construction (1949–1978) and the post-1979 reform in the PRC, with China's economy now the second largest in the world. At the same time, the Jianghan dikes also changed from being unstable to stable. In the Qing and the Republic, these dikes were frequently breached; since 1954, however, all the major river dikes in the Jianghan area have remained secure (with a few exceptions). It is true that in the Qing and the Republic any particular dike might remain sound for a relatively long time, but for all the major river dikes to remain safe for decades simply never happened until the PRC era. This book explores the reasons behind this remarkable accomplishment.

Departing from popular criticism of the Qing government for mismanaging (or entirely failing to manage) water-control projects, Jane Kate Leonard (writing on the Grand Canal) and Randall Dodgen (on the management of the Yellow River) have defended the Qing central state—particularly during the Daoguang reign (1821–1850)—by highlighting the positive side of Qing governance and arguing that frequent water calamities were mostly caused by factors beyond human control.[21] I concur that it is unfair to put all the blame on the corruption and incompetence of the Qing government, especially

21 Leonard 1996; Dodgen 2001.

since charges of malfeasance were sometimes leveled by fellow officials whose motives may not always have been pure. In the long run, however, it is undeniable that there was widespread corruption in the management of water-control projects, including in the Daoguang reign.

It is also not entirely clear whether all officials responsible for river administration in the mid-nineteenth century were professional technical bureaucrats, or were mostly officials with practical experience, or were "hybrid literati-technologist[s],"[22] and not necessarily trained hydraulic technicians. In fact, none of them were trained in hydraulic technology; instead, they were recruited through China's rigorous civil examination system in which the candidates did not have to demonstrate any knowledge of technology. In county and prefectural yamens (i.e., the offices of public officials or departments), private secretaries (*muyou*) who served magistrates and prefects were the true professionals. Since it took time for officials responsible for water conservancy to learn the details of the local situation, it was suggested in 1763 that it would be better not to frequently change the personnel in charge of high-water control and annual repairs of dangerous dike segments.[23]

To argue that the frequent breaches of Jianghan's dikes were mainly caused by mismanagement does not mean that all the officials in charge were always irresponsible nor that all of Jianghan's dikes were always mismanaged. On the contrary, since the dikes were all-important to local society and economy, many Qing officials took or had to take their responsibilities very seriously. And indeed, some dikes or dike segments remained secure for a considerable amount of time. The people of Mianyang Department, for instance, even built a temple to worship a prefect (when he was still alive) for his contribution to water control,[24] a rare honor for an imperial bureaucrat. The record also shows that there were other officials who recognized they were appointed to "promote the beneficial and abolish the harmful" for the local people,[25] a moral concept later fully adopted by the Chinese Communist Party (CCP).

But not all the personnel responsible for water management, particularly yamen clerks, fulfilled their duties. Thus, it was common that once a dike broke the official in charge would try to evade responsibility by claiming there was so much water that taming it was simply impossible. If an official was in fact held responsible, he could lose his position or even be jailed. In a pre-industrial society, it is possible that a major water calamity or dike rupture could indeed

---

22 Dodgen 2001: 146.

23 Chen Zhenhan et al. 1989, vol. 2: 407.

24 *Mianyang zhou zhi* 1894, vol. 7: 77a.

25 *Mianyang zhou zhi* 1894, vol. 3: 31a.

be caused by a "natural disaster" such as heavy rains or huge floods that were beyond human control, but to use that to explain frequent dike ruptures was a feeble excuse. As we will see, there were differences in the response to several historic floods on the Jianghan Plain in 1788, 1870, 1931, 1935, 1954, and 1998. The human factor, particularly good dike management, was always crucial in the reduction of the frequency of dike ruptures and their harmfulness. Numerous authors in the Qing and the Republic indeed explicitly stated that the frequent ruptures of Jianghan's dikes could not be attributed to nature, but instead were man-made (i.e., the result of mismanagement).[26]

3 Water Control, Environment, and the State

Understanding China requires understanding water control. Among Western scholars who have studied water control in China, one of the most influential was Karl August Wittfogel (1896–1988), whose theory of oriental despotism holds that the despotic power of the imperial Chinese state was based on its total domination of water control.[27] This idea has been generally criticized as not in accord with China's historical reality. But if one goes deeper, one will find that the object of Wittfogel's attack was not imperial China but contemporary socialist China. According to him, although "hydraulic states" such as China and polities in the Near East escaped colonialization, they faced huge pressure from the industrialized West, and after World War II they had ample opportunity to change, but in China all these opportunities "were blocked by the Communists."[28] His major target was the Soviet Union. According to him, tsarist Russia was also an "oriental" or hydraulic state, although the conditions of water control of tsarist Russia were far different from those of imperial China. Addressing China, Wittfogel claimed that after World War II, "Mao Tse-tung and his followers established an agrarian despotism" that "bore a close resemblance to the great despotic regimes of China's past," and after the CCP took power, "the basic trend toward the crystallization of a totalitarian system of power, economy, and class structure is unmistakable."[29] That is to say, his concern was not really hydraulic societies, but totalitarianism: his true goal was to attack contemporary Communist China. His magnum opus,

26 Hu Zuhe 1999 [1838]: 157; Yu Changlie 1999 [1840]: 75; *Songzi xian zhi* 1982 [1937]: 73; *Zhongxiang xian zhi* 1937, vol. 3: 29a.

27 Wittfogel 1981 [1957].

28 Wittfogel 1981 [1957]: 433–36.

29 Wittfogel 1981 [1957]: 441, 443.

Oriental Despotism: A Comparative Study of Total Power (1957), is essentially a theory-driven polemic that avoids discussing the concrete. From the perspective of water control, for example, precisely how was contemporary Communist China despotic or totalitarian? In fact, he did not even mention the construction of water-control projects in the early PRC, the core part of an "oriental" or hydraulic society. Furthermore, according to Friedman, Pickowicz, and Selden, "to see China as totalitarian fails to capture the significance of the political strife among leaders committed to reform socialism, traditional socialism, and communist fundamentalism."[30]

While it is true that Wittfogel metamorphosed from being a member of the German Communist Party into a fervent anti-Communist, and that his life and academic work came to be driven by his anti-Communist mission, that in itself does not mean that waterworks were not of crucial importance to Chinese governance. In fact, he correctly highlighted the decisive role played by the state in the management of water-control projects. He was also correct in pointing out that it was necessary for the central state to manage water control, but he nonetheless exaggerated the role of the central imperial state. The Qing central state in fact did not directly manage every important dike, and, moreover, intentionally left the management of most dikes in the Jianghan area to the local people.[31] Obviously, such a state does not meet the basic condition for Wittfogel's "oriental despotism." Subsequently, almost all disputes over water use and dike affairs on the Jianghan Plain in the Qing and the Republic involved mobilization and organization by the locals themselves.[32] Furthermore, Wittfogel focused on dry regions, but the Jianghan Plain has plenty of water. The priority there has been flood-control and drainage, which is completely different from the North China Plain where the priority has been irrigation. Therefore, in this research I will explore whether and how failures to control the major rivers impacted local waterworks.

Simply put, Wittfogel asserted that the creation and management of large waterworks were the fundamental cause of the rise of highly centralized and bureaucratized despotic states, but this oversimplified view conceals the complex interaction among various elements such as the central government, local governments, local society, and the environment. It should also be noted that the dike systems, particularly the polder dikes, in the Jianghan area were and remain of a different order and scale, and that the local governments were and still are involved more than the central government, and local governments

30 Friedman, Pickowicz, and Selden 1991: 268.

31 Jiayan Zhang 2014.

32 Jiayan Zhang 2013b.

and societies more than the centralized bureaucracy. This makes for a picture much more complex than centralized "despotism."

Water-control projects are public goods. According to the free-rider theory, people—driven by their selfish instincts—wish to benefit from such projects for free. Hence, historically the state had to become involved in their management. Also, water-control projects usually spread across more than one administrative jurisdiction, and thus when disputes arose among these jurisdictions, finding a solution required the involvement of the upper-level administration. It is therefore logical that the central government would be involved in the management of water-control projects that were important to the whole nation. This has nothing to do with ethnicity, civilization, level of development, or the political system or ideological background.[33] Hence the need for a triangular, multidimensional framework of analysis.

Based on the Jianghan experience, it can be seen that, from the perspective of water-control management, the first three decades of the PRC, the period of collectivization or the Mao era, came the closest to what Wittfogel claimed was oriental despotism. During that period, the state penetrated deeply into and controlled rural society, down to the household level, and at the same time engaged in an unprecedented construction of water-control projects across the nation, likely the largest in China's history. The period from the 1950s to the 1970s is thus widely regarded as the golden age of water control. These projects had many problems, but their achievements cannot be denied. In fact, water control has been regarded as one of the "three great successes" of the Mao era—the other two being agricultural extension and national defense. Based on the achievements of agricultural extension in the early PRC, some have argued that "the Mao era was not the simple picture of totalitarian oppression and ecological disaster that is presented in many accounts available in

33 A good example is the dispute between the states of Georgia, Alabama, and Florida and the U.S. Army Corps of Engineers mainly over the use of water from Lake Lanier, a man-made reservoir created by the building of the Bulford Dam on the Chattahoochee River in 1956. As the city of Atlanta has grown, it has needed more water from the lake, but Alabama challenged the U.S. Army Corps of Engineers' reallocation of more water to Atlanta by arguing that Alabama also needs more water to expand and maintain a nuclear plant, and downstream Florida argued that allocating more water to Atlanta would endanger some aquatic species and that Florida needs more water for its fishery industry. The dispute began in 1990, and since it could not be solved at the state level, it eventually went to a national-level authority—the Supreme Court (LaManna 2010). What the American government at the national level did is similar to what the central government of imperial China did, particularly with the management of the Yellow River and the Grand Canal and their dikes.

English."[34] The fight against the plague and snail fever in the PRC, for example, has been a remarkable success.[35]

Some Western scholars, however, believe water control was the chief culprit behind totalitarian oppression and ecological disaster in the Mao era. For example, they assert that most land reclamation in the early PRC was ordered—against "the wishes of local peasants"—by the central government.[36] But that is not the case in the Jianghan area, where many state-led large-scale reclamation projects were actually overwhelmingly welcomed by the local peasants, who had tried to reclaim land earlier but failed. Moreover, many small polders were enclosed by local peasants without government permission.

To correct excessively idealistic notions about Maoist China—green mountains, blue water, pure air—in the 1970s, Vaclav Smil, using official publications, captured the destructive impact on the environment caused by "socialist construction": "To describe it unequivocally is impossible: what a mixture of some excellent intentions and notable achievements with much casual neglect, astonishing irresponsibility, and staggering outright destruction!"[37] His corrective was so influential that this image of a deteriorating Chinese environment has become common sense in the West. But this is certainly not the whole picture of socialist construction in China. Even Judith Shapiro, who coined the term "Mao's war against nature," admitted that environmental destruction in the post-Mao era has been worse than in the Mao era.[38] For example, in my childhood in the 1970s, it was common and easy to catch many fish in the rice paddy fields in the early spring when fish swim against the water flow to spawn. But with the increasing application of fertilizers and pesticides since the 1980s, that possibility has gradually disappeared. Although water-control projects in the early PRC had many problems, these projects actually contributed greatly to the economic development of the PRC. As a village boy, a peasant's son, a graduate of an agricultural college, and a former agricultural technician with two township governments, in this book I explore China's water control from a more comprehensive yet practical angle.

---

34 Schmalzer 2016: 2, 236n49.

35 Cao Shuji and Li Yushang 2006; Gross 2016.

36 Smil 1984: 67.

37 Smil 1984: xi, 8.

38 Shapiro 2012: xx.

4 A Historical and Environmental Perspective

Even before the popularity of environmental history, scholars had already begun to pay close attention to the relationship between the environment and socioeconomic change in rural China.[39] For example, in his classic study of rural development in the Yangzi delta, *The Peasant Family and Rural Development in the Yangzi Delta, 1350–1988* (1990), Philip Huang starts with a detailed discussion of the region's "ecosystem." He explains that his work explores "the interdependence of natural environment, social-economic structure, and human agency. Each aspect, it will be shown, interacted with the other parts of the entire 'system.'"[40] Although elaborating on the ecosystem is not the focus of Huang's book, at its heart is the notion that the ecosystem should be understood as organic, as if it were a part of biology or an interactive entity in life sciences, rather than as a one-way push-pull mechanism as seen from the perspective of inorganic physical sciences.[41] In other words, Huang's book stands out from a great many works in social science and history that lack a sense of humanity and have treated people as objects. My research is in the same vein: it treats peasants as real people, rather than as abstract figures such as rebels bursting with revolutionary zeal as portrayed in Marxist writings or as exemplars of rational economic man posited by market theory.

Environmental history has gradually become a hot topic in recent years. A growing number of scholars are trying to relate the subjects they study to environmental change, or to discuss their subjects from an environmental perspective, such as agrarian development, river and water management, war and plague, and frontier zones.[42] My research, however, should not be misunderstood to be intentionally designed to "swim in the mainstream"; rather, it finds itself in the mainstream because water control is a crucial part of environmental history. In addition, it makes perfect sense to study the Jianghan dikes within the framework of environmental history, as the never-ending building and rebuilding of dikes on the Jianghan Plain has greatly altered the local environment. Finally, since studies of environmental history are more valuable when they take a long-term perspective, this research puts the Qing,

39 Jiayan Zhang 2014: 3–5.
40 Philip C.C. Huang 1990: 21.
41 Philip C.C. Huang and Yuan Gao 2015.
42 For example, Marks 1998; Jiayan Zhang 2014; Pietz 2015; Muscolino 2015; Bello 2016; Gross 2016; Ling Zhang 2016; Schlesinger 2017; Kinzley 2018; Lavelle 2020; Yan Gao 2022; Xiangli Ding 2024.

the Republic, and the PRC together to show the changes and continuities in dike management on the Jianghan Plain.

To say that there is a close relationship and mutual influence between human activities and the environment is to state the obvious. There are a great many scholars nowadays who believe that human activities have had such profound impact upon the earth that the current epoch should be called "the Anthropocene" (human epoch), though others have doubted the role that mankind has played in geological change.[43] Human activities no doubt have caused great damage to the environment. In fact, the history of mankind is a history of environmental destruction, but the human being also has the ability to rebuild the environment. The landscape of the Netherlands, for example, is largely man-made, and Dutch agriculture is among the most impressive in the world. It would be ideal if the original ecology could be maintained everywhere, but this possibility has been shrinking in today's world. The issue, therefore, is not really whether human activities alter their environment, but whether the human being is able or willing to rebuild a better environment.

On the Jianghan Plain in the Ming-Qing era, environmental changes in the upstream mountains clearly influenced the downstream plains. The reclamation of mountainous lands caused serious erosion of topsoil, and rainwater carried away these soils and deposited them in the downstream rivers. This in turn raised the riverbeds and subsequently forced the local people to build ever-higher dikes, which eventually turned the rivers into hanging rivers and increased the possibility of dike ruptures. But silt from upstream land could also be used to enrich downstream land. One reason the people in the Jianghan Plain strove to reclaim river beaches, even though doing so was illegal, is because the soil was fertile—so fertile, in fact, that it was not unusual to intentionally break open a dike to use the silt carried by floodwater as fertilizer to enrich the land. The silt carried by floodwater also could be deposited in lakes and gradually make a part of them reclaimable. Millions of acres of farmland were reclaimed from silted lakes in central China alone in past centuries, a feat that substantially reduced local population pressure.

Some environmentalists, however, have failed to take population pressure into account; others have considered humans to be the biggest environmental problem. We certainly should not attribute all environmental problems to population pressure, but we also should not turn a blind eye to the basic condition of China—a very large population with limited per capita arable land. As early as the Republic, some hydraulic experts had argued that lakeside

43 Davies 2016.

wasteland should not only be used to discharge floodwater, but also should be used as farmland to ease population pressure.[44] It is generally true that the PRC government in the 1950s–1970s was able to ensure that its orders and prohibitions would be strictly enforced, but it also felt quite powerless to stop the reclamation of river beaches/islets and lakes. A major reason was the mounting population pressure.

Ironically, whether in the imperial era or in more recent times, those who criticized polder enclosure and dike building the harshest usually were not engaged in agriculture nor relied on farming to make a living. Peasants themselves as well as elites or officials who had intimate knowledge of agriculture held a very different view of dike building. For example, everyone knew that building more dikes would block waterways and compete for space with water, but according to Yu Changlie, a county-level Qing official, that was the only way "to protect farmland and households."[45] Practical minded county-level officials frequently clashed with high-ranking bureaucrats over this issue. Influenced by Wang Baixin, a prominent late-Qing scholar who suggested that dikes in the middle Yangzi River valley be abandoned, a magistrate of Jiangling in the Daoguang reign once suggested to his higher-ups that some branch river dikes be abandoned, but his proposal was rejected.[46] Even in Jianli, Wang's native county, some officials emphasized the importance of preserving and protecting dikes.[47]

Jia Rang, a hydraulic official-engineer who flourished around 100 BCE, is famous for his three strategies for managing the Yellow River. The best strategy according to Jia was to let the river run naturally; the second-best was to build channels to discharge water; and the third, and worst, was to build dikes.[48] Scholar-officials in later dynasties frequently mentioned him when they discussed water control, particularly dike affairs. Modern environmentalists would concur with his first strategy. According to Jia's thinking, the Jianghan people had chosen the worst strategy by building a great number of dikes. But pragmatically speaking, and also according to many local officials, this was the only available option. The compiler of a local gazetteer of the early Qing argued that since both national revenue and the people's destiny (*guofu minming*) relied on the protection afforded by the dike systems, even the people who built the

---

44 Sun Fushi 1939: 75.

45 Yu Changlie 1999 [1840]: 12.

46 *Jiangling xian zhi* 1876, vol. 8: 10a.

47 *Jianli xian zhi* 1872, vol. 3: 4b.

48 Zhou Kuiyi et al. 1990: 30–33.

dikes were not afraid of the hardship involved.[49] Ni Wenwei indeed disagreed with Jia's assessment. According to Ni, with the increase of population and the reclamation of low-lying land, there was no choice but to build dikes.[50] And according to the compilers of a late Qing local gazetteer, even the third strategy, or the worst method in Jia Rang's eyes, that is, building dikes, was not easily implemented.[51] In the Republic, responding to suggestions to abandon dikes, hydraulic technicians argued that, to guard against the floodwater of the Han River in its lower reaches, diking was the only possible solution.[52]

In the first half of the twentieth century, damming was still regarded as a major means of reducing and protecting against inundation. During the second half of the century, however, people increasingly realized that dams had adverse side effects, and since the early 1980s some environmentalists and human rights activists have launched a global anti-dam campaign, advocated the destruction of large dams, and successfully killed many proposals for large dams worldwide.[53] One of their targets was China's Three Gorges Dam. It is probably true that many dams should not have been built in the first place, but the dike systems on the Jianghan Plain are different. Many places in the world may be fine or even better off without dams, but it is impossible for the Jianghan people to abandon their dikes.

Studies of Chinese environmental history present many contrasting arguments. One key issue is whether China's development is sustainable. On the one hand, some studies have highly evaluated traditional Chinese agriculture as it maintained an ecological equilibrium and was sustainable. On the other hand, some have criticized traditional Chinese agriculture for causing serious ecological deterioration due to its exploitive use of land, soil erosion, and deforestation, thus rendering development unsustainable. Subsequently, a common theme in the literature has been China's environmental deterioration stretching over many centuries and environmental unsustainability in China today. In this telling, China's environment is now on the brink of collapse. The concept of sustainable development per se, however, is a reaction to the side effects of industrialization in the West. In contrast to the West, China remained a rural economy until the second half of the twentieth century and for centuries it faced a shortage of natural resources. One must take this historical context into account to understand the choices available to the people of China.

49 *Jingzhou fu zhi* 1685, vol. 8: 1a.
50 Ni Wenwei 2002 [1885]: 6.
51 *Songzi xian zhi* 1869, vol. 4: 20a–21b.
52 Song Xishang 1954: 74.
53 McNeill 2000: 157–82; McCully 1996.

The most influential proponent of the argument that China's development has been unsustainable was the late Mark Elvin. One of the main pieces of evidence Elvin offered is the continual investment throughout Chinese history in the maintenance of water-control projects. To ensure the security of these projects people had to pour in a great deal of manpower and capital year after year. A famous example is the Dujiang Weir project in Sichuan.[54] However, Elvin might have overstated his case here, since no water-control project, or any engineering project for that matter, can operate well without constant maintenance. In other words, following Elvin's logic, not a single water-control project in human history was sustainable. But in fact, according to many Chinese scholars, the Dujiang Weir is one of the best examples of a sustainable water-control project, as it has operated for more than two millennia and still functions well today.

The traditional agriculture of the Jianghan Plain, protected by the dike systems, was actually sustainable.[55] For this reason, in this book I will not try to offer a deeper analysis of the Jianghan economy. Also, since there were no alternatives to diking, I do not focus on the economic benefits of the dikes, but instead focus on how political change influenced the dike systems, how the dike systems influenced the local economy and society, and especially how the dike systems were involved in and influenced the daily life of the Jianghan people.

A common criticism of the Chinese tradition of protecting nature is that it existed only on paper, and that, hence, in reality it failed to prevent environmental degradation or deforestation. Although this claim may appear to be valid, it does not tell the whole story. This divorce between words or thoughts and deeds has to be understood in a more practical way. The core of the traditional Chinese economy was agriculture, and most Chinese up until the end of the twentieth century were peasants. As Rhoads Murphey pointed out long ago, elites/scholars who advocated environmental protection only accounted for five percent of China's population, and they usually did not participate in farming; those who actually cleared forests and reclaimed land, actions that "damaged" the environment, were usually commoners, who had no option to do otherwise, since their very survival required that land be brought under cultivation.[56] But their behavior should not overshadow the important influence of the ideas of elites/scholars. Even the common people, in many cases, followed the elites' exhortation to protect nature and the environment. A new

54 Elvin 1993.

55 Jiayan Zhang 2014.

56 Murphey 1983: 116–17.

study reveals that in southern China, from 1000 to 1600, without much governmental supervision, the local people constantly planted/replanted a great number of forests to replace natural forests.[57] The true environmental chaos only started about three hundred years ago when China's population doubled within one century. Even during this period, people in Central China's mountainous areas intentionally replanted forests to meet the market demand for wood.[58] In the pre-1949 Jianghan area, in many cases local communities, villages or clans, set up regulations to close hillsides and ancestral cemeteries to facilitate afforestation and to close lakes to prevent overfishing.[59]

In his study of the Yellow River and "the problem of water" in modern China, David Pietz argued that to understand the current problems of China we have to go back and explore China's history.[60] I would add, that also applies to the Jianghan dikes. Any student of Chinese history knows that it was normal for new dynasties to rebuild water-control projects devastated by warfare, to reclaim abandoned farmland, and to open new land. What the PRC did in the 1950s is merely the continuation of this age-old imperial tradition, even though the PRC embraced a very different political ideology. However, in many respects the early PRC government clearly did a better job of water control than many imperial governments. Frequent dike ruptures on the Jianghan Plain, which had plagued the region for centuries, gradually disappeared. At the same time, the powerful Communist state mobilized millions of laborers and opened millions of acres of farmland that supported hundreds of thousands of families, just as in the Ming-Qing era but on a far larger scale and in a far shorter time. These reclamations did have side effects: they reduced the area in which floodwater could be discharged and increased the pressure on the dikes. Indeed, in today's scholarship on China's environmental history, both at home and abroad, a main theme is the unanimous criticism of over-reclamation in China, particularly in the Mao era. Many of these criticisms are justified, but they are also ahistorical, as most peasants at that time faced a life-or-death struggle for sheer survival. It does not make much sense to ask peasants to protect the environment while their stomachs are empty. Without bringing more land under cultivation, how could a huge and growing population be fed? The

57 Miller 2020.

58 Meng Zhang 2021.

59 *Wuchang xian zhi* 1989: 232; Jin Feng 1996: 248; *Honghu xian zhi* 1992: 163; *Hanchuan xian zhi* 1992: 176. According to these new gazetteers, wealthy lake owners closed a portion of their large lakes and banned fishing for a time, and then reopened them and doubled the charge for anyone who wanted to fish. Although this was ideologically interpreted as exploitation, it is actually a good way of avoiding overfishing.

60 Pietz 2015.

reclamation of wasteland was one of the best means to feed the growing population. Realizing this fact, millions of acres of wasteland were opened in the early PRC on the Jianghan Plain alone, which supported both the local people and immigrants who came from all over the country.

According to historian Robert Marks, China before 1949 underwent not only a socioeconomic crisis but also an environmental crisis, and thus, "the Chinese Communists inherited a seriously degraded natural environment."[61] After the Communist Party took power, it made great efforts to improve this "seriously degraded natural environment." In the Jianghan area that included the reinforcement of the dike systems and the reduction of water calamities, which not only won the support of the local people but also deepened their sense of national identity and their acceptance of the CCP's rule as legitimate.

Because of their ideological stance, Marxist scholars in China have usually portrayed the Republic under the rule of the Nationalists (Guomindang, or GMD) negatively, but recently there has been a trend toward painting this period with rosy colors. Although the period is not as dark or negative as Marxist scholars had portrayed, it was also not as positive as some revisionist scholars would like us to believe. To arrive at a balanced assessment, I will emphasize the central features of the transition from the Republic to the PRC, particularly regarding dike management.

Chinese scholars also usually make a clear-cut distinction between "Old China" (from 1919 to 1949, or from 1840 to 1949) and "New China" (after 1949), not just because of the different ideologies in each period, but also because their research tradition: in China's academic world it is taken for granted that scholars who study the PRC will not or should not study the Qing, and vice versa. The Qing dynasty, the Republic, and the PRC are usually studied separately, particularly by those who study the PRC, for whom 1949 is the watershed between "New China" and "Old China." Western scholars, however, tend to stress China's historical continuity and oppose artificially carving up China's history. If we look at dike management on the Jianghan Plain, what happened after 1949 certainly cannot be severed from what transpired before 1949—the importance of dikes, the situation in which human beings contended with water for land, and the use of manual laborers to patrol dikes as an essential part of high-water control all continued after 1949. Yet that there were great differences in dike management before and after 1949 cannot be denied. At the end of a lengthy essay on water control in the Hubei basin (i.e., the Jianghan Plain), Pierre-Etienne Will mentioned that, regarding the relationship between

---

61 Marks 1998: 333; Marks 2017: 298.

the central government and local power as seen in water control, "the continuities between late imperial and modern times would deserve closer study."[62] He seems to have implied that the PRC did a better job of water control than its predecessor regimes. Peter Perdue—who studied the Lake Dongting Plain—implied the same.[63] But neither of them delved deeper into water control in central China in the PRC. It is precisely this subject—water control in central China in the PRC—that will be one of the focal points of this book.

In sum, past scholarship has mostly emphasized the role of the central government in water control. But that overlooks the fact that government in China, even today when the state at times seems all powerful, is not monolithic. It also ignores the reality that effective water control calls for different techniques and measures in different natural environments. And, finally, it turns a blind eye to the fact that governments at the prefectural and county levels on the one hand and at the provincial and central levels on the other have different or, in some cases, even completely opposite interests. Furthermore, historically grassroots government officials were not mere cogs in the state's huge bureaucratic machine, nor were they all corrupt and incompetent. This research constitutes a corrective to these partial and limited interpretations by employing a multilateral interactive approach, including a triangular, multidimensional analytical framework consisting of the central state, the local state, and local society, and their interaction with the natural environment. In the past, scholarly analysis was mostly limited to two of these dimensions. An example would be Wittfogel's research on water control and the central state (or the overall governance system), as well as countless works that deal with simply the two dimensions of government and environment. Few researchers have considered the role of initiatives of the local governments and local society or have paid sufficient attention to issues surrounding the competence and integrity of local government/bureaucrats, as well as the effects of population pressure on local society and the active participation of the people. Here too this book departs from past works that focus on the state by inviting the reader to look at things from the perspective of the peasants whose lives literally depended on the integrity of the dike systems, not from the perspective of bystanders and outsiders. The analytical framework of this research can be summarized as the interaction between environment and government (both central and local) and society and economy (especially demographic factors) and their diachronic evolution.

62 Will 1985: 347.

63 Perdue 1987.

Since the structure and operation of the government is not the main concern of this book, the terms "state" and "government" are used interchangeably. Moreover, governments at all levels are usually represented by their chief bureaucrats, including governor-generals, governors, prefects, and county magistrates in the Qing dynasty and their counterparts in the Republic and the PRC. Since provincial-level bureaucrats are officials at the same level as ministers, they usually look at issues such as water control from a national perspective. Therefore, the provincial government can be conceptually included in the central government. Local government mainly refers to the state at the prefecture and county level in the Qing dynasty, and also includes governments below the county level during the Republic and the PRC.[64]

5 Synopsis of Research and Sources

Apart from the Introduction and Conclusion, this book includes ten thematic chapters, each with sections organized both thematically and chronologically. This may make it easier for readers to discern what has changed, what has not changed, and the hows and whys of those changes, over the past three centuries. Another advantage of this arrangement is that if a reader is interested only in one topic/chapter, that reader can read and comprehend that topic/chapter without perusing other topics/chapters.

Chapter 1 examines aspects of the relationship between the dike systems and administration that past scholarship has overlooked or has not discussed in sufficient detail, such as the composition and nature of the relevant management organizations and changes in those units, the role and responsibilities of administrators (including hydraulic technicians), the rewards and punishments involved in dike management, and the politicization of water conservancy in the PRC.

Chapters 2 and 3 explore the practical steps and methods used in both high-water control and annual repairs in the Jianghan Plain over the past three centuries. Generally speaking, high-water control was more important than annual repairs in the Qing and the Republic, though some officials in the late Qing had already begun to argue that annual repairs should take priority. In the PRC, both high-water control and annual repairs have been treated seriously, but it is clear that the official strategy is to prevent floods, and thus high-water

64 As an administrative unit, *shi* (city) in the PRC may be a megacity such as Wuhan, or it may be a prefecture-level or county-level administrative area. To avoid confusion, the text includes a mention of the corresponding level of each "city," such as prefecture or county.

control and annual repairs gradually became part of government's routine throughout the year, not just a seasonal job as in the past. Chapter 2 focuses on high-water control, and Chapter 3 on annual repairs.

Chapter 4 reviews the annual repairs and high-water control agencies (*xiufangchu*) of people's polders (*minyuan*, polders paid for, built, and managed by local communities but registered with the government), particularly in the 1930s–1940s. These agencies have long been portrayed very negatively by PRC publications. Based on newly available archival materials, I provide a more complete and rounded view of their structure, operations, and leadership, as well as their relationship with the rural government. Polder agencies could not function well without the involvement of local government. This reflects the increasing penetration of state power into rural society in the Republic and foreshadowed a transition from the state's indirect involvement in dike management in the Republic to direct involvement in the PRC.

Chapter 5 analyzes the relationship between dike management and water deity worship and its vicissitudes from the Qing dynasty to the PRC. Responding to frequent floods, the Jianghan people built temples to worship water deities and put iron oxen statues on dikes as talismans to ward off inundations. Before 1949, governmental officials stressed the importance of the worship of water deities in dike management—or at least they recognized its psychological function. In the early PRC, the worship of deities was criticized as a superstition, and the state revived the ancient idea that "human beings can overcome nature." In any event, with the decrease in the number of dike ruptures, the worship of water deities seems to have come to an end.

Chapter 6 discusses the changing relationship between polders and lakes, including the reclamation of lakes and enclosure of polders as a way of dealing with the pressure of population growth; the decline of polders in the PRC due to the increasing integrity of river dikes; the shrinking of lakes and the changes in their functions; and the changes in the purpose of lake enclosures in the PRC—mainly to enlarge farmland to support a growing population and to prevent the spread of snail fever.

Chapter 7 probes the reclamation of private polders from the Qing to the PRC. These were always banned because they blocked waterways and caused water calamities. Before 1949, however, local government officials usually did not enforce the ban partly because of population pressure. Even the new government after 1949 failed to stop reclamation because of the need to feed a growing population. With the economic growth since the 1980s, most young villagers have moved away and found non-farming jobs and have lost any desire to reclaim land. Furthermore, low-lying land in many places has been returned to lakes for receiving and holding floodwater and supporting fishing, which is now more profitable than farming.

Chapter 8 investigates the military functions of the Jianghan dikes and their relationship with military affairs, a topic long overlooked by past scholarship. Dikes have been used not just to control the flow of water, but since ancient times they also have had a very close relationship with military affairs. Since dikes can be used as defensive works, armed forces were widely involved in dike affairs. Moreover, war had a profound impact on dikes, particularly in the first half of the twentieth century.

Chapter 9 scrutinizes the relationship between the Three Gorges Dam project and high-water control on the Jianghan Plain. One of the major concerns of critics of the Three Gorges project is that it required the relocation of more than a million people. However, compared to generating electricity, controlling high-water in the Jianghan area, and dealing with complicated environmental consequences, population resettlement has been a relatively easy challenge for the unique party-state system of the PRC to deal with.

Chapter 10 assesses the effects of the changes in the dike systems on the socioeconomic life of the local people. Although Jianghan society benefited from the protection afforded by dikes, diking was also exhausting work. With the building of still more river dikes in the PRC, however, many polder dikes were abandoned or turned into farmland. The increasing security of the dike systems, particularly after the completion of the Three Gorges Dam in 2009, brought dramatic changes to dike management and had a profound impact on the Jianghan people, whose daily lives had been interwoven with dikes for centuries. These changes included a reduction in the frequency of disputes over dikes, changes in crop choices, the introduction of new house-building materials and changes in the location where houses were built, and an increasing propensity to take the integrity of dikes for granted.

My research is based on three major types of sources. The first consists of dike-related materials from the Qing dynasty, including memorials and reports by imperial officials at different levels (county, prefecture, province), the gazetteer of the Wancheng Dike, and local water-conservancy treatises.

The second is local gazetteers, or local histories, including traditional gazetteers compiled before 1949, mostly during the Qing and the 1930s, and new gazetteers compiled after 1949, mostly in the 1980s–1990s and after. The latter have rarely been systematically used by researchers. Traditional gazetteers contain a wealth of information on the local economy before 1949, and are, in fact, the most important source of information for analyzing local rural economic conditions. But records on dike affairs in traditional gazetteers—most of which are county gazetteers—are usually very brief. After 1949, Chinese governments at various levels devoted a great deal of effort to compiling and publishing new gazetteers, including a great number of township and even some village gazetteers in the Jianghan area, many of which include records on dike affairs. This makes the new gazetteers a much richer mine of materials than traditional gazetteers.

Records in the new gazetteers usually emphasize the positive, and pay little attention, if any, to the negative side. But the subjects in gazetteers of different levels and departments are not necessarily the same. For example, provincial gazetteers may not mention much about the side effects of water conservancy projects, while county and township gazetteers may contain more detailed information; the general gazetteer of a county or a prefecture might not record the number of deaths due to illness related to water-control projects, but public health gazetteers might highlight the government's achievements by recording numbers of patients treated. Therefore, careful use of gazetteers makes it possible to find nuanced information with which to establish and assess historical facts.

The third and the most important sources are archival materials, particularly county archival materials, that have only become available recently. Local gazetteers are local encyclopedias, and hence their records are brief and generalized. Archival materials, particularly unedited original documents, on the other hand, are much more detailed, specific, and closer to reality. Influenced by their ideological disposition, for example, most new gazetteers portray dike management in the 1930s–1940s very negatively. They almost unanimously describe the directors of annual repairs and high-water control agencies of people's polders as "evil" gentry who seized the position and were only interested in bullying and exploiting the villagers and embezzling dike fees. Archival materials in the Songzi City Archives, however, present a completely different picture—most agency directors in the 1940s were normal villagers (with experience working on dikes) elected by their peers; most of them received no payment (except for some subsidies); financial policies left little room for embezzlement of dike fees; and in fact it was hard for the directors to collect dike fees and some personally advanced funds to pay for dike expenditures. By the same token, new gazetteers also portray dike management in the PRC positively. They generally describe villagers after 1949 as very happy to go to dike worksites, but archival records clearly indicate that some peasants in the years of 1950–1953 were actually loath to go to the dike worksites, mainly because the workload was heavy and unequally and often arbitrarily distributed.

Many of these materials—from reports and memorials by Qing officials to records in the PRC archives—are official documents. The new gazetteers are mostly based on official documents as well. One should be cautious about accepting at face value claims of achievements recorded in these official documents, but the problems they reported or recorded—such as corruption among officials in the Qing and hardships on dike worksites in the early PRC—were indeed real.

Many scholars who study the Republic and the PRC have their own ideological and political stances, which unavoidably prejudices their research. The result is that discourse and reality are often inconsistent. Furthermore, official publications of the PRC have accentuated the positive aspects of water control in the People's Republic, while Western research has emphasized the negative aspects. Here archival materials can be of great help in bringing one closer to what really happened. In this book, I seek to avoid these kinds of prejudices and shortcomings and let the archives speak for themselves.

Since environmental history is an interdisciplinary field, researchers ought to have some scientific background or at least a good understanding of the science involved. I have the advantage of being trained in science in college. Moreover, I have firsthand experience of diking in the area. In the 1970s and 1980s, I often participated in both the annual repairs and high-water control of local dikes. In the past two decades, I also toured the Jianghan plain, by car, by bicycle, and on foot, and investigated many dikes, and toured some dike segments multiple times—including the Great Jingjiang Dike, the major dikes along the southern bank of the Jing River, the Han River dikes, the Dongjing River dike, the Ouchi River dike, the Songzi River dike, the Tiaoxian River dike, and many polder dikes, as well as numerous riverside parks. I had an opportunity in 2017 to board a small plane to get a bird's eye view of the Jingzhou-Shashi segment of the Great Jingjiang Dike. All this has been useful in understanding the current situation with various dikes throughout the region, something that one cannot get from written materials. For example, on my visits I found that, contrary to the claims in many documents, illegal construction on dikes and farming dikes still exists. This underscores the fact that one cannot rely solely on written materials, whether they be gazetteers or documents in archives.

Over the years, I have conducted many field surveys—at family gatherings, in visits with governmental officials, along dikes, in the fields, in local parks, and so on, but I did not interview people in a formal way. Rather, the interviews were informal, and mostly causal conversations to verify information I had gleaned from local gazetteers and archives. I did not take notes or otherwise record these conversations since that would almost certainly have made informants uncomfortable or caused them to say what they guessed I might want to hear. For this reason, although I have greatly benefited from the information they shared, I have tried to rely on such information as little as possible. In general, my principle in using source materials is to rely on them in this order: archival materials first, local gazetteers second, and other sources third.

CHAPTER 1

Dikes: Systems and Administration

In their treatment of the different roles played by government personnel, rural gentry or elites, and peasants in the management of the Jianghan dikes in the Qing and the Republic and the category of river dikes and polder dikes, scholars have generally contended that the state was only responsible for a few important segments of river dikes, leaving the management of the majority of river dikes and all polder dikes in the hands of the local people, albeit with an important role for the gentry.[1] In the PRC, however, the gentry as a social class no longer existed, and polders were no longer privately owned by clans or residents. Thus, this chapter discusses aspects of the relationship between the dike systems and administration that past scholarship overlooked or did not treat in detail, such as the management units and their vicissitudes, the role and responsibilities of administrators (including hydraulic technicians), the rewards and punishments entailed in dike management, and the politicization of water conservancy in the PRC.

1 The Dikes of Jianghan: Categories and Terminology

Depending on different criteria, various terms have been used to describe the Jianghan dikes. Based on their fundamental function, in the Qing and the Republic, there were only two general categories—river dikes and polder dikes, which also included lake dikes in a few places. In the PRC, channel dikes and reservoir dikes were added to the list. Dikes have also been identified by the river where they are located: the Yangzi River dike, the Han River dike, the Dongjing River dike, and so on. They can also be described by their function: flood-storage dikes, flood-control dikes, city-protection dikes, and street dikes. They can also be named after their location, such as the south dike, north dike, west dike, upper dike, lower dike, inner dike, outer dike, the Wancheng Dike, and the Distance Dike. Finally, they can be identified by the materials from which they are built: earth dikes and stone dikes.

Before 1949, dikes could be identified by the administrative jurisdiction in which they were located: for instance, *Jingdi* (the dikes of Jingzhou), *fudi* (dikes

1 Ts'ui-jung Liu 1970; Will 1985; Jiayan Zhang 2014.

 | DOI:10.1163/9789004739482_003

belonging to a prefecture), and *xiandi* (dikes belonging to a county). They could also be named in honor of the person who led the building or repair a certain dike, such as the Revered Mr. Zhou dike and the Revered Mr. Zhang dike. To satirize the poor quality of some dikes, people in the Republic described them as *doufu di* (tofu dikes), *douzha di* (dregs of soybean curd dikes), or *cailan di* (vegetable basket dikes).

Although in most cases river dikes are found on both banks of a section of a river, their importance is vastly different. The Great Jingjiang Dike, for example, lies on the northern bank of the Jing River and is the most important dike on the Jianghan Plain; the dike along the southern bank of the Jing River is less important. The same is true of the Distance Dike (Yaodi), which is situated 20 li or about 10,000 meters from the eastern bank of the Han River and is as important as the Great Jingjiang Dike, but the dike on the other side of the Han River along its western bank is far less important.

According to the sources of funds and legal status, in the Qing dynasty dikes could be classified as official dikes, imperial dikes, and ministry dikes (which were mostly important segments of river dikes), military dikes (*jundi*, managed by *wei* or grain transport stations), people's dikes (built by the local people, with no government funding, but officially registered with the government), and illegal or private dikes (which were not supposed to exist). The legal status of any particular dike, however, was changeable. It could be changed from a people's dike to an official dike, or vice versa, and a polder dike also could be turned into a river dike.

Although dikes in the Qing dynasty came in many different types and were known by a variety of names, there was no official classification of dikes. To decide which level of government or administrative unit should be responsible for a particular dike, and which dike should receive government investment and how much, the Hubei government in the 1920s began to officially classify all dikes into two categories: main dikes (*gandi*) and people's dikes (*mindi*). However, the classification criteria were always debatable and controversial, and the financial support for different categories of dikes was hugely different.

The official classification started in 1926 when many dikes ruptured during that year's high-water season and had to be repaired in the following dry season. To make better use of the limited funds available, the Hubei's Bureau of Water Conservancy decided all dikes along the Yangzi and the Han Rivers were "main dikes," and the provincial government was responsible for their repair; all other dikes were classified as "people's dikes," and repairs were the responsibility of the local governments. Applying this simple classification was easy in theory, but very problematic in reality, since some polder dikes or people's dikes were higher or longer or more important than some segments of river dikes. In 1931, the government of Hubei decided that all high and uninterrupted river dikes were main dikes, and all the rest were people's dikes. In 1937, all dikes along the Yangzi and the Han Rivers that protected more than 20,000 mu (1 mu = 1/6 acre or 667 square meters) of farmland were classified as main dikes, and all other river dikes were considered people's dikes, with one exception: any former people's dike that protected more than 20,000 mu of farmland and was the same height and thickness as the nearby main dike was now reclassified as a main dike. In 1945 main dikes were further divided into three categories: first-grade main dikes, which protected 200,000 or more mu of farmland in total, or 4,000 or more mu of farmland per kilometer of dike; second-grade main dikes, which protected 30,000 or more mu of farmland, or 1,500 or less mu of farmland per kilometer of dike; and third-grade main dikes, which protected less than 30,000 mu of farmland.[2] Since the category to which a particular dike or dike segment was assigned had a direct bearing on the availability of funds for dike work, it is understandable that local governments frequently wanted the rank of dikes under their jurisdiction to be upgraded. This was rarely successful.

Disputes over the classification of dikes continued into the PRC, as the amount of state funds allocated for a particular dike was closely related to its categorization. According to a provincial gazetteer, in the PRC the Jianghan dikes were generally classified into three categories: main dikes (major dikes along the Yangzi and the Han Rivers), branch dikes (dikes along river branches or tributaries), and people's dikes.[3] In this gazetteer, people's dikes referred to polder dikes; but in prefectural and county gazetteers, the nature of people's dikes could be very different. In the gazetteer of Xiaonan District (which was separated from the county of Xiaogan in 1993), for example, a "people's dike" could be a lake dike, a polder dike, or a segment of the dike along a rivulet.[4]

According to local archives and gazetteers, aside from these three officially recognized categories, there have been some locally recognized categories, such as *weidi* (enclosed or embanked dikes) and *badi* (clinging dikes).[5] The

2 *Hubei sheng zhi shuili* 1995: 23.

3 *Hubei sheng zhi shuili* 1995: 250.

4 *Xiaonan qu zhi* 2015: 187. In Yingcheng, according to a 1947 survey, even the dike of one polder could be subject to the management of different parties. For example, the dike of Yujia polder was managed by four parties: one segment was managed by households that lived adjacent to the dike, a second segment was co-managed by two clans, a third by one clan, and a fourth by residents of the whole polder (Yingcheng 49-1-1 [Note: since the category numbers of the Yingcheng and Jiangxia (Wuchang) archives are the same, I have added the names of the counties for clarity]).

5 *Shishou shi zhi* 2014: 175, 177; *Hubei shuili zhi* 2000: 527.

hierarchical ranking of these dikes, however, does not always reflect their size or even their importance, as, for example, some weidi may be larger than a branch river dike or people's dike. In my field survey, I was told that according to regulations, neither weidi, embanked dikes or polder dikes that are mostly managed by the local county, nor badi, dikes of a tiny polder (*bayuan*) enclosed outside a weidi, should exist since they were illegal. But the local government did not ban them, particularly weidi. With the agreement of the upper-level government, some weidi could be reclassified as a people's dike, simply because many people had already lived there, and it would be too costly to destroy it. Quite often local governments asked upper-level governments to legalize these dikes based on their current situation. Sometimes upper-level governments deemed it necessary to accept such *faits accomplis*.

The status and names of different segments of a dike can also differ based on their relative importance or what could be called their application time—that is, the time the local government asked the upper-level government to recognize a particular dike. One segment could be considered part of a main dike, while the next segment could be classified as a branch dike or people's dike. The dike along the Yangzi River in Jianli County, for example, is divided into two segments: one (47.5 kilometers) is part of the Great Jingjiang Dike, while the other (96.45 kilometers) is part of "the main dike of the Yangzi River."[6] In Xiaonan District, a segment of the dike along a rivulet is termed a "branch dike," but another segment is called a "people's dike."[7]

The Great Jingjiang Dike was the most important main dike on the Jianghan Plain in the Qing and the Republic. In 1954, it was further classified by the PRC state as one of the dikes that "must-be-protected" (i.e., under no circumstances would it be allowed to rupture)[8] and was usually listed separately from other main dikes to highlight its uniqueness and crucial importance. The reason is simple: this dike is located along the upper reaches of the Jing River, where the water level is more than 10 meters higher than the ground behind the Great Jingjiang Dike.[9] If the dike broke, floodwater would engulf most of the Jianghan Plain and the city of Wuhan, and the Yangzi River would change its course—such a disaster would be of such monumental proportions as to be unthinkable. That is why when Zhou Enlai, the premier of the State Council of

6 *Jianli difang zhi* 1991: 75.

7 *Xiaonan qu zhi* 2015: 187.

8 *Hubei shuili zhi* 2000: 495.

9 In 1998 along the Haoxue section of the Great Jingjiang Dike, the highest water level of the Yangzi River was more than 15 meters higher than the ground on its inner side (*Jiangling xian zhi* 2015: 24).

the PRC from 1949 to 1976, visited an endangered segment of this dike in the town of Haoxue in 1958, he felt he was "trembling with fear, as if standing on the edge of an abyss and treading on thin ice."[10] Almost half a century later Wang Shengtie, a former leader of Jingzhou, expressed the same feeling.[11] Thus, the PRC state from the very beginning invested a huge amount of money and manpower in harnessing the Jing River and reinforcing the Great Jingjiang Dike. In the Qing dynasty, emperors frequently visited water conservancy projects, particularly the Yellow River dikes and the Sea Wall along the southeast coast, but none of them ever visited the Yangzi River dikes in its middle reaches. In the PRC, however, many of the nation's top leaders—Mao Zedong, Zhou Enlai, Dong Biwu, Li Xiannian, Li Peng, Jiang Zemin, Zhu Rongji, Wen Jiabao, and Xi Jinping—visited the Great Jingjiang Dike one or more times, a fact that underscores the dike's importance.[12]

It should be kept in mind that river dikes are not necessarily higher or longer than polder dikes. In fact, some branch dikes may be relatively small while the dikes of some large polders may be far higher and longer than river dikes. After decades of water conservancy construction in the PRC, almost all major rivers on the Jianghan Plain are now sandwiched between two dikes, one on either bank, and thus most former polder dikes located behind the river dikes (on the inner side) became useless and were turned into farmland, except for some polder dikes that served to protect land from lake water. Therefore, by the end of the twentieth century, the Hubei Department of Water Conservancy categorized the Jianghan dikes as consisting of ten dike systems, with the vast majority located along the Yangzi and Han Rivers and their tributaries.[13]

2 Dike Management

During the Qing dynasty, the government assigned officials to the job of managing dike affairs, but not many government agencies were established to specifically handle dike management. In the early Republic, the engineering bureau responsible for new construction was separated from the high-water control bureau, which existed only during the high-water season. In other words, in

10 *'98 Jiangling kanghong zhi* 1999: 3.

11 *Jingjiang difang zhi* 2012: Wang's preface.

12 Mao did not actually walk on the Great Jingjiang Dike, but he passed by on a ship twice and summoned local leaders to the ship to report to him on the situation with the dike (Luo Xianqun 2013).

13 *Hubei shuili zhi* 2000: 495–528.

Hubei, there was no single official responsible for dikes year-round.[14] In both the Qing and the Republic, the state was only responsible for river dikes; all polder dikes were locally managed. This changed in the PRC. The management of dikes was put in the hands of government agencies at different levels; dike management was institutionalizing by incorporating those agencies into the formal administrative system; and dike management was routinizing or formalizing by making the responsible agencies permanent units with assigned staff employed all year round.

The management of the Great Jingjiang Dike is a good example of the changing role of government in dike management. Before 1788, no prefecture-level official was assigned the task of taking care of this important dike. After 1788, a prefecture-level official was charged with managing the dike, and a few troops were assigned the task of patrolling and maintaining some important segments of the dike. In the Republic, the General Bureau of the Wancheng Dike, later replaced by the Bureau of Dike Work on the Jing River (jingjiang digongju), was established as a permanent government unit to manage the Wancheng Dike. In the PRC, the Great Jingjiang Dike was first managed by the Jingzhou Water Conservancy Branch, and then, since 1951, the General Section of Management of Jiangling.[15] In 1999, management of the dike was assigned to several administrative units: in Jiangling it was managed by the Jiangling Sub-Bureau of the Bureau of the Yangzi River Course Management in Jingzhou.[16]

In general, dike maintenance became more institutionalized after 1949. In the early PRC the goal was "the localization of [dike] maintenance,"[17] and in response specialized governmental dike management agencies were established at the county level and above. Below the county level, dikes were managed mostly by maintenance teams (*guanyangzu*) or maintenance points (*guanyangdian*), quasi-government units. Although these teams or points were official units at the lowest level in the hierarchy of dike administration in the PRC, many of their personnel were not formal government employees, but instead were informal (*yigong yinong* ["both worker and peasant"]) dike keepers. They worked full-time at the maintenance teams or points, but remained peasants who did not receive a regular salary, but instead earned "workpoints," a measure of work done during the collective era, from their production teams.

14 *Gailiang digong yijian shu* n.d.: 4.

15 *Jiangling xian zhi* 1990: 352.

16 *Jingjiang difang zhi* 2012: 469–73.

17 D35-1-148a.

In all of Hubei, there were 1,371 such dike keepers in 1965, 3,487 in 1985, and 4,254 in 1995.[18]

The dikes they maintained were state-managed major dikes. Polder dikes, however, were traditionally managed by the local people in the Qing and the Republic, although in the 1930s and 1940s many polders established agencies to manage annual repairs and high-water control. These organizations, however, were non-government units (see Chapter 4). In the PRC, on the other hand, even polder dikes were managed by the government. For example, in Songzi County (renamed Songzi City in 1996), on the southwestern edge of the Jianghan Plain, the old agencies were replaced in 1950 by maintenance departments, each of which had a staff that ranged from seven employees (including four full-timers) for large polders to three to five (three full-timers) for small polders.[19] These departments were different from the old agencies in three significant ways. First, the staff members were appointed by the state, not elected by the people. Second, most of them worked full-time. And third, they were a part of the state system and no longer remained independent. Two years later, these maintenance departments were also abolished, and the management of polder dikes was fully incorporated into the government's administrative system.[20]

In the 1950s, great efforts were made to implement the "localization of [dike] maintenance," which included the recruitment of local residents as dike keepers. In 1955, each township along the Great Jingjiang Dike set up a dike maintenance committee. There were a total of 41 maintenance teams under these committees, each with one full-time dike keeper. In 1961, there were 15 maintenance stations along the Great Jingjiang Dike, each with a director and one or two dike keepers.[21] At the end of 1963, along the Han River dikes, there were 38 dike maintenance stations and 193 dike maintenance teams, with a total of 678 personnel, of which 58 were full-timers and 620 were part-timers, all recruited from nearby production brigades.[22]

This kind of mobilization had an immediate effect. In Jingmen County, for example, cadres were no longer frustrated in their attempts to drive cattle and pigs off the dikes. Instead, it was reported, after the establishment of a local dike maintenance unit in 1957, the number of pigs on dikes was drastically

18 *Hubei sheng zhi shuili* 1995: 260; *Hubei shuili zhi* 2000: 553.

19 33-1-2.

20 33-1-16.

21 SZ113-2-274-003.

22 SZ113-2-274-009.

reduced.[23] In the early PRC, due to a shortage of funding, not only were the local people mobilized to maintain dikes, but also the maintenance teams were encouraged to be self-supporting, and thus dike keepers also farmed, and raised silkworms and lotus (valued for its edible roots) as well as livestock (pigs, cattle, sheep, and fish), and planted commercial trees.[24] By 1979, there were 860 or more kilometers of dikes in the Yangzi River valley in Jingzhou, with an exclusion zone (i.e., land along the foot of dikes reserved for maintenance and protection of the dikes) that covered 69,100 mu. Trees could be planted in the exclusion zone, which dike keepers used to generate income, but farming was prohibited. Thus, in Gong'an County, for example, dike keepers, almost half of whom were former production team leaders, enjoyed a higher income than village schoolteachers.[25]

Another major change in the PRC has been the politicization of dike affairs, a subject also discussed in other chapters. In the Qing, the political reliability of dike managers and dike maintainers was never a concern, but it became a serious matter in the PRC, particularly in its early years. In the 1950s, very often militias were deployed to dike worksites to maintain order and surveil landlords, counterrevolutionaries, and other "troublemakers" (see Chapter 8 for more details). The emphasis on political reliability reached its zenith during the Cultural Revolution. In 1969 in Zhijiang County, for example, a special unit was established to strengthen the management of a dike and its sluice gate. A report on this unit explained why a certain Yu was chosen to be a leader of this unit: because he came from a peasant family, was a party member and a demobilized soldier, had served in many positions in the brigade committee, and was loyal to Mao.[26] In a word, he was politically reliable and thus qualified for the job.

From the Qing to the PRC, although the organizations responsible for dike management frequently changed over time, one thing remained constant: the chief military and civil leaders at any level of administration were responsible for high-water control in the area under their jurisdiction. In both the Qing and the Republic, the county magistrate oversaw high-water control in his county; in the PRC, the county head was assigned that job. Simply put, from the Qing through the PRC the top administrator in any area oversaw high-water control in his or her jurisdiction. This was a sign of the importance of high-water control.

23 D35-1-148a.

24 SZ113-2-311-006; SZ113-3-231-008; 40-1-180; 46-1-15ab.

25 40-1-39-1.

26 45-1-61.

After 1788, at the provincial level, the governor-general of Hunan and Hubei and the governor of Hubei were required to go in person to Jingzhou each year to direct local officials in handling high-water control. If the governor-general or governor was too busy to go, he was to appoint the prefect of Jingzhou to be his representative, a step that required the approval of the emperor. At the end of the nineteenth century, this became the norm—the prefect of Jingzhou was always asked to represent the governors in supervising high-water control.[27] At the prefectural level, the sub-prefect of water control (*shuili tongzhi*) was assigned the task of specifying how the Wancheng Dike would be managed. At the county level, it was the county magistrate who was to lead high-water control in his county. The residents of an area protected by a polder dike, however, were responsible for high-water control of their polders.

This system continued in the Republic, with the county head and the commander of locally stationing military forces (if any) responsible for leading high-water control in the county. Hubei suffered another huge flood in 1931, but there was no specific high-water control organization before the flood. The government of Hubei established some organizations during the high-water season, but since they were of limited effectiveness due to poor coordination, they were disbanded after the high-water season. The provincial government only began to set up an organization specifically for high-water control before the coming of the high-water season in 1932.[28] According to this organization's 1936 regulations, members were to come from many government departments, and the date the organization would begin working and finish working would depend on the high-water situation.[29] At the county level, there was also an organization, a high-water control committee, that was responsible for high-water control, with the county head as its director. Township leaders were required to supervise the gentry in their townships to take care of the high-water control of polder dikes.[30] These were all temporary organizations.

The management of people's polders was slightly different. For instance, it seems the five annual repairs and high-water control agencies established in Jianli in 1935 were permanent organizations, as they had to work all year round. But since they only received operating funds for half a year, it is likely they too were temporary units since no one would work for half a year without pay.[31] Most people's polder agencies in Songzi, for example, only operated for several

27 *Zaixu xingshui jinjian* 2004: 663–768.

28 *Hubei shuili zhi* 2000: 228.

29 LS1-5-639.

30 LS1-5-640; *Mianyang xian zhi* 1989: 147.

31 LS31-4-347.

months, but in a few cases, some staff received pay for twelve months, implying they worked all year round.[32]

In the PRC, the basic principle of high-water control has been "centralized and unified leadership, with each level [of government] bearing a different responsibility."[33] There is a temporary high-water control organization at every level of administration, from the central government to the provincial, prefectural, and county governments. Each high-water control organization at the county level unites all affairs related to high-water control within that county, but it is also under the leadership of the immediately superior high-water control organization. At the county level and above, there is a high-water control organization at each level. Each of these organizations consists of members from different governmental departments, just as in the Republic under Nationalist rule. The similarities and continuities in political and governmental organization between the GMD and the CCP in part were the result of the fact that both adopted freely from the USSR model. Nonetheless, organization in the PRC has clearly been more comprehensive and stronger than in the Republic. The evolution of organizations of this kind reflects the "centralized minimalism" in the Chinese governmental system, characterized by semiformal governance by quasi-officials from the Qing to the PRC.[34]

In 1983, the Hubei provincial office of high-water control became a permanent governmental unit, operating year-round, not just during the high-water season.[35] In the following year, the offices of high-water control at the prefectural and county levels also became permanent units.[36] This constituted a dramatic change in dike management.

3 The Role and Responsibility of Administrators

Given the importance of water conservancy, administrators have borne a huge responsibility, a tradition that can be traced back to the legendary Yu the Great, who was said to have lived in the third millennium BCE. It is generally believed that it was the Yellow River that Yu the Great tamed, but a recent

32 E.g., 100-8-97: 150.

33 *Jingjiang dadi zhi* 1989: 217.

34 Philip C.C. Huang 2008.

35 *Hubei shuili zhi* 2000: 228.

36 *Mianyang xian zhi* 1989: 148; *Jingjiang dadi zhi* 1989: 218. Below the county level, each township with dikes also has its own high-water control organization, and even villages have such organizations (*Songzi xian zhi* 1986: 374; *Mianyang xian zhi* 1989: 147). They are, however, mostly temporary.

study argues that it might also have included the Yangzi River, particularly the Jiangling-Shashi section in Hubei.[37] Yu was widely praised as so devoted to the harnessing of the Yellow River and/or the Yangzi River that he passed by his house without walking in to see his family. The true reason, however, is that he dared not do so because of his profound responsibility. His father, Gun, had been tasked with taming the river and had built imposing dikes, but they had failed. For this, Gun was executed. Yu thus had to be extremely cautious.[38] In the Han dynasty (202 BCE–220 CE), a minister even committed suicide over his dereliction of his duty to manage the Yellow River.[39]

The most important water conservancy projects on the Jianghan Plain are the dike systems. By the early Qing dynasty, dike safety in this region had already been regarded as more important than the prevention of thievery or robbery, since those crimes might harm only a few or at most tens of families, whereas a dike rupture could have disastrous consequences for hundreds or thousands of families over a vast area. Also, both tax revenue and military supplies relied on a stable dike system.[40] A local gentryman in 1802 further explained that "even if there were thieves or robbers, they were only a few loafers who filched like rats and snatched like dogs and dared not show their face in the light of day. But if a breached dike was not repaired, that would turn a small thief to a big robber, and a big robber into a rebellious bandit."[41] The head of Hanchuan County in 1946 also claimed that high-water control was more important than guarding against bandits.[42]

The safety of dikes thus became the first priority of local officials, particularly prefects and magistrates. For example, in Gong'an during the Qing, it was said that the top priority of governance was dike management.[43] As for Jianli, according to a magistrate in the early 1870s, "the difficulty in ruling Jianli consists not in the governance of the people, but the management of the dikes."[44] In Jingzhou, it was said that when two of its prefects in the late Qing assumed office during the high-water season, one's beard and hair turned white within a month, and the other, who witnessed the grave danger facing the Wancheng Dike, vomited several liters of blood within a couple of months.[45] And an

37 Yin Lingling 2019.
38 Zhang Fang 2007: 2.
39 Zhou Kuiyi et al. 1990: 26.
40 *Mulingshu* 1990 [1848], vol. 9: 40b–41a.
41 *Xiangdi cheng'an* 1969, vol. 1: 29b.
42 LS31-4-60A.
43 *Gong'an xian zhi* 1874, vol. 3: 37a.
44 *Jianli xian zhi* 1872, Lin's preface: 3a.
45 *Jingjiang di zhi* 1937, vol. 4: 15.

official assigned the job of blocking a breach in the Han River dike in 1936 even planned to commit suicide if he failed.[46] All these anecdotes illustrate the tremendous responsibility and pressure that officials bore.[47]

To explain the frequent water calamities on the Jianghan Plain during the Qing and the Republic, past scholarship pointed to the mismanagement of the dike systems as one of the main culprits. Mismanagement entailed, among other things, corruption among governmental personnel, particularly informal ones such as yamen clerks and runners.[48] That, however, should not lead one to assume that all government personnel were corrupt and incapable, and neglected their responsibility. On the contrary, many of them did fulfill their duties. For example, while inspecting high-water control of the Han River dikes in 1838, Lin Zexu, the governor-general of Hunan and Hubei (and later the famous anti-opium hero in the First Opium War [1839–1842]), witnessed the hard work of some minor water-control officials on the dike sites in Mianyang and Jingzhou.[49] Some of them, particularly formal governmental officials, were model administrators. Here I will briefly discuss three of them.

The first is Wang Zhiyi, the governor-general of Hunan and Hubei from 1806 to 1810. He assumed office in winter 1806. To understand the situation in flooded areas in spring 1807, he used his own money to hire a small boat, with one servant and one bodyguard, and traveled through the heart of the Jianghan Plain. The boat was so inconspicuous, without any of the trappings usually associated with officialdom, that anyone standing on a riverbank would have no idea that it carried a very important official. Moreover, there were no welcoming parties hosted by local officials either. Wang traveled widely and collected a trove of information—even from residents of remote areas, insofar as he felt their opinions were worthwhile. Mainly based on his surveys, he proposed a plan to harness the waterways of the Jianghan Plain, particularly in the Han River valley, without favoring any party. The key to his proposal was that if there were problems in the upper reaches, then the solution was to block the openings of the river and build dikes; if there were problems in the lower reaches, the answer was dredging the river and discharging floodwater. That is because the old course of the tributaries had already silted up, and it would be too costly to dredge them. But if they were not dredged, an opening

46 Chen Zhangguan 1982.

47 A gazetteer mentions that a native of Mianyang in the Qing dynasty resorted to bribery to avoiding serving in the Yellow River administration, even though the post was very lucrative. His concern was simple: the post was extremely perilous since he would be executed if the dike ruptured (*Miancheng zhi* 2000: 214).

48 Peng Yuxin and Zhang Jianmin 1993: 235–38; Jiayan Zhang 2014: 64–72.

49 Lin Zexu 1935 [1885], vol. 5: 139–40.

in the upstream dikes would submerge a huge area of the downstream valley. Therefore, the best solution was to close the tributaries. However, for the Han River, since its course in the lower valley could easily silt up (as the water slowed down), the raising of its riverbed increased the threat to dikes, so it needed to be dredged regularly.[50] Because Wang was familiar with the situation on the ground and was able to propose an unbiased yet practical solution, dikes were built and reinforced and rivers were dredged. The Han River valley benefited from his work for decades.

The second is Yu Changlie, a county official in the early nineteenth century. One stormy night in summer 1827, rising river water overflowed the dike outside the county seat of Jianli, and clerks and laborers ran away. At that moment, Yu walked to the dike and, reportedly, built small dikes of bricks atop the old dike, and appealed to people to add dirt. By dawn, the floodwater was blocked, the dike was saved, and Yu was acclaimed.[51] On another day, in summer 1840, after it had rained for a month, the water of the Yangzi River rose and reached the Wancheng Dike, and with rain still falling, a segment of the dike collapsed.

> All villagers were running helter-skelter, howling, and desperate to escape. Yu … handwrote a farewell note to his wife: "If some unexpected things happened, [that means] I died at my post." … He urgently recruited people to explore the real circumstances of the collapse, [and then] poured in soybeans and cotton quilts to block [the water], stopping the inner seepage, [and then] dirt was added above, and the dike held. [After that] Yu, with the summer rain still falling, walked to examine every spot. He blocked about a hundred such leaks in person, and the dike eventually held. At that moment, tens of thousands of clerks, laborers, and bystanders had gathered, and they all prostrated themselves before Yu and declared, "You saved our life!"[52]

These two examples reveal that Yu Changlie was not only dedicated to dike affairs but was also an official of great talent.

The third is Ding Fengwu, the second deputy magistrate (*zhubu*) of Haoxue early in the nineteenth century. One day in the high-water season of 1813, the Yangzi River water rose suddenly in the segment in Haoxue and threatened to unleash a catastrophe. Most government officials fled, but from noon to midnight, Ding stood at the most dangerous spot on the dike to lead emergency

50 Hu Zuhe 1999 [1838]: 165–66; *Xiangdi cheng'an* 1969, vol. 2: 1a–10b.
51 Yu Changlie 1999 [1840]: 9.
52 Yu Changlie 1999 [1840]: 7.

repairs. The water gradually rose above his knees, but he did not move, swearing that he would rather die at his post. People were inspired, worked even harder, and saved the dike. After Ding died, the local people set up a wooden statue of him and put it in a local temple. After that, whenever there was a threat to the dike, the local people would carry Ding's statue out, inspiring laborers, who then did their best to fend off the danger.[53] Unlike Wang Zhiyi, who surveyed the flooded areas and offered practical solutions, and Yu Changlie, who had a rich knowledge of dike affairs and knew how to deal with threats to dikes, here the local people were actually inspired and motivated by Ding's high sense of duty, although they did not necessarily benefit from his knowledge of dike management. In these accounts, which were released in the PRC, it is unclear how Ding died. According to a memoir published in Taiwan, Ding jumped into the river to pray for help from the water deity.[54] Understandably, PRC accounts would not want to highlight this kind of "superstition."

These kinds of models who sacrificed themselves for others have been particularly welcomed in the revolutionary discourse of the PRC. In 1954, Yuan Youfa, the head of Shishou County, jumped into a boat to block a dike rupture and was engulfed by rapid-flowing floodwater and killed. He was posthumously recognized as a revolutionary martyr.[55] In the same year, Li Dahan, a deputy county head of Jiangling, together with several laborers, jumped into floodwater to block a hole in the Great Jingjiang Dike. That was a very dangerous action—they could easily have been swallowed by floodwater. More laborers quickly jumped into the water, and others filled up the hole. National newspapers praised Li as a hero.[56] In 1998, sixteen CCP members in Wuhan signed a "life and death plate," pledging to safeguard the dike with their lives.[57] Although the expressions of respecting the high-water control heroes were different in different times, they undoubtedly played a role in motivating people to unite and fight against the floods in their respective times.

A more typical yet influential local PRC official of this kind, however, was Rao Mintai (1909–1979). Rao was born to a poor peasant family and joined the Communist guerrilla forces when he was young. In the struggle against bandits

53 *Haoxue zhen zhi* 1986: 37–38; Fang Ming and Dong Yumei 2003: 60.

54 Luo Senlin 1970. There was a similar story about the Yellow River in the Qing dynasty. If the dike was in danger, the responsible official might jump into the water as a sacrifice to the water deity. It was said that if the river dike ruptured, these officials would be punished, but if they jumped into the water, it is possible that they would be posthumously worshiped as a river deity (Dong Longkai 2001: 499).

55 *Shishou xian zhi* 1990: 143; *Jingjiang difang zhi* 2012: 788.

56 Anonymous 1991: 77.

57 *Jianghan qu zhi* 2007: 8.

in 1949, he was appointed the first Communist to serve as the head of Songzi County. In 1952 he led his fellow residents in participating in the construction of a massive flood-control scheme—the Jingjiang flood diversion project—on the southern bank of the Jingzhou section of the Yangzi River. In one case, when the rapid flow of water made it very difficult and dangerous to complete the closure of a dam in a river, Rao jumped in a boat loaded with stone to be used to complete the closure, and shouted: "[If there is going to be] death, I'm the first one to die—you follow me!" Many brave laborers followed him, the closure was completed, and Rao became a model. He was soon promoted to the deputy prefect of Jingzhou and worked on water conservancy in Jingzhou for the rest of his life. After his death, to honor his dedication the local government buried him atop a hill along a reservoir that was built under his leadership. As a local chief cadre, he wore a straw hat and straw shoes, he carried dirt with a shoulder pole along with other laborers, and he did not hesitate in the face of danger.[58] All of this was normal for cadres of that generation. However, it appears that their spirit of sacrifice is rare among CCP cadres in more recent decades, except at sites of high-water control. In the huge flood in the Yangzi River valley in 1998, for example, hundreds of thousands of governmental officials stayed on the dikes and were exposed to the sun and rain for weeks or even months. As a result, their faces became deeply tanned. In my field survey, even the commoner folk, who always seem to complain about governmental employees, admitted that at that time those officials behaved "like a Communist."

When it came to government officials' involvement in dike affairs in the Qing and Republic, a common theme was their malpractice. The sources make clear that the arrival of the PRC did not immediately change things in this regard. In 1963, for example, the State Council issued a report on the widespread misuse of water conservancy funds involving such offenses as waste, misappropriation, corruption, theft, and falsified reporting in some provinces.[59] The report did not mention Hubei, and it is unclear how widespread these kinds of misconduct were in Hubei. But misconduct certainly existed in Hubei. In 1954, in Songzi, for example, many village cadres embezzled funds that were to be used to pay laborers; embezzlement and jerry-building were also found in that year's repair of a ruptured dike in the county of Xiaogan; in 1955 in the prefecture of Xiaogan, there were twenty-three cases of corruption and fourteen of embezzlement, as well as some other forms of misconduct in repair projects; and in 1960, several hundred cooks and managers in 141 mess halls on

58 Jingzhou shi shuiliju et al. 2008: 2–3, 20–23, 177–86; Xin Zhiying 2005: 207.

59 SZ113-3-393-002.

Jingzhou's dike worksites embezzled money that was to pay for meals.[60] Yet, overall, the money involved in such offenses were small. In 1982, many financial irregularities in the Bureau of Water Conservancy of Jiangling were uncovered. These included embezzlement, corruption, theft of public property, and using state funds to build private houses. It is unclear how much money was involved in these cases, but in one instance the "borrowed" money amounted to 10,000 yuan (or roughly 3,817 USD),[61] a huge sum in 1982.

According to the historian Huaiyin Li, however, compared to China before 1949 and to the post-Mao era, "the grassroots cadres in the Maoist era faced the strongest political constraints and had the least room for corruption and abuse."[62] That was certainly the case with dike management. Given the importance of the dike systems of the Jianghan Plain and the strict discipline of the CCP, although embezzlement never disappeared, it was uncommon before the 1990s for cadres or governmental employees to dare to embezzle dike funds. But corruption within the government, including in dike management, has spread in recent decades. Not many cases related to theft and misuse of dike funds have been officially reported, but some interviewees told me that due to institutional reasons, in the 1990s and 2000s, it was considered not bad if 70 to 80 percent of the funds for water conservancy were actually used for such projects. Even the funds issued by the central government after the 1998 flood to aid relocatees were very often diverted to build hotels, restaurants, and official buildings, and also were distributed as wages.

The continuing malpractice in dike affairs can best be seen in the corruption surrounding boats that shipped stone. To reinforce some important and dangerous segments of the Yangzi River dikes, large stones were annually dumped into water at the foot of targeted dike segments. But in the Qing, it was common that the amount of stone dumped was less than half of what was requested.[63] According to Ni Wenwei, it was easy to check the amount of stone loaded in boats, but the supervisor could be fooled if he was not diligent, or he might even have been an accomplice.[64] The same sort of malpractice occurred in the Republic. In a case from 1931, for example, a director of an annual repairs and high-water control agency in Jianli falsely reported that a project had used 6,000 fang (more than 22,000 cubic meters) of stone, while in fact only 1,900

60 D35-1-50c; 50-1-22; Wuchang 49-1-4b; D35-1-207a.
61 48-1-342.
62 Huaiyin Li 2009: 135–36.
63 Yu Changlie 1999 [1840]: 138.
64 Ni Wenwei 2002 [1885]: 125.

fang were used.[65] Even after 1949, since many stone-delivery boats were privately owned, their owners sometimes charged more and falsely reported the amount of stone they shipped, or shipped subpar stones. According to a 1959 report, when loading stones on board, inflating the volume of stone by 25 percent was actually a guild regulation.[66] But even so, an inflation of 50 percent or more cannot be justified.

These kinds of malpractices became increasingly difficult to perpetrate in the PRC, since most private stone-delivery boats were dispensed with and replaced by state-run boats. A 1963 summary of dike work reported that supervisors were sent to quarries to ensure the stone met the required standards and that the boats were fully loaded.[67] But malpractice never completely disappeared and reared its head again in the reform era. In the year 2000, a reporter found shocking wrongdoing in a project to protect the Yangzi River dike. The stone-delivery boats were only half full, many stones were substandard small rocks or weathered stones, both the amount and cost of the stone were inflated, and all this was ignored by the on-site supervisors.[68] This sounds no different from what happened in the Qing and the Republic. I asked a local water conservancy official about this and was told that, putting aside the question of corruption, it was not bad if half of the planned volume of large stones had been dumped into the river. The reason is that no one involved in the project—those who quarried the stone, those who moved the stone from quarries to boats, and those who shipped the stone—wanted to report the real volume. The people who did the quarrying would report nine fang as ten fang, and the people in charge of transporting the stone to the boats would do likewise. Those who did the shipping were even worse: they piled up large stones on four sides of the boat but left the center empty. Although the supervisors knew the truth, they just recorded whatever they had been told, because if they checked and recorded the real amount, they could not find anyone to do these jobs. They felt they had to turn a blind eye.

---

65 Hubei sheng shuiliting and Hubei sheng dang'anju 1999: 413–14. According to a contemporary report, one fang equals 3.7 cubic meters (National Flood Relief Commission 1933: 84n1).

66 D35-1-170.

67 00046-001-0046; D35-1-285; SZ73-05-0215-014.

68 Wu Xiaofeng 2000.

4 Technicians and Dike Administration

Technology and technicians were an important part of dike administration. In imperial times, those who were assigned positions in river administration or were responsible for dike management were not selected because they were trained as hydraulic technicians, but because they were governmental officials. In the Qing, most of them got their job by passing the civil service examination, which tested applicants' knowledge of some of the most treasured of China's ancient classics, skill at composing poetry, and ability to write essays in a rigidly prescribed style. Few officials had technical expertise or experience, and thus whatever practical knowledge they needed to carry out their duties they had to acquire on the job. Those who served in the Yellow River Administration were called "Confucian engineers," or more precisely, "hybrid literati-technologist[s]."[69]

In the Republic, as a part of its nation-building project, the government established several new institutions, such as the Huai River Conservancy Commission,[70] and new schools such as the Hohai College of Engineering, formerly the Hohai Civil Engineering School, to train hydraulic technicians. In 1922, the government of Hubei selected more than thirty young people from counties with dikes and sent them to Hohai to be trained as hydraulic technicians.[71]

In the PRC, water conservancy technicians were included in the government administrative system; that is, they were "state cadres."[72] But almost all government departments in the new state lacked professionals—including hydraulic technicians. In 1950 in Gong'an, after a dike breach, it was found that technical workers were not well organized; a summary report on the 1951 annual repairs of the Dongjing River dikes recognized that there was a shortage of technicians everywhere, and called for augmenting their numbers; in Zhijiang in 1954 technical knowledge could not keep pace with the on-site realities of high-water control; and in 1957 in Jingshan County, it was widely admitted that there were too few hydraulic technicians (only seven, whereas there were five thousand laborers), and the technical level at diking sites was low.[73]

Thus, in the early PRC, although "old" technicians, that is, those trained before 1949, might have been politically unreliable, the new state still needed

69 Dodgen 2001: 146.

70 Pietz 2002.

71 *Hubei shuili zhi* 2000: 492.

72 In the 1950s, physicians on diking sites were also categorized as technicians (50-1-24). But my discussion focuses on hydraulic engineers or water conservancy engineers and technicians.

73 38-1-4a; 40-1-13; 45-1-8; 00046-001-0032.

their expertise. In 1949 in Mianyang, for instance, three to six dike engineers with a high level of knowledge were chosen from each dike project to receive double the food allowance; that is, two times the food allowance that the Communist cadres received.[74] Although these privileged technicians would likely have worked in the bureau of dike engineering when the Nationalists were in power, in 1949, as we see, the Communists nevertheless treated them with respect and temporarily put class struggle ideology aside; after all, technology was crucial in dike work. The locals, however, did not necessarily appreciate that. For example, in Jianli in 1951, technical guidance lagged the work of laborers.[75] Anti-old-technician sentiment was high at the time. Many people considered technicians to be "villains" and ridiculed them as "dogs." This understandably tended to deter technicians from fully playing their role.[76] In response, local leaders argued out that since diking technology cannot be ignored, the old technicians should not be cast aside. They went on to argue that being "anti-local-bullies and anti-old-dike-managers does not mean [we should] reject the old technicians, let alone [their] technical guidance."[77]

But over the years, some technicians from the Nationalist era still favored technology over politics, and thus the Communists, particularly during the Cultural Revolution, called on them to spend more time on "political learning."[78] One cadre even suggested that if any of them already met the criteria for membership in the Communist Party or the Communist Youth League and wanted to become a member, they should be admitted.[79] Even during the Cultural Revolution, "old" technicians, though of questionable political reliability, were still needed and were treated fairly, as long as they did not conceal their background and did not engage in any sabotage after 1949.[80]

All of this shows the importance of technology and technicians in dike affairs. Numerous reports in local archives confirm that local authorities recognized this fact. For example, in Mianyang in 1950, on-site cadres were asked to pay attention to and learn technology.[81] In Qianjiang County in the same year, cadres were called on to rely on and respect technology, and laborers were asked to follow technicians' guidance.[82] A 1951 report from Jingzhou

74 Yingcheng 49-1-2b.
75 1-1-029-022.
76 22-1-013-010.
77 SZ1-5-59A.
78 SZ34-2-414; 54-1-108.
79 50-1-28b.
80 00046-001-0117.
81 22-1-006-016.
82 40-1-10a.

claimed that technicians won the trust of cadres and the masses by using their knowledge to effect emergency repairs.[83] In a 1952 report also from Jingzhou, technical guidance was acknowledged to be an important factor in the success of high-water control.[84] Finally, a 1954 report from Jingzhou noted that because of a shortage of technicians, and because most cadres had no technical proficiency, some knowledgeable "reform-through-labor" prisoners were used on high-water control sites.[85]

Thus, at the very beginning of the PRC, the new state also began to train its own technicians. In 1955, the government of Hubei planned to train twenty thousand peasant technicians, including twenty-six hundred in Jingzhou, where some young peasants were selected to be trained as water conservancy technicians. The goal was to prepare them to take charge of the water conservancy work of their units.[86] They were trained on water conservancy project sites, but in Jingshan only 37 percent of them passed the exam; many had worked hard yet still failed because of their poor educational background.[87] In 1956, when the Yingcheng County government also organized a class to train rural water conservancy technicians, thirty-seven of all 174 attendees were illiterate; naturally, some of them did not learn what they were supposed to learn.[88] Xiaogan in 1957 addressed this problem by requiring that potential hydraulic technician candidates be under the age of thirty, be in good health, have an elementary school education and experience with dikes, and be endorsed by their fellow villagers.[89]

By 1957, the province of Hubei had trained more than thirty thousand peasant hydraulic technicians, all of whom were required to work on water conservancy projects.[90] In Zhongxiang County in that year, the major technical force consisted of peasant technicians who had been trained in recent years.[91] In 1958, to support the upsurge of water conservancy construction, the government of Jiangling established a class to train peasant water conservancy technicians. The candidates were required to be male, healthy, between the ages of 18 and 35, politically reliable, and to have had some basic education. Eventually, 632 people—99 cadres and 533 peasants—attended the training

83 D35-1-4a.
84 D35-1-38b.
85 40-1-57b.
86 00046-001-0007.
87 00046-001-0017.
88 Yingcheng 49-1-45.
89 81-1-53.
90 Wuchang 49-2-14.
91 54-1-35a.

class. Although the students varied in their educational background and the length of time they attended the class (some were enrolled late), many admitted they learned a great deal.[92] Also, using water conservancy worksites as classrooms and water conservancy projects as textbooks, more than three hundred peasant hydraulic technicians were trained in Hanyang County in 1960.[93] The most famous local peasant technician on dike affairs is Luo Heshan, who studied how to control insects in breakwater forests (or wavebreak forests, trees planted outside a dike [facing the water] to protect it) and earned national fame in the 1970s.[94]

Technicians were supposed to concentrate on their technical tasks. But this was a luxury for rural cadres, many of whom had to fulfill multiple duties, which was particularly the case during the Cultural Revolution. In 1974 in Zhongxiang, for example, some dike technicians were temporarily assigned other jobs, which had a serious impact on dike work during the high-water control period.[95] The shortage of technicians also became a common problem in Jingzhou in that year. In the Han River valley in Jingzhou Prefecture, 255 cadres worked on dike administration, but only 83 of them were technical cadres. Moreover, the technicians had little practical experience.[96] At the same time, the numbers of qualified technicians could shrink, as was reported in Tianmen in 1975.[97]

Over time, particularly after the 1979 rural reform—a monumental shift from collectivization and a return of production decisions made by individual peasant households—the situation gradually improved. Hubei established several hydraulic colleges and technical secondary schools to train hydraulic technicians. In 1950, the province had only eighty such technicians, fourteen of whom were qualified engineers; but by 1995 the number of technicians had skyrocketed to 17,509.[98] Statistics from various counties reveal the growth in the numbers of technicians. Qianjiang, for example, had no hydraulic technicians at all before 1949, but by 1995 it had 168.[99] Similarly, the dike management system in Hanchuan had a staff of 36 in the 1950s, only a few of whom were technicians. But by 2005, it had 153, including 55 college and technical

92 48-1-54ab.

93 30-1-24a.

94 42-1-188. See Schmalzer (2016) for a detailed discussion of peasant agricultural technicians in the Mao era.

95 42-1-155.

96 42-1-158b.

97 42-1-172.

98 *Hubei shuili zhi* 2000: 1275, 1327.

99 *Qianjiang shuili zhi* 1997: 270–71.

secondary school graduates.[100] In still another case, by the year 2007 Honghu had 396 hydraulic technicians, 235 of whom had graduated from college or technical secondary schools.[101]

5 Rewards and Punishments

Because of the crucial importance of the dike systems, local officials were subject to punishment if dikes under their jurisdiction broke. An early example was Yin Zhongkan, a prefect of Jingzhou in the Jin dynasty (265–420), who was demoted because the Yangzi River washed away thousands of houses in Jiangling due to mismanagement of the river dike.[102]

In the Qing dynasty, the state established regulations on the responsibility of officials, and those who failed to meet the standards were subject to punishment.[103] Demotion was one of the most common punishments. Most officials who were demoted were reduced in rank by one grade, although some were demoted two grades, or even four grades.[104] Another common punishment was deposition. It was often the case that a deposed official could be restored to his former position if he repaired a broken dike well or did a good job the following year. And some officials also were fined to cover the cost for the repair of broken dikes. For example, after the 1788 flood, the fines for officials at different levels ranged from 5,000 to 40,000 taels of silver, heavy penalties inasmuch as the annual salary of a county magistrate in the Qing dynasty was normally about 1,000 taels of silver.[105] That is to say, the fines reported here ranged from five to forty times a magistrate's annual salary. Forfeiting of salary was also a form of punishment. And under rare circumstances, officials also be imprisoned or even could be executed.[106]

In most cases, an official usually suffered only one of these punishments, but it was not unusual for some to receive multiple punishments: deposition first, followed by demotion, and a fine. The magistrate of Jingshan, for instance, was fired and also fined 20,061 taels of silver because of repeated ruptures of the Han River dike at Wangjiaying in 1824 and 1825. And in 1841, the magistrate of

100 *Hanchuan difang zhi* 2007: 222.
101 *Honghu shi zhi* 2014: 219.
102 Zhang Fang 2007: 78–79.
103 Jiayan Zhang 2014: 49–51.
104 *Zaixu xingshui jinjian* 2004: 286–94.
105 Ni Wenwei 2002 [1885]: 210–12; Jin Feng 1996: 105.
106 *Zaixu xingshui jinjian* 2004: 294, 594.

Jiangling not only lost his position but also had to pay the cost of a failed dike project due to his "stupidity and incompetence."[107]

Most officials who were punished were either prefects or county magistrates, although some were as high as governor-generals, and others as low as sub-county magistrates. They were mainly punished because of dike ruptures caused by their mismanagement: carelessness in supervising diking, failure to manage high-water control and emergency repairs, and so on.[108] These lapses all involved river dikes, since polder dikes were maintained by the local people, and rewards and punishments did not extend down to them.

Understandably, not everyone admitted responsibility when a dike ruptured. In many cases, people argued that dikes were "over-flooded," that is, there was simply so much floodwater that it was beyond the capability of humans to control, and hence the responsible officials should not be blamed. Some officials indeed used this as an excuse to evade blame, such as Shu Chang, the governor-general of Hunan and Hubei. In 1788, immediately after floodwater engulfed the city of Jingzhou, Shu explained in a memorial that the disaster was mainly caused by such a huge volume of floodwater that coping with it was "beyond what any human could do."[109] The emperor rejected this excuse, and Shu and twenty-two other officials were punished for this incident.[110] In 1824, the magistrate of Jingshan was dismissed for one dike rupture. He also had argued that the Han River water suddenly rose 14 or more chi (1 Qing chi = 31.1 cm) and then 8 chi, "plus there was wind and rain, and humans had no power to do anything about it," thus the dike in Wangjiaying broke.[111] But according to the Qing regulations, if the inundation was indeed caused by nature, that is, the floodwater suddenly rose and over-flooded the dike, then responsible personnel could not be blamed, and hence would be exempted from punishment.[112] In 1860, the volume of the Yangzi River floodwater in Yichang reached a record high level (of 92,500 m^3/s),[113] and thus the governor-general of Hunan and Hubei reported that most magistrates were exempted from punishment, but he still blamed the prefect of Mianyang for not doing his best to protect the Han River dike.[114] Although officials frequently used "too much water" as an

107 *Zaixu xingshui jinjian* 2004: 27, 81, 269.
108 *Zaixu xingshui jinjian* 2004: 22–786.
109 Shuili dianlibu 1991: 481.
110 Ni Wenwei 2002 [1885]: 210–14.
111 Shuili dianlibu 1991: 670.
112 Zhang Youyu and Gao Chao 1991: 852.
113 *Hubei shuili zhi* 2000: 181.
114 Shuili dianlibu 1991: 926.

excuse, this defense was often flatly rejected by their superiors and the officials were punished.

This included some well-known people. As the hardworking governor-general of Hunan and Hubei, Lin Zexu did much to improve water conservancy in Hubei, particularly in the Jianghan area, but he was also punished by being demoted two grades because of the 1838 flood. Yu Changlie, a very diligent county official, was punished in 1851 for dike ruptures in his county when he was the magistrate. The Qing court considered dike ruptures intolerable, and its regulations were not mere formalities. Indeed, in 1840, because of his carelessness, the magistrate of Jiangxia County was deposed "as a warning to those who did not take dike work seriously."[115] In thirteen of the thirty years of the Daoguang reign (1821–1850), officials in the Jianghan area were punished for dereliction of their duty to ensure the effectiveness of dike work.[116] This high frequency of punishment also demonstrates the high frequency of water calamities in the reign of Daoguang, or as the locals put it, "Daoguang Daoguang, nine years out of ten are famine years."

In the Republic, after a dike ruptured the responsible officials were also punished by being removed from office, or demoted, or assigned a demerit (a new form of minor punishment). The official position on the cataclysmic 1931 flood was that it was "one of the most disastrous natural calamities which the world has witnessed," and was beyond human control—that is to say, the GMD government disclaimed responsible for it, while in the PRC sources the GMD has usually held responsible.[117] However, the disaster was exacerbated by problems entirely within human control, such as a shortage of materials, poor organization, and irresponsible high-water control.[118] Thus, after the 1931 flood, twenty-four or so governmental officials, including twenty-one county heads, were punished: some were removed from office, and others were demoted to a lower rank, or hit with a salary cut.[119] After the 1935 flood, Lei Xiaoqing, the head of Jiangling, was dismissed; Xu Guorui, the director of the Bureau of Dike Work on the Jing River, received a demerit; and two other minor officials were permanently kicked out of the government.[120]

115 *Zaixu xingshui jinjian* 2004: 252, 253, 338.

116 *Zaixu xingshui jinjian* 2004: 22–294.

117 National Flood Relief Commission 1933: foreword; Xie Qianmao 1931; *Wuhan difang zhi* 1986: 18.

118 Xie Qianmao 1931: 40, 88–89.

119 LS1-4-452.

120 *Shashi shi zhi 1* 1992: 173.

Although in the Qing dynasty, it was almost entirely government personnel who were punished for failing to prevent floods,[121] this began to change in the Republic. A 1928 regulation provided that any resident who did not obey the orders of the county head and refused to assist in dike repair would be fined 10 to 500 yuan.[122] Although this regulation extended penalties to the common people, at least in theory, it is unclear how it worked in reality, or how many villagers (if any) were fined. According to a 1939 temporary regulation on the management of people's polders in Mianyang, if a member of the staff of a people's polder annual repairs and high-water control agency engaged in any kind of wrongdoing, that person would receive the same punishment as a government functionary,[123] even though the staff members of people's polders agencies were not government personnel. In 1947 in Peixing polder in Songzi, there was a rule that each household had to provide one laborer for high-water control, and if anyone refused to obey this rule or left without permission, that person would be punished.[124] Again, it is unclear what the punishment was and how widely it was applied. This polder in any case was managed by the residents themselves, not by the state.

Although, as we have seen, dikes frequently ruptured and many responsible officials were punished in the Qing and the Republic, few of them forfeited their lives. In 1947 the governor of Hubei issued an order to apply military law to high-water control, which included the possibility of on-site executions,[125] but it is unclear if anyone had really been killed. In summer 1950, however, when a segment of the Dongjing River dike broke, the responsible engineering technician was accused of serious dereliction of duty and was sentenced to death.[126] This was a rare case, as there is no evidence that other PRC cadres lost their lives due to their failure in carrying out dike work. However, they could be fired or jailed. For example, in the 1953 annual repairs in Xiaogan, eleven people were punished, including two who were dismissed, four who were removed from office, and one who was sentenced to prison.[127] In 1954 in Shishou, a former religious leader was sentenced to three years in prison because he agitated against a rational approach to fighting floods, and instead claimed that the

121 In Qianjiang, it was said that in the Qing dynasty, dike administrators would be punished if a dike ruptured due to carelessness (*Qianjiang shuili zhi* 1997: 142). They were not government officials, and it is unclear if any of them were actually punished.

122 *Hubei shuili zhi* 2000: 407.

123 LS36-2-15.

124 100-8-108: 52–53.

125 *Xiantao shuili zhi* 2008: 91.

126 *Honghu xian zhi* 1992: 18.

127 Wuchang 49-2-8.

right way—which the GMD followed, according to him—was to kill pigs and sheep and throw them into water, at which point the flood would instantly retreat.[128] In 1958, when a segment of the Wuhan dike was in danger, a factory cadre did not lead people to make emergency repairs, but instead he hurried to save his own luggage. The CCP considered this shameful and warranted harsh punishment, but the cadre was only placed on probation within the party for one year, since he "had been wounded for the cause of the people's revolution" before 1949.[129] In 1959, to get water from the Yangzi River to combat the severe drought that year, some cadres in Shishou opened a branch river dike without approval from their superiors, but did not close it in time, and the dike ruptured. Six people were punished for this incident by being either expelled from the party or deposed.[130] In Shashi during the Cultural Revolution, one person was fired from public employment and two people were sentenced to seven and fifteen years respectively, because they embezzled dike fees.[131]

Although neither the death penalty nor imprisonment because of failure to control water have been common in the PRC, punishment does seem to have been harsher than before. As early as 1951, the government of Hubei made it clear that no failure in high-water control would be tolerated.[132] In other words, any dike ruptures would be attributed to man-made factors, and someone would be held accountable. The responsible personnel could be punished for dereliction of duty, refusing to heed commands from above, leaving their posts during high-water control, and violating regulations.

As in the Qing and the Republic, in the PRC many responsible officials have also been removed from their official position or discharged from the government after dike ruptures, but it seems there have been no cases of demotion. In fact, the most common punishment in the PRC has not been administrative, but disciplinary, which includes expulsion from the party, receiving a reprimand from the party, being assigned a demerit, and being publicly criticized (e.g., in the newspapers). In the PRC's party-state system, those who have been expelled from the party or received a disciplinary warning within the party might keep their position in government, but usually they have had no hope for promotion, and their political career most likely ended. To be publicly criticized in a newspaper, though humiliating, is a much lighter punishment.

128 39-1-15-3.

129 XX000129-WS01-0043-0015.

130 1-1-556-67.

131 *Shashi shuili difang zhi* 1994: 160.

132 33-1-5 (1)a.

As in the Republic, punishment in the PRC has continued to extend down to nongovernmental personnel who live in the countryside. For example, in Zhongxiang three rural cadres—two in 1954 and one in 1964—lost their positions due to their negligent handling of dike work.[133] But it is difficult to apply this kind of punishment to people who have no position in government. In Yingcheng in 1959, after the rupture of a polder dike due to the carelessness of the on-site leaders, the county headquarters for high-water control sent people to investigate the case. One of the on-site leaders, a certain Wang, responded: "Investigate what? Accountability? [Can anyone] annul my status as a commune member?"[134] At that time, village leaders were still peasants, not government employees, and they did not receive a salary from the state. Thus, to dismiss them from their positions did not really "punish" them much. In fact, their dismissal was more like a warning to other people to handle dike work carefully. It was also a symbolic demonstration of the state's control of rural society. In another case, during the 1998 Yangzi River flood, 137 people were punished (by being dismissed, expelled from the party, assigned a demerit, etc.) in Jingzhou alone: among them, 76 were rural party members and village cadres.[135] But none of them had a post above the county level. A possible explanation is that no major dike had ruptured.

Although generally speaking in the Qing there were more punishments than rewards, some officials were rewarded for their dike work. A common reward was promotion. For example, Tang Shuyi, the magistrate of Jiangxia, was promoted one grade in 1833 because of his contribution to annual repairs and high-water control.[136] In the late Qing, the emperor would send a plaque to local governments as a kind of reward for keeping the dikes safe. But the security of dikes was routinely recognized as mandatory for the responsible officials, and thus they could not expect to be rewarded just because they did what they were supposed to do (see Chapter 5 for details). For the common people, it was suggested in the early nineteenth century that anyone—no matter who that person might be—finding and reporting a leak would be rewarded with 10 min (or 10,000 wen, equivalent to about 10 taels of silver), which was to be publicly announced before the high-water season.[137] It would appear that this

---

133 *Zhongxiang shuili zhi* 1998: 189.

134 Yingcheng 49-1-75.

135 *'98 Jingzhou kanghong zhi* 1999: 110. According to another source, in total 119 cadres (54 state cadres and 65 village cadres) were punished in Jingzhou (*Jingjiang difang zhi* 2012: 591).

136 *Zaixu xingshui jinjian* 2004: 171.

137 Ni Wenwei 2002 [1885]: 154. The *wen* (usually called "cash" in English-language works) were typically copper alloy coins with a square hole in the center. They were the chief units of currency in China from the fourth century BCE to the early twentieth century.

was a bounty available to the public, but it is unclear how many, if any, commoners ever received such a reward. The Republic had regulations stipulating that county governments should reward local residents who had contributed to dike safety by pointing out problems.[138] In 1948, the head of Jianli recommended that three villagers be rewarded for their outstanding contribution to that year's high-water control.[139]

Many officials in the PRC also got promoted because of their outstanding achievements in water conservancy, particularly their work on the dike systems on the Jianghan Plain. Rao Mintai, for example, was promoted from county head to deputy prefect in the early 1950s. Song Yuying, a native of Henan who had relocated to Zhongxiang, impressed her superiors during the high-water control of 1970 by jumping into the rapids to stop floodwater. For this, she was promoted from her position as the party secretary of her village to a leading county-level post, and then to a prestigious provincial-level position.[140]

Just as with punishments, rewards also have been extended to the villagers in the PRC. To encourage people who participated in high-water control to fulfill their duties, in 1950 the General Headquarters of Flood Control of the Central Government issued temporary rules on punishments and rewards, which targeted both laborers and cadres.[141] The government of Hubei issued the same rules in the same year, and again in 1954.[142] The Regulations on Rewards and Punishments of Cadres and Laborers of Water Conservancy Projects in Jingzhou, promulgated in 1954, specified that both cadres and laborers were equally subject to rewards and punishments.[143] The city of Wuhan issued similar regulations for high-water control in 1954, and the counties of Jianli and Hanyang did the same in 1959 and 1960, respectively.[144] In all these cases, what was most important was not how people were rewarded and punished, but that both cadres and laborers were treated equally. And in terms of rewards, nonmaterial rewards such as being recognized as a model laborer and being praised in the newspapers were considered more valuable than material rewards. In Mianyang in 1949, model laborers were awarded a plate, and "diking heroes" were selected from among model laborers and were awarded a flag and a towel, shoes, and farm implements.[145] To bolster the morale and enthusiasm of peasant laborers, some model laborers in Zhongxiang in 1951 were

---

138 LS31-4-1170.

139 100-10-482: 46.

140 *Chaihu zhen zhi* 2013: 462–64.

141 22-1-006-010.

142 *Jianli difang zhi* 1991: 335–36; 30-1-1.

143 40-1-55c.

144 XX000129-WS01-0060-0006; 1-1-271-013; 30-1-24b.

145 Yingcheng 49-1-2b.

chosen through a process of public appraisal and were awarded a hand towel, soap, socks, and a shovel.[146] In 1952, Xin Zhiying, a national model laborer in Songzi, was even awarded two cattle and a suite of farm implements for her outstanding contribution to the Jingjiang flood diversion project.[147] In 1983 in Jingzhou, "advanced individuals" in high-water control were awarded 50 RMB (close to 20 USD),[148] about one month's wages at that time. In Hanchuan in the same year, a sub-township party secretary, four village party secretaries, and several villagers jumped into floodwater and successfully blocked a leak. All were rewarded. The cadre received a pay increase, and all others were awarded a Yongjiu ("Forever") bicycle—famous, reliable, Shanghai-made, and hard-to-get.[149] Owning such a bicycle in the countryside in the 1980s was a matter of pride.

In Tianmen in 1961, all the villagers were mobilized to participate in dike maintenance and worked under a mutual supervision system. Those who did a good job were awarded some workpoints, and those who damaged dikes were punished.[150] This approach was even incorporated into village regulations. For example, Madian Village, Jiangling County, ruled that all villagers had to take part in water conservancy work.[151] All these cases represent clear signs that state power had deeply penetrated the fabric of rural society. From the Qing, to the Republic, and then to the PRC, state power penetrated more and more deeply into rural society. Concomitantly, dike management gradually transitioned from the Qing practice whereby dikes were mainly managed by the locals themselves, to the Republic, when the state began to deepen its intervention, and to the PRC, when the state directly managed all dikes through grassroots organizations.

---

146 54-1-3a.

147 Xin Zhiying 2005: 209.

148 54-1-232.

149 *Fenshui zhen zhi* 1998: 177–78.

150 SZ113-2-195A.

151 *Madian cun zhi* 2004: 30.

CHAPTER 2

High-Water Control

The major tasks of dike management are high-water control in the summer-autumn season and annual repairs in the winter-spring season. Here I will elaborate on the practical steps and methods used in both high-water control and annual repairs on the Jianghan Plain in the past three centuries. Generally speaking, in the Qing and the Republic high-water control was considered more important than annual repairs, though some officials in the late Qing had already begun to argue that annual repairs should be prioritized over high-water control. In the PRC, both high-water control and annual repairs have been considered essential, but the official strategy emphasizes flood prevention, and thus high-water control and annual repairs have gradually become a governmental routine that is practiced throughout the year, not just a seasonal task as before. This chapter explores high-water control; annual repairs will be discussed in the next chapter.

1 The Specifics of High-Water Control

The elements of high-water control discussed here include the temporal parameters, the criteria for determining the specific water level needed to trigger high-water control, and the role of laborers.

Qing sources are usually unclear about the temporal boundaries of high-water control. A new gazetteer of Mianyang indicates in the Qing, high-water control in Mianyang extended from the sixth day of the fifth lunar month to the Hanlu day;[1] that is, roughly four months. The high-water control period became clearer in the Republic but nonetheless varied in different times and areas. In 1933 the military authorities declared that the period of high-water control on the Yangzi and Han River dikes was to extend from June 15 to September 15, subject to adjustment according to circumstances.[2] In 1936 the Hubei government determined that the period of high-water control was to be from June 1 to the end of August, but according to a regulation issued in 1946,

1 *Mianyang xian zhi* 1989: 148. Hanlu ("cold dew") is the seventeenth solar term in the traditional Chinese calendar (which divides the year into 24 solar terms); the Hanlu day usually falls on October 8 or 9.

2 LS1-5-614.

 | DOI:10.1163/9789004739482_004

the period on the major river dikes in Hubei was from June 16 to October 15, also subject to adjustment according to circumstances.[3] The period was from June 1 to September 30 in Qianjiang and Songzi, from June 15 to September 15 in Jiangling, and from June 1 to August 31 (from 1926 to 1945) then from June 6 to October 15 (after 1945) in Wuchang.[4] That is to say, the period of high-water control generally lasted three to four months in the Republic.

In the PRC, the government of Hubei determined that the period of high-water control for the whole province would extend from May 1 to October 15, but also made clear that each region could adjust the period to suit local conditions.[5] Accordingly, the period runs from May 1 to October 15 in Mianyang and May 1 to October 20 in Tianmen, but from April 15 to October 15 in Xiaogan, Jingzhou, and along the Dongjing River.[6] That is about half a year—two to three months longer than in the Qing and the Republic.

During the high-water control period, it is important for the dike managers or responsible officials to have information on the water regimen (*shuiqing*),[7] or hydrological data, in a timely fashion. This, however, was not easy in the Qing and the Republic. Hydrological data and orders were mostly delivered by hand by couriers on foot or on horseback. In the Qing in Jingzhou, for example, during high-water control of the Wancheng Dike, after floodwater overflowed the river beaches, "circulating labels" were used to report the water regimen. These were bamboo boards (each "three chi long, five cun [1 Qing cun = 3.11 cm] wide, and half a cun thick"), which were passed from one station to the next, with each station noting the water level of the Yangzi River. The records were reported to the local government once every three days, or were transmitted immediately if floodwater rose or dropped dramatically.[8] Red flags, smoke and bonfires, and gongs and drums were also used to transmit information—most likely to send an alarm out when an emergency arose.[9] Based on this reporting system, the responsible officials had no idea about the water regimen nor whether it had changed in the upper and lower reaches, and hence no idea

3 *Tianmen xian zhi* 1989: 142; *Hubei shuili zhi* 2000: 235.

4 *Qianjiang xian zhi* 1990: 292; 100-8-87: 35; *Jiangling difang zhi* 1984: 96; Jin Feng 1996: 149.

5 Hubei sheng shuiliting and Hubei sheng fangxun kanghan zhihuibu bangongshi 2000: 142.

6 *Xiantao shuili zhi* 2008: 94; *Tianmen shi zhi* 2013: 95; *Xiaogan diqu shuili zhi* 1996: 58; *Jingzhou diqu zhi* 1996: 146; *Dongjinghe difang zhi* 1994: 3.

7 I use "water regimen" purely to refer to such hydrological data as flow and water level, which is very different from Yan Gao's "water regime". For Gao, "water regime" includes almost everything related to water, such as rivers, lakes, wetlands, dikes, animals, human activities, etc. (Yan Gao 2022: 1).

8 *Jingjiang dadi zhi* 1989: 225.

9 *Qianjiang shuili zhi* 1997: 146.

how long the high water would last.[10] This made quick and appropriate decisions virtually impossible.

One source mentions that telegraph was used in China to transmit information about the water regimen in 1889, and the telephone was used for the same purpose along the middle and lower reaches of the Yellow River and the Yangzi River in 1899, and post offices and radio stations along the Yangzi River were used to transmit such information without charge in 1933.[11] All these were likely only occasionally used by state agencies at that time, as one archival source indicates that in 1933 information about the water regimen along the Han River in Hubei was still mainly delivered by couriers on foot.[12] A gazetteer of Shashi—one of the most advanced places in the Jianghan area—indicates that postal service and telegraph only became available after the Anti-Japanese War, or the Second Sino-Japanese War (1937–1945).[13] Even in 1950, in many places, such as a sub-county high-water control office in Gong'an, there were still no phones.[14] The use of telegraph and the telephone only gradually spread in the PRC, and "any telegraph or phone call that related to the water regimen during the period of high-water control was treated with dispatch."[15] This may be true in theory, but it was not always the case in reality. In an emergency on the Yangzi River dike at Tianjiakou in Honghu in 1969, the people on the site repeatedly tried to call the high-water control headquarters of a nearby state farm and the high-water control headquarters of Honghu County to seek help, but they could not get through because when they tried to call, the line was busy: a news broadcast was being transmitted over the phone line. The dike soon collapsed.[16] The source apparently believed that this tragedy could have been easily avoided if the line had not been busy; a 1975 note from Jingzhou indeed spelled out that the high-water control phone line could only be used for high-water control.[17] In 1972, a high-water control report on Zhijiang particularly complained that there were not enough telephone lines along the Zhuzhang River.[18] In the following decades, the communication system along the Jianghan dikes gradually improved; in 1979, when the government of Jingzhou distributed that year's high-water control funds to the dike

---

10 *Mianyang xian zhi* 1989: 148.
11 Gu Hao 2006: 161.
12 LS1-5-658.
13 *Shashi shuili difang zhi* 1994: 102.
14 38-1-4b.
15 *Jingjiang dadi zhi* 1989: 226.
16 *Jingjiang difang zhi* 2012: 617.
17 40-1-24-1.
18 45-1-104.

maintenance units, it made clear that a portion of the funds had to be used to improve the communication system.[19] Since the end of the twentieth century, both computers (connected via the internet) and cellphones have been more and more widely used to process and transmit hydrological data, and GPS has been used to monitor high-water control infrastructure.[20] This has greatly helped officials to collect, share, and analyze data and make informed decisions.

During the Qing it seems there were no clear criteria of what constituted "high water." And in the Republic, only a few places in the entire Jianghan Plain had some rough criteria. In Qianjiang, for example, "the initial dangerous stage" was when floodwater reached the foot of the dike, "the dangerous stage" when floodwater rose to two meters below the top of the dike, and "the most dangerous stage" when floodwater reached the top of the dike. In each stage, different and increasing numbers of laborers were required to go to the dike to patrol and watch over it, and carry out emergency repairs.[21] In another example, each stage in Songzi was determined by the distance from the flood to the top of the dike: five meters was the "flood-control stage," three meters the "stress stage," and one and a half meters the "dangerous stage."[22]

In the PRC, there have usually been three threshold water levels, or in a term used locally, "three lines" (*sanxian*), each marking a different stage of high-water control. These three have varied substantially from dike to dike, but the principle has been the same. The first, or "starting," line begins when floodwater has overflowed the river beach and reached the dike: this marks the start of high-water control. The second line is the "alerting" level: the point at which floodwater has risen to a level on the dike such that an emergency could occur. And the third line is the "certified" level: the point at which floodwater has reached the highest water level the dike was designed to contain or the highest historical water level.[23]

Each of these "lines" entailed regulations on who should be sent to work on high-water control sites. In general, before 2005, when the system of compulsory labor for water conservancy projects was abolished, those who lived closest to a dike were the first-line laborers; those who lived a little farther from the dike were the second-line laborers; and those who lived still farther were the third-line laborers. In Jiangling, these distances were five li (first line), five

19 42-1-216.

20 *Hubei shuili zhi* 2000: 1062; *Jingjiang difang zhi* 2012: 551; *Shishou shi zhi* 2014: 174; *Wuhan shi zhi 1* 2006: 340.

21 *Qianjiang xian zhi* 1990: 292.

22 100-2-52: 80.

23 *Tianmen xian zhi* 1989: 142; Hubei sheng shuiliting and Hubei sheng fangxun kanghan zhihuibu bangongshi 2000: 44.

to eight li (second line) and eight or more li (third line).[24] At different levels (lines), different numbers of laborers were required to patrol and watch over the dike. On the Great Jingjiang Dike, for instance, the first line required 6 to 10 laborers per kilometer, the second required 30 to 50 laborers per kilometer, and the third 80 to 100 per kilometer.[25] In 1964, on average there were 735 laborers on every kilometer of the Dongjing River dike.[26] In other words, there was one high-water control laborer for every 1.36 meters of dike. The numbers were far less on branch river dikes. In Songzi, for example, the numbers required at the different levels (lines) were only 10, 20, and 40 laborers per kilometer respectively.[27] In Qianjiang there were four water levels: starting, warning, certified, and "strive for" (i.e., strive for the safety of the dike), and the required laborers for each level after 1964 were 8, 60, 210, and 350 per kilometer respectively.[28] In cases of extreme danger, more laborers were needed. In Jianli in 1998, for instance, about 150,000 laborers were stationed on the county's dikes—one person per meter on some important dike segments.[29] In 1954, around 200,000 laborers were stationed on the Great Jingjiang Dike, about one person for each 0.9 meter of dike.[30] As a "must-be-protected" dike, under no circumstances would the government withhold resources and allow the Great Jingjiang Dike to rupture.

2 Materials and Equipment

Other than laborers, high-water control also requires a huge quantity of materials and equipment. In the Qing, these were mostly provided by the local people, whereas in the Republic the state started to provide some materials, and in the PRC it increasingly supplied materials and equipment.

As for the specific materials need for high-water control, records from the Qing dynasty show that, for example, people responsible for the maintenance of the Wancheng Dike were required to provide "earth oxen" (*tuniu*, earth piled up for high-water control) before the high-water season. A Qing source indicates that each earth oxen was approximately two chi wide at the top, one zhang (ten chi, or about 10.2 ft) wide at the bottom, two zhang long, and four

24 D35-1-4a.
25 *Jingjiang dadi zhi* 1989: 219.
26 *Qianjiang shuili zhi* 1997: 144–45.
27 *Songzi xian zhi* 1986: 374.
28 *Qianjiang xian zhi* 1990: 292.
29 *Jianli xian zhi* 2013: 193.
30 Li Shouhe 2001.

chi high,[31] or, in other words, about 13 cubic meters. As for other materials, Qing sources on the Wancheng Dike, for example, mention stone, bricks, wooden stakes (long and short) and boards, grass or straw, firewood, reeds and reed mats, tree branches, straw mats, baskets, sacks, ragged cotton quilts, and broken woks and vats. In addition, the equipment and tools required included stone rammers, shovels, dustpans, wooden buckets, shoulder poles, lamps and lampstands, torches, candles, umbrellas, bamboo hats, palm-bark rain capes, gongs, wooden clappers, and circulating labels.[32] In this case, the sources do not mention how many of each item was needed. But in a few instances, sources did provide numbers: "One sign board, … enough palm-bark rain capes and bamboo hats, two lanterns, two patrol boards, enough torches, two gongs, two hoes and two shovels, ten to twenty pairs of carrying baskets with bamboo shoulder poles, two hammers, two rammers, two iron woks, several cotton-padded jackets and quilts, and ten sacks."[33] The required items and their numbers were slightly different in Tianmen and likely in other places as well.[34] In any case, the local people were responsible for providing all these materials and equipment.

Most of the required materials and equipment remained the same in the Republic, but some new items, such as wire and raincoats, were added to the list. State agencies such as the Bureau of Engineering Work on the Yangzi and the Han Rivers (Jianghan gongchengju) started to purchase some of the items needed for high-water control on major river dikes. If there was a shortage, materials and equipment would be requisitioned.[35] But most items were still collected from or provided by the locals. It is certain that they had to provide all the equipment and tools, but in most cases it is unclear who was responsible for what kind of materials. A few sources, however, do mention who was required to provide tools and equipment, and in what quantities, in various locales. For example, in Honghu in the Qing and the Republic each household had to provide one log, ten wooden stakes, and three bundles of reeds or firewood.[36] In Songzi, the annual repairs and high-water control agency

31 Wang Fengsheng 1832, vol. 2: 45a. But according to another source, the volume of each tuniu was about 4.8 fang (*Jingjiang dadi zhi* 1989: 264), or 17.76 cubic meters.

32 *Jingjiang dadi zhi* 1989: 220; *Jianli difang zhi* 1991: 234. According to an order from Zhongxiang in 1956, each torch was "6 chi long and 1 chi in circumference" (54-1-27) (roughly, 180 cm by 30 cm). Torches were usually used for lighting, but also could be used for signaling an alarm (*Hanjiang difang zhi* 2015: 80).

33 Yu Changlie 1999 [1840]: 127.

34 *Tianmen xian zhi* 1989: 143; *Zaoshi zhen zhi* 2009: 126.

35 *Hubei shuili zhi* 2000: 235–36; LS19-2-2295 [1]; LS1-5-608.

36 *Honghu xian zhi* 1992: 123.

of Yonghe polder in 1945 was also responsible for all materials for high-water control in this polder. In Futai polder in 1946, laborers were only required to bring one spade, a pair of carrying baskets, and a shoulder pole; the polder's annual repairs and high-water control agency provided all other high-water control materials.[37] In Peixing polder in 1947, each laborer had to bring one bamboo hat, one shoulder pole, and a pair of carrying baskets.[38]

In the early PRC, the locals continued to responsible for a large share of these materials and equipment, with some supplementary help from the state. However, over time the state gradually took on more of the burden.[39] The basic principle was that for high-water control of all river dikes, the state would provide the most important materials, with the locals responsible for materials that were readily available locally. For high-water control of polder dikes, the locals were responsible for providing all materials.[40] If the locals were unable to supply the requisite materials, various workarounds were used. For example, the Zhijiang County government in 1954 required people who lived outside the polder area to help out by providing one straw bag per household to those who lived in the polder area.[41] In Jianli in 1964, every household was required to provide wooden stakes, rice straw, wheat straw, and broken bricks in such quantities that the needs of that year's high-water control were met.[42] In Yunmeng County in 1972, each household had to provide one shed made of tree twigs and one straw bag, and each person had to provide 1 jin of rice straw and 0.0025 cubic meters of brick slag.[43] In 1995 in Jiangxia District, each household had to turn in three polypropylene knitted bags and one stake to the village for high-water control of village reservoirs and polder dikes.[44]

In determining the quantities of materials needed, the principle has been to err on the side of caution. Clearly, ending with materials left over might be a problem, but the consequences of having too few materials could be disastrous.[45] A 1954 document warned that one should not underestimate the

---

37 100-10-186: 66; 100-8-42: 54.

38 100-8-108: 53.

39 *Jianli xian zhi* 1994: 175; *Shashi shuili difang zhi* 1994: 101.

40 *Xiaogan diqu shuili zhi* 1996: 57–58.

41 45-1-8. Another document in the Zhijiang City Archives reports that during the 1955 high-water control period, 10,000 jin (catties) of rice straw were made into 1,500 straw bags (45-1-11b), or roughly 6.67 jin of rice straw per bag. It was difficult for an area where cotton was the dominant crop to come up with enough rice straw (45-1-104).

42 1-1-502-009.

43 46-1-13.

44 *Jiangxia qu zhi* 2007: 175.

45 D35-1-41C; *Shashi shuili difang zhi* 1994: 101.

importance of even small items such as light bulbs or lampshades, since when they are needed, they could be crucial.[46] At the same time, more materials such as iron wire, kerosene, gasoline, diesel oil, dynamite, and plastic sheeting have been added to the list in the PRC, with the state providing all of these new items as well as a portion of some regular materials such as straw bags, reed mats, and bamboo. Auxiliary items such as straw, broken woks, ragged cotton-padded quilts, and carrying baskets and poles are still provided by the locals.[47] But now they are not necessarily provided by individual laborers or households; instead, they are furnished by their villages. In Gong'an, for instance, each village is required to prepare 1,000 kilograms of straw, 1,000 polypropylene knitted bags, 100 kilograms of cotton fiber, ten wooden stakes, and ten lanterns.[48] In Jiangxia, the government has been responsible for all high-water control materials for the Yangzi River dikes since 1980, freeing the locals from any obligation to provide materials.[49]

The quantity of materials used in high-water control is staggering. In 1998 in Honghu, the materials consumed in that year's high-water control included "11.35 million polypropylene knitted bags, 135,000 straw bags, 120,000 meters of plastic sheeting, 840,000 wooden stakes, 500 tons of reeds, 40 tons of rice straw, 430,000 fang of gravel, 200 tons of steel, 50 tons of iron wire, and 116,200 life jackets."[50] In all of Hubei, "531,700 moso bamboo poles, 3.5 million tons of gravel, and 175 million emergency bags of all kinds were used."[51] Many of these objects, such as polypropylene knitted bags, plastic sheeting, and steel, were not available in the Qing and the Republic. In Shishou, at the peak of the 1998 high-water control (August 7), all major and branch dikes—259 kilometers in total—were lit by electricity,[52] a great improvement that could not have been imagined in the Qing and the Republic, or even in the early PRC.

From the Qing to the PRC, the state required that the materials for high-water control be transported to and stored in assigned places along dikes. Traditionally these materials were transported by human porters or pack animals. The first-time motor vehicles were used in high-water control in Shashi was after the Anti-Japanese War.[53] In 1954, the state ordered that in the high-water control season, all means of transportation be used to transport

46 50-1-15b.
47 *Jingjiang dadi zhi* 1989: 220; *Maliang zhen zhi* 2013: 124.
48 *Gong'an xian zhi* 2010: 105.
49 *Jiangxia qu zhi* 2007: 175.
50 *Honghu shi zhi* 2014: 203.
51 *Hubei shuili zhi* 2000: 293.
52 *Shishou shi zhi* 2014: 12.
53 *Shashi shuili difang zhi* 1994: 102.

emergency materials.[54] In August of that year, the State Council even sent airplanes to airdrop 26,000 sacks at an important segment of the Distance Dike in Zhongxiang,[55] a method that had never been used before. The use of motorized transport has become much more common in recent decades.

Gathering and storing sufficient materials before the high-water season has been essential to ensuring the safety of dikes. In 1931, in dealing with a dike emergency, Xu Guorui recalled that three times it was Heaven's intercession alone that prevented the Jing River dike from collapsing. In one case, he needed a huge volume of beans. Normally it would take ten or more hours to transport them from Shashi, but miraculously a team of thirty-six packhorses—loaded with wheat and beans—suddenly showed up. All the goods they carried were purchased on the spot and used to fill holes in the dike. In another case, Xu needed some lime, which would take at least ten hours if shipped from Shashi, but a ship passing by happened to be loaded with lime. In a third case, Xu needed some stone. It would take half a month to get what he needed if he ordered it from a stone yard, but a boat carrying stone suddenly emerged. Xu regarded all these as miracles.[56] Little wonder that he became an even more devout Buddhist after this high-water season (see Chapter 5). A shortage of high-water control materials indeed contributed to the collapse of the Dongjing River dike in a spot in Qianjiang in 1950 and the rupture of the Yangzi River dike at Tianjiakou in Honghu in 1969.[57]

During the collective era, when most materials were controlled by the state, local governments usually issued official notices specifying which counties were to receive what high-water control materials, such as straw bags and reed mats. For electrical equipment and fuel, the notices even spelled out where materials were to be obtained: for example, light bulbs and nails from the hardware store, searchlights from the supply and marketing cooperative (*gongxiaoshe*), batteries from the department store, and gasoline and kerosene from the oil company.[58] In some places local supply and marketing cooperatives were required to stock a sufficient quantity of materials essential for high-water control such as sacks, straw bags, and kerosene—this was considered an "important political task."[59] In the CCP's lexicon, a "political task" is one that is of the very highest priority and must be carried out at all costs.

54 SZ68-2-102.
55 *Tianmen xian zhi* 1989: 146.
56 Hubei sheng shuiliting and Hubei sheng dang'anju 1999: 401–02.
57 SZ113-2-11; *Jingjiang difang zhi* 2012: 617.
58 42-1-183.
59 *Pengxing xiang zhi* 2011: 125.

As we have seen, most of the materials and equipment used for high-water control remained unchanged from the Qing to the PRC, a remarkable continuity. But in the PRC, some new industrial products such as polypropylene knitted bags, plastic sheeting, steel, gasoline, and dynamite were introduced, enhancing the local people's ability to ensure the integrity of dikes. At the same time, we also find that in the PRC, most materials used in high-water control have been provided by the state, with the local people only needing to supply some auxiliary materials that are mostly locally available, which undoubtedly has largely reduced their financial burden.

3 Methods and Importance of High-Water Control

Ensuring that sufficient materials and equipment are gathered is one of the most important tasks before the high-water season. In the Qing, once the high-water season on major river dikes started, laborers were divided into groups, with each group responsible for taking care of a dike section. These groups were required to carefully patrol and inspect the section assigned to them day and night, looking for any potential danger and immediately attending to it.[60] This has not changed much from the Qing to the PRC. However, aside from some basic principles or even detailed rules of high-water control,[61] we do not have much information on precisely how high-water control was carried out in the Qing, particularly regarding labor arrangements.

Information of this sort became more available in the Republic. In 1946, able-bodied residents who lived near major river dikes were selected and organized into groups to patrol the dikes in rotation or together, depending on the situation with the dikes. They were required to bring their own tools and rain gear.[62] High-water control on polder dikes in some cases simply followed the high-water control regulations for the major river dikes,[63] but in other instances there were slight differences. In 1947, in Xida polder in Songzi, for example, there were two high-water control groups, each with thirty people. Patrols, rotated between the two groups, were carried out both day and night. For the night patrol, twenty people were assigned to walk along the foot of the

60 *Jingjiang dadi zhi* 1989: 222.

61 Wang Fengsheng 1832, vol. 2: 49a–53b; Yu Changlie 1999 [1840]: 126–28; Ni Wenwei 2002 [1885]: 151–60.

62 *Hubei shuili zhi* 2000: 235.

63 100-10-393: 28.

dike, and if they discovered a dangerous problem, they were to beat a gong, calling more laborers to the dike.[64]

In the PRC, information on high-water control is even richer and more detailed. According to a high-water control manual, laborers who will be patrolling and inspecting dikes must be responsible, capable, and knowledgeable (and be acquainted with the dike), and must have participated in high-water control before. They should be middle-aged men from families with two or more laborers, they should be arranged into groups, and the groups should be fixed (i.e., there should be no change in their composition) within the year.[65] While patrolling and inspecting dikes, laborers have to bring equipment with them, which, in Xiantao (known as Mianyang until 1986), included a shovel, rain gear, a flashlight or lantern, a gong or whistle, some red and white flags (to mark endangered spots), and a notebook (to record dangers), etc.[66] In the 1990s, in Tuoshi Town in Tianmen, forty-eight militia members from six villages formed six teams for high-water control and emergency repairs. Each team member had "one camouflage suit, one pair of Liberation shoes, one shovel, one flashlight, one pair of shoulder baskets, and one polypropylene knitted bag."[67]

In patrolling dikes, laborers are required to fully use their eyes, by carefully inspecting the dike from the top to the bottom and scrutinizing the nearby water surface for any sign of potential danger such as cracks, water leakage, pop-up sand (that is, sand on the water's surface), and small swirls; their ears, by listening to the sound of the water; their feet, by walking barefoot and feeling for potential dangers, such as soft soil (a sign of potential collapse) and cool water (a sign of river water leakage); and their hands, by checking potential problem spots by feeling them.[68] In Hanyang in 1956, regulations specified that of the two people who patrolled the side of dike facing the water, at least one should be able to swim.[69] In the 1998 high-water control, the image of barefooted people with sticks and flashlights in hand patrolling dikes in groups was imprinted in the mind of millions of Chinese via national media, particularly television and newspapers.

Since high-water control requires many laborers and the high-water control season usually lasts four to six months, it is understandable that laborers—who

---

64 100-30-15: 4.

65 Zhou Hongbo and Gong Tianzhao 1999: 176.

66 *Xiantao shuili zhi* 2008: 109.

67 *Tuoshi zhen zhi* 2008: 500. Liberation shoes are green rubber-soled canvas footwear worn by soldiers.

68 Zhou Hongbo and Gong Tianzhao 1999: 178; *Jingjiang dadi zhi* 1989: 223.

69 30-1-14.

are mostly peasants and have to attend to work on the farm—cannot stay at a dike for long. Thus, in many cases, some will return home and be replaced by a new group of laborers. To help laborers become familiar with a dike and the potential threats to it, it is ideal to replace or relieve them as infrequently as possible, particularly when floodwater is near its peak. It is possible to arrange for laborers to be relieved between two peaks, but group leaders should not be relieved.[70] Each group is responsible for a section of a dike.

Patrolling and inspecting dikes is a tough job, especially since laborers are required to walk along dikes and inspect them every hour or two, day and night, rain or shine. In fact, they are required to be especially cautious during mealtimes, break time, at dusk and dawn, when it is windy or rainy, and when the water level rises or drops drastically, since it is during these times that people are liable to relax their vigilance. Thus, the saying "Dikes won't rupture in the dangerous sections," since people are well acquainted with these sections and hence are likely to patrol and inspect them carefully, thus ensuring their integrity.[71]

Once a threat has been spotted, laborers must immediately report it to their superiors, and it must be immediately rectified. One author recalled that in summer 1962, as he and his colleagues were patrolling and inspecting a dike section late at night, they spotted some small waves in a nearby lotus pond. When they checked, they found some surging muddy water, which they immediately reported. Technicians thought it was a leakage of river water and it was very dangerous. No time was wasted in acting and fixing the problem.[72] This is a good example of the standard procedures for emergency repairs.

At moments like that, materials stored ahead of time can be invaluable. According to a high-water control regulation in the Qing, if water was leaking from a square or round-shaped hole, one should first use a wok (face down) to block the mouth, then put dirt on the wok, thus blocking the leak. If water was leaking from an oblique or rectangular-shaped hole and one wok was not big enough to cover it, then one should first stuff cotton quilts and the like into the outlet, then put dirt over that to block the leak. If all methods failed and the dike collapsed, people on the site should not panic at that moment, but instead first use wood stakes and firewood to build a fence on the river side to reduce the waterflow, and then shovel dirt into the opening, and at the same time, have some soldiers and/or laborers stand in the middle of the opening, hand in hand, with their eyes closed, and jump up and down while other people

---

70 *Jingjiang dadi zhi* 1989: 224.

71 *Jianli difang zhi* 1991: 248–49; Zhou Hongbo and Gong Tianzhao 1999: 178.

72 Xiang Xin 1998.

throw dirt over their heads. This would gradually compact the dirt and block the opening.[73]

These sorts of rules should not be thought of as just something that could be found only in books. In fact, they were frequently applied. In 1837, Lin Zexu witnessed in person how a leak on the Wancheng Dike was blocked by covering it with a wok.[74] In 1840, another leak in this dike was immediately blocked by stuffing cotton wadding and beans into it.[75] In 1946, a threat to the Great Jingjiang Dike was overcome by the use of soybeans and other grains and dirt-sacks.[76] In 1981, a brigade accountant found some leaking holes near a river dike, and he immediately stuffed in a cotton quilt. But the quilt was quickly sucked in. He shouted for help, people rushed in and blocked the leak with fifteen quilts and twenty straw bags.[77] In many emergencies, people resorted to blocking floodwater with their bodies. In 1954, a "human dike"—that is, people standing together as a dike—was used in Ouchi in Shishou to protect its dikes in that year's huge flood. The same happened in Wuhan where, whenever danger emerged, soldiers from the Chinese army (the People's Liberation Army, or PLA) would jump into water to form a "human wall." And on the night of July 25, 1983, five cadres and several patrol laborers jumped into the water and used cotton wadding to block a leak. In 1998, when one segment of an auxiliary dike (*zidi*) in Tiaoguan, Shishou, collapsed, thirty or so PLA officers and soldiers jumped into the water to once again form a "human wall" so that other people could quickly rebuild the collapsed section behind them.[78] All these examples mark a clear continuation of the use of the same materials and methods in high-water control from the Qing to the PRC.

The use of broad beans and miscellaneous grains in an emergency repair to stop a leakage in Qianjiang in 1951, however, was criticized as wasteful and something that should have been avoided.[79] This case illustrates a dilemma in high-water control. Since the purpose of emergency repairs is to avert a disaster, at that moment any available materials could and should be used. But in the early PRC, the state was still financially weak and lacked resources, and thus it encouraged laborers to be frugal in the use of high-water-control materials. In 1963, the central government even issued a notice that in high-water

---

73 Wang Fengsheng 1832, vol. 2: 52b–53b.
74 Shuili dianlibu 1991: 798.
75 Ni Wenwei 2002 [1885]: 247.
76 *Jiangling xian shuili zhi* 1984: 58.
77 45-1-310a.
78 *Ouchi zhen zhi* 1985: 62; XX000065-WS01-0156-0001; *Xiaogan shi zhi* 2013: 218–19; *'98 Jingzhou kanghong zhi* 1999: 9.
79 D35-1-4a.

control, people everywhere should use local materials, such as firewood and straw bags, insofar as possible, but should use jute sacks—particularly fine jute sacks—as little as possible or not at all.[80] Using local materials as much as possible in high-water control was in fact a basic principle in the early PRC in this region. That, however, soon began to change. In 1987 in Wuhan, many high-water control materials, including straw bags, for major river dikes could be purchased with state funds.[81]

By the same token, unused materials generally should be stored for high-water control in the following year, but it is not unusual for them to be used for different purposes. In Yingcheng in 1956, for example, 14,661 straw bags were left over after that year's high-water control, but later it was found that 3,840 of them were missing. It turns out they had been used in building houses, pigsties, and privies, and were also put to use as grain bags, rain capes, sleeping mats, and kindling.[82] All this was another sign of material shortages in the early PRC.

In most cases, regular materials stored ahead of need should be sufficient. But anything can happen in an emergency, and the people responsible for high-water control might use whatever is available—grain, clothing, cars, and even boats—to block a leak. Therefore, the leaders responsible for on-site emergency repairs act the same as military commanders in a battle: they have authority to requisition any useful objects.

The discussion above also illustrates how important the responsibility of laborers and their leaders was, since the slightest negligence could cause a disaster. Many a time Qing officials were punished for their negligence, and there were numerous similar examples in the Republic. In 1926, for instance, after a segment of the Great Jingjiang Dike in Jianli ruptured, it was later found that when this occurred the dike manager was gambling in town, and consequently failed to take any action to promptly plug a leak.[83] In another example, after many dikes ruptured in the 1931 flood, responsible officials were punished for their negligent handling of annual repairs and high-water control.[84] And in still another example, in 1935, when a small breach in the Great Jingjiang Dike was spotted, the people on the site could do nothing since the director, Wu Jingtang, had absconded. Moreover, no materials were available. After a while the breach became bigger and wider, and the people at the site had no

80 XX000065-WS01-1391-0019.

81 27-2-2.

82 Yingcheng 49-1-51b.

83 LS1-3-566.

84 LS1-4-452.

choice but to flee. As a result, the dike ruptured. For his dereliction of duty, Wu was punished.[85]

The Great Jingjiang Dike has certainly not been the only victim of carelessness. For example, in 1935, when the dike of a polder mostly located in Tianmen broke, the people who lived nearby were not eager to prevent inundation because they wanted to benefit from the silt carried by the floodwater. When warning gongs were beaten, only around ten people answered the call. A security team then forced some people to come to the dike, but they fled, giving the excuse that they did not bring any tools. The dike eventually collapsed.[86]

In the early PRC, on the other hand, there were numerous examples of cadres who worked diligently on high-water control. In 1952 in Zhijiang, for example, township leaders went without sleep when patrolling the dikes on windy and rainy nights, which, it was said, greatly inspired other cadres and laborers.[87] At the same time, there were, unsurprisingly, also incompetent cadres. In 1954, a dike in Honghu was breached mainly because of the carelessness of those who were responsible for local high-water control. This dispirited other laborers, and they wanted to go home.[88] In another case, after the dike of Mengxi polder in Gong'an broke in 1998, it was later discovered that, first, the people who were responsible for a section of the dike did not carefully patrol and inspect it. In fact, two endangered places were spotted, but by schoolchildren. Second, after the first danger was discovered, it took five weeks for the dike to rupture, but when it did, those who were responsible for that dike section were off duty and failed to prepare enough materials for an emergency repair. And third, when the situation deteriorated, no laborers and materials were on the site. It is no wonder that eleven people were punished after this incident.[89]

Most cadres in the early PRC worked diligently on high-water control not because it was regarded as necessary for protecting the environment, but rather because it was essential to the national economy. For instance, in 1950, when the Jianli County government issued an order to repair and rebuild the dikes, it made clear that "dike building and water conservancy are basic demands of our country's peasants."[90] Similarly, the people of Zhongxiang had also suffered from water calamities and wanted to overcome them, thus the first priority of the county government in 1951 was to build dikes and irrigation

85 Li Zinan 2001.

86 *Jianghe xiufang jiyao* 1971 [1936]: 256.

87 45-1-2.

88 SZ1-02-0216-007.

89 *'98 Jingzhou kanghong zhi* 1999: 110, 252.

90 1-1-010-004.

channels to develop agriculture.[91] Here the county government regarded high-water control as a "great political task" and one that had to be well organized and prepared.[92] In other words, improving high-water control was not just the government's policy, but also something demanded by the people.

In an emergency high-water control meeting in 1958, the Hubei party secretary emphasized that if a dike collapsed, not only would the economic loss be painful, but "more importantly, the political fallout will be bad."[93] According to Zhao Fulin, the party secretary of Jingzhou in the early 1980s, in the 1960s, because of serious water calamities, the first priority of governments at all levels in Jingzhou was water conservancy construction and high-water control.[94]

To emphasize the importance of high-water control, in 1955 a Hanchuan County leader remarked that the development of industry required grain and other raw materials. Bumper harvests thus were needed to support the development of industry, while bumper harvests in turn required secure dikes.[95] County leaders of Shishou concurred. They contended that the integrity of Shishou's dikes greatly contributed to the bumper harvest in 1955, which in turn "strongly supported socialist construction nationwide."[96] That is to say, these local leaders believed that successful high-water control and the security of dikes were important not only to the local economy, but also to the development of industry and the national economy. This logic instantly raised dike safety in their counties to a level of national importance.

4 Surveying and Reinforcing Dikes before the Coming of the Annual High Water

High-water control requires surveying dikes to uncover problems and then resolving them before the coming of the annual high water. Many improper activities can damage dikes: planting crops or building houses, pigsties, or cattle pens on dikes or in dike exclusion zones, using soil from dikes to make bricks, opening dikes for irrigation, removing grass from dikes, and dumping garbage and debris on dikes. There is only piecemeal information about these activities and the damage they caused in the Qing and the Republic, but such

91 54-1-3a.
92 54-1-5b.
93 40-1-156.
94 Jingzhou shi shuiliju et al. 2008: Zhao's preface.
95 50-1-28a.
96 39-1-19-3.

information became widely available in the PRC, particularly in its early years. This section focuses on that period.

In 1949 the Hubei government issued a temporary regulation stating that people should treat the major dikes along the Yangzi and Han Rivers as if they were as precious as their own lives and do nothing to harm them.[97] In the following year, Hubei issued additional regulations on the maintenance of dikes along the Yangzi River and its important tributaries, which included a ban on cultivating crops, planting trees, and animal grazing on dikes.[98] In the 1950s, it was said that people along the Dongjing River regarded dikes as their "granaries" and "lives," thus protecting the dikes was equivalent to protecting their granaries and lives. Hence the importance of dike maintenance.[99] Key areas protected by dikes were also important locations where the state purchased grain.[100] In short, ensuring the safety of these dikes was indispensable for the survival of the new state as well as for the sustenance of the local people.

Thus, all the above-mentioned irresponsible activities were banned in the early PRC. Yet they nonetheless continued, particularly the cultivation of dikes and their exclusion zones.[101] For instance, around 1952, some cadres in Jiangling were reluctant to transform some fertile land into dike exclusion zones, while others thought that giving up farmland would interfere with the development of their jurisdictions. These views were criticized as one-sided and evidence of "selfish departmentalist" thinking.[102] Nonetheless, such thinking was widespread. To cite a few examples: according to a report, before 1956 the people of Mianyang left no land for dike exclusion zones at all.[103] Another document reports that in 1957 some people saw no problem with farming dikes, such as in Yingcheng, where there was a "tradition" of growing crops on dikes.[104] In Qianjiang in 1961, the locals made no secret of their disgust with the notion of leaving land unfarmed and some were unhappy that so much of their land was

97 Yingcheng 49-1-4b.

98 SZ34-02-0112-028.

99 SZ113-2-144-009.

100 SZ34-2-406.

101 In 1950, some PLA troops dug out grass from dikes and reclaimed dikes to grow crops. The reclamation of wasteland for farming was a commendable PLA tradition, but cultivating dikes seriously damaged them. Thus, an order was issued to ban this activity, otherwise the responsible people would be arrested and punished (22-1-006-008). This episode suggests that the troops, who likely did not understand the importance of dikes in the region, failed to realize that they were damaging the dikes.

102 48-1-13.

103 SZ113-2-311-006.

104 Yingcheng 49-1-51a.

designated as a dike exclusion zone. They farmed much of this land anyway.[105] People in Shishou also thought it was wasteful to turn large amounts of land into river dike exclusion zones. From the perspective of a province or a prefecture, the reduction in cultivated land may not have seemed significant, but some villages along the dikes in Shishou could lose 3 percent of their land. The county government thus had to take this into account and consequently requested that the width of the dike exclusion zone be narrowed.[106] The same happened in Songzi, where the local government in 1961 narrowed the dike exclusion zone to comply with local traditions.[107]

To solve the serious problem of food shortages during the Great Leap Forward (GLF, 1958–1961), also known as "Three Years of Natural Disasters" or "Three Years of Difficulties" in the official CCP discourse, the state encouraged people to "make use of every bit of space" (*jianfeng chazhen*) to grow crops. Responding to this call, in 1959 residents of Gong'an started "to root up grass from dikes in order to plant pumpkins, rape, broad beans, wheat, and vegetables." In September 1960 the county began to clean up its dikes, and "removed 47 houses, 3,363 fences, 2,327 privies, 551 cesspits, 119 pigsties and cattle pens, and 32,000 pumpkin patches." It was found by spring 1961 that 430 kilometers of dike had been damaged by cultivating crops.[108] In Jianli in 1960, some cadres openly asked the local villagers to plow all dike exclusion zones and grow vegetables, and a leader of a fish farm, who cultivated 100 meters of a dike, even threatened to tie up those who dared to try to stop him.[109]

A report on dike management and maintenance highlighted three major reasons behind the penchant for tilling dikes in Shishou around 1960. First, poor leadership: local cadres either did not pay enough attention to the regulations on dike maintenance or neglected dike management. Second, carelessness: local cadres did not think there was a problem and even encouraged people to farm dikes. Third, ignorance: the government did not spend enough time in disseminating information on the importance of dikes, and peasants did not realize farming on dikes was wrong.[110] According to another report, at the end of 1960, about 6,420 mu of dike surface in Gong'an had been damaged by the growth of broad beans and rape. The report complained that the locals did not even dare to use a stake to tie their cattle when grazing them on dikes

105 40-1-234.
106 1-1-543-46.
107 33-1-63.
108 *Gong'an xian zhi* 1990: 106.
109 22-1-329-005.
110 39-1-38-3.

in 1954, but they turned to farming dikes around 1958, particularly in 1960. The report attributed this to the carelessness of local cadres, some of whom openly endorsed the cultivation of dikes.[111] Both reports, however, ignored the serious shortage of food grain at that time. Local residents simply felt that if they did not grow crops anywhere land was available, they might go hungry.

Damage to dikes was also widespread in Wuhan. In 1961, some villagers not only cultivated dikes, but even erected fences to enclose them as if they were their own private property. When dike keepers came to stop them, these villagers pretended to acquiesce, but continued to cultivate the dikes after the keepers left. Even commune leaders did not want to intervene.[112] With this kind of resistance, the acreage under crops on the dikes of Wuhan reached an astonishing 3,000 mu in 1962.[113]

In urban areas, some people also followed the principle of *jianfeng chazhen* and grew vegetables on dikes, such as in Wuhan in 1962. They were criticized for misunderstanding the principle and were ordered to immediately pull up the vegetables.[114] But it was not easy to enforce the ban. This was illustrated by a certain Lin, a retired worker, who since 1955 had been constantly reclaiming and cultivating the dike in Qiaokou District, Wuhan. By 1962, he worked as much as 1 mu. The responsible dike keepers had repeatedly tried to persuade him to stop but always failed. His excuse was his life was hard. He declared he would defend his crops to the death. An investigation found that his life in fact was not that hard. After much public criticism and private cajoling, he agreed in 1963 to eradicate the crops he planted in the dike exclusion zone.[115]

The government admitted in March 1962 that grain was indeed in short supply in Wuchang, and thus crops should be planted on all available land. But it also recognized that the cultivation of the ramps and tops of dikes was wrong and dangerous, and thus should be prohibited. Although the cultivation of the dike exclusion zone was understandable, following the best principles of dike management, the government ordered that after the harvest of the crops in the exclusion zone, no more new cultivation would be permitted.[116]

This proved to be easier said than done. In 1963, it was found that on the 94.5 kilometers of dikes in Dongxihu District, 1,200 mu of vegetables, wheat, beans, and sorghum were still grown on dikes and their exclusion zones. In addition,

111 SZ113-2-195B.

112 XX000065-WS01-1138-0034.

113 XX000065-WS01-1273-0010.

114 XX000065-WS01-1281-0020.

115 XX000065-WS01-1281-0062; XX000065-WS01-1488-0005.

116 XX000065-WS01-1281-0030; XX000065-WS01-1281-0036.

many houses, manure pits, pigsties, and cattle pens also remained on dikes or their exclusion zones.[117] This led to frequent conflicts between dike keepers and local residents. In a village in Wuhan in 1963, some villagers reclaimed two plots in a dike exclusion zone to grow paddy rice, and when dike keepers talked to the commune's secretary, he responded: "[We are] now making great efforts to develop agriculture and grain. What's the big deal about transplanting some rice seedlings and harvesting some rice? Talk to me after the rice has ripened."[118] This is a perfect example of how local cadres undercut dike safety using the excuse that agricultural development required expanding the acreage under plow, forgetting that if their dikes failed, their crops could be washed away. Some peasants in 1964 even criticized dike keepers for "undermining production" by rooting up crops planted on dikes.[119] Although it was absolutely forbidden to root out grass from dikes, in spring 1965 in Wuhan, many villagers nonetheless did so. When dike keepers went to the local brigade secretary for help, he complained that these dike keepers were paid wages, but did nothing but cause trouble for local residents. The peasants, on the other hand, had to earn workpoints to survive and had no "idle time to deal with unimportant matters."[120]

The Han River dikes suffered from the same problems. An investigation in April 1961 found that 43 percent of the Han River dikes in Hanchuan had been damaged by improper activities. The dikes were covered with 1,835 vegetable gardens, 3,300-plus pumpkin patches, 24 pigsties, 18 cattle pens, 357 privies, and 240 mu of soybeans and wheat, etc. The situation with branch dikes and polder dikes was even worse.[121] One reason for this was a lack of vigilance. By the end of the 1950s, with the beginning of the very large Danjiangkou Reservoir project (at the time, one of the largest in Asia) and the completion of the Dujiatai flood diversion zone, people living along the lower reaches of the Han River no longer worried about dike ruptures. Consequently, they farmed the dikes, dug up grass, and built houses, privies, cattle pens, pigsties, and chicken and duck sheds, and even tombs, on the dikes. In 1961, they were ordered to restore the dikes.[122]

117 XX000065-WS01-1273-0004.
118 XX000065-WS01-1386-0011.
119 XX000065-WS01-1439-0007.
120 XX000065-WS01-1954-0016.
121 50-1-87.
122 SZ113-2-195-005.

After the GLF, the cultivation of dikes and dike exclusion zones gradually declined but never disappeared, and many other improper activities continued as well. In 1964, it was found that the cultivation of dikes, the building of houses, pigsties, and privies, and especially the grazing of livestock on dikes and their exclusion zones were still common in Wuhan.[123] In 1981, residents of the state livestock farm of Mianyang grew rape and wheat on an 800-meter segment of the dike, potentially threatening more than seventy thousand people and two hundred thousand mu of land of Hannan Farm. The latter called on the former to stop misusing the dike, but the former argued that the dike belonged to them, and thus they had the right to farm it. They also contended that if they had to destroy the crops on the dike, they were entitled to compensation.[124] This was unreasonable and was purely driven by selfishness, since in 1981, unlike in the GLF, there was no great shortage of grain. Another problem, particularly in urban areas, was the use of river beaches to store trash and building materials, which blocked waterways and increased the threat to nearby dikes.[125]

But all these and various other threatening activities gradually decreased following the 1979 rural reform, since people no longer felt the need to misuse the dikes for tiny gains. In the early PRC, when people dug up grass from dikes, they offered various justifications, such as they lacked firewood, or the grass was a readily available source of much-needed fertilizer.[126] But both explanations have lost ground in the reform era, with rural residents increasingly relying on natural gas and electricity for cooking and chemical fertilizers for farming. In recent decades, most villagers have even lost interest in farming regular land, let alone dikes and their exclusion zones.[127]

Moreover, since 2002, the use of rural compulsory labor has been reduced (and abolished since 2005), and in line with this, the surveying and reinforcement of the Han River dikes before the coming of the annual high water has gradually ended.[128] This represents another historic change in dike management.

---

123 XX000065-WS01-1489-0008.

124 XX000084-WS03-0083-0030.

125 27-2-8.

126 50-1-50a; Wuchang 49-2-21.

127 But dikes and their exclusion zones are still cultivated. In particular vegetables are grown along some minor dikes. As one drives over Jianghan's dikes, one will see sporadic small plots of vegetables along the dikes.

128 *Hanjiang difang zhi* 2015: 240.

5 Laborers and Funding

From the Qing to the PRC, the basic principle for recruiting laborers for high-water control remained unchanged: those who benefited from a nearby dike should take the responsibility for high-water control on that dike. But there were some variations on this theme, particularly in the PRC.

Since land was privately owned in the Qing, those who had land protected by a dike were required to provide laborers for high-water control in proportion to the area of land they owned. These laborers, who were to patrol and inspect the dike during the high-water control period, might have come from nearby villages.[129] The same principle—those who had land protected by a dike were required to provide laborers in proportion to the land they owned—was followed in the Republic as well, as recorded in Zhongxiang and Jiangxia.[130] During the 1947 high-water control period, each household in Peixing polder in Songzi had to provide at least one laborer, although, an archival source notes, "when necessary, everyone must be mobilized."[131] Nonetheless, in some cases, it was very difficult to mobilize residents for high-water control.[132]

Although the principle that those who benefit from a dike should be responsible for its high-water control has mostly survived in the PRC, the situation has become more complicated since the PRC has a political and socioeconomic system that is very different from that of the previous eras, particularly after the state abolished private landownership. First, who were the laborers? They were mostly working-age villagers. But the scope of "working-age" has varied slightly from place to place and time to time. In the early 1950s, for instance, in Gong'an it was men from age 18 to 55 and women from 20 to 45; in Songzi, men from 17 to 50 and women from 18 to 45; in Qianjiang, Tianmen and Xiantao, men from 18 to 50 and women from 18 to 45.[133] In Zhongxiang, the working-age range changed almost annually: in 1951, 1952, and 1955, it was 18–50, 18–55, and 18–45 for men, and 20–40, 20–45, and 18–40 for women, respectively.[134] In Hanyang in 1951, all men from 17 to 55 years old were required to participate in high-water control, while most women aged 18 to 35 voluntarily participated.[135] Under some extreme circumstances, the age range

129 *Zaixu xingshui jinjian* 2004: 11.

130 *Zhongxiang shuili zhi* 1998: 185; Jin Feng 1996: 150.

131 100-8-108: 52.

132 100-8-99: 41.

133 D35-1-12; D35-1-37; 40-1-113; *Tianmen shuili zhi* 1999: 125; *Xiantao shuili zhi* 2008: 457.

134 54-1-3a; 54-1-5a; 54-1-15.

135 30-1-5. Judged by these cases, it was quite normal for my production team to send me, a seventeen-year-old high school graduate, to a dike to participate in high-water control

could be very different. During the peak of high-water control in 1998, when a record-breaking high-water raged across the region, a township in Shishou required almost all its residents—men from 16 to 60 and women from 16 to 55—to go to the dikes for high-water control.[136] In general, high-water control laborers were expected to be healthy and robust. Thus, in Qianjiang in 1952, the old, weak, disabled, and pregnant women on the site of high-water control were asked to go home. Alternatively, only with the express permission of their leaders, were such people allowed to stay on the dikes to do some light jobs.[137]

Women usually did not participate in high-water control in the Qing and the Republic. In 1945 a polder in Songzi adopted a regulation that "the elderly, the weak, women, and children" could not serve as high-water control laborers, since such laborers bore a huge responsibility.[138] But in the PRC, women gradually began to engage in activities beyond the family compound, including high-water control. In Zhongxiang's high-water control of 1951, one woman jumped into water to block a leak, and another woman patrolled the dikes in rainy nights.[139] In Songzi in 1952, women even organized teams to participate in annual repairs and high-water control, as they "wanted to enhance their status at working"; in the First Ward, 120 women "voluntarily joined the emergency team."[140] Pregnant women, however, were absolutely prohibited from engaging in high-water control.[141] Since the high-water control season was also the busy farming season, women in Hanchuan in 1952 were mobilized to harvest the autumn crops, leaving the men to attend to high-water control.[142] The same was true in a township in Qianjiang in 1955, where the women not only farmed the land but also transported materials to the high-water control site. Thus, a summary report on high-water control clearly stated that women's high-water control achievements "cannot be obliterated."[143]

In the 1954 flood, some women who joined in high-water control even became role models.[144] In 1958 in Hanchuan, fifteen women organized an

---

in the early 1980s. In fact, in some Jianghan counties, a seventeen-year-old young man was considered a "full laborer." In the collective era, men aged 16 to 55 and women 16 to 50 were categorized as "full laborers" in Zhongxiang and Qianjiang (*Zhongxiang xian zhi* 1990: 288; *Qianjiang xian zhi* 1990: 224).

136 *Shishou shi zhi* 2014: 177.

137 40-1-35d.

138 100-8-49: 33.

139 54-1-3b.

140 33-1-8; 33-1-10.

141 40-1-113.

142 SZ113-2-35B.

143 40-1-93.

144 Alley 1956: 62–65.

emergency repair team named after Mu Guiying, a legendary female general in the Song dynasty (960–1279), to participate in that year's high-water control. It was reported that they worked as hard as or even harder than their male peers.[145] The prominent role of women was repeated in Zhongkou Farm in Qianjiang in 1974, where eight educated young women joined in "the Iron Girl Fighting Team" to engage in high-water control.[146] But even in the PRC, asking all working-age women to participate in high-water control has been rare. When women have been called on, that means the situation must have been critical and an infusion of laborers was desperately needed. That was exactly the case in high-water control in Jiangling in 1998: when more laborers were needed, women were also sent to participate in high-water control. They could be found everywhere—patrolling dikes, making emergency repairs, and providing other crucial services—on the high-water control site. In such instances, they truly "held up half the sky."[147]

Second, where did the men and women who fought to protect the dikes come from? In theory, as we have noted, they came from areas that were protected by the dikes in question. In the PRC era, however, more and more people who were not peasants, such as government employees and urbanites, also benefited from the protection afforded by dikes. Thus, in 1954, grassroots cadres in Wuhan were assigned the "political task" of "going to the dike" to participate in high-water control. These cadres were usually role models for others.[148] Just as with Yu the Great, the idolization of such "models" was commonplace. In a case in Zhijiang, even after his son had died, a militia team leader stayed on at a dike to lead an emergency repair and declared that high-water control was more important than his personal bereavement. Another person paid no attention to the death of his mother but rushed to deal with an emergency.[149] In Wuchang, a township head stayed on a dike for high-water control without going home even after his mother passed away, his house collapsed, and his son died.[150] In Qianjiang in 1957, everyone, even full-time fishermen, woodcutters, duck raisers, and porters and construction workers, were obliged to work on high-water control, but they were exempt from providing high-water control materials.[151] In Jianli before 1966, government employees and urbanites also had to share in high-water control work, although that task was gradually

145 50-1-65.
146 40-1-413.
147 Liu Qiansheng n.d.: 117–21.
148 Tu Mingxin 2014.
149 45-1-8.
150 XX000129-WS01-0100-0001.
151 40-1-148a.

shifted to rural residents alone.[152] In Qianjiang from the 1960s to the 1980s, cadres, teachers and students, and employees of state-owned enterprises were normally not responsible for high-water control, but in an emergency, they too had to answer the call.[153] If an emergency was too grave to be handled by the local government and the people alone, upon request, troops would be sent to help (see Chapter 8).

Third, how were laborers assigned or apportioned? There is no short answer since the recruitment and allocation of workers has undergone intricate twists and turns over the years. In Tianmen since 1953, laborers were apportioned based on output (one laborer had to be provided for every 800 jin of unhusked rice) and the size of the labor force (one laborer for each working-age person). Beginning in 1963, there were two categories of apportionment: for people who lived inside dikes but had land outside dikes (that is, facing water), laborers were apportioned based on the size of the workforce; for those who lived outside dikes but had land inside dikes, laborers were apportioned based on the area of their land. After 1985 there was only one category: laborers were apportioned according to the size of the workforce.[154] In Jianli, since the mid-1960s, in areas that benefited from a dike, laborers responsible for the dike were apportioned according to the whole population (30 percent) and the total acreage of land (70 percent); after 1981, laborers were assigned to individual households but counted by villages.[155] In Jingzhou, in 1955 laborers were apportioned based on output (every 500 jin of unhusked rice was counted as one unit), workforce (each working-age person was one unit), and land (with the exception of beachland, wasteland reclaimed within the last five years, and abandoned land reclaimed within past three years). In 1956 the task of providing laborers was assigned to the collective, but in 1981 it was shifted to individual households.[156] Almost everywhere the vast majority of laborers assigned to high-water control were peasants, but in Shashi, urban workers also shared the burden.[157]

Urban workers in Wuhan also participated in high-water control. A major difference in the Wuhan area was the emphasis on laborers' political reliability. In Wuhan's Wuchang District in 1956, people patrolling dikes had to be politically reliable and the core of the patrol team had to be CCP and/or Communist Youth League activists.[158] In 1962 in Qingshan District, people who

152 *Jianli difang zhi* 1991: 367.

153 *Qianjiang shuili zhi* 1997: 267–68.

154 *Tianmen xian zhi* 1989: 143; *Tianmen shuili zhi* 1999: 125.

155 *Jianli difang zhi* 1991: 367.

156 *Dongjinghe difang zhi* 1994: 312–13.

157 *Shashi shuili difang zhi* 1994: 99.

158 Wuchang 49-1-5.

were chosen to patrol dikes not only had to be physically strong, but also had to pass a strict political investigation and be "politically absolutely reliable."[159] In 1963 in Qiaokou District, members of emergency teams had to be politically reliable, and on endangered dike sections, when floodwater reached a certain level, armed militia were deployed to watch over the dike day and night.[160]

One principle, however, remained unchanged over time: when called upon, participating in high-water control was compulsory. Moreover, no one was paid, aside from a few rare cases where laborers might receive a small subsidy, such as 100 wen a day per laborer in the late Qing, or a deduction of one fang of earth per laborer in Songzi in the 1940s, or a payment of 0.3 yuan per laborer per day in Xiaogan in the 1950s.[161] But after 1963, laborers in Xiaogan were paid nothing.[162] In fact, the government of Hubei had made it clear in 1962 that all high-water control work assignments were compulsory and came without any wages or payments.[163] This rule, however, did not seem to apply to urban participants. In Wuhan, urban units paid their workers participating in high-water control several subsidies, such as meals and what were known as "cooling beverage fees," or payments for working on hot summer days.[164] Most rural residents or peasants who participated in high-water control certainly did not get these sorts of payments.

The unequal treatment of peasants and workers on high-water control sites reflects the systematic rural-urban divide in China. In 1954 it was found that even when peasants, soldiers, and urban workers worked side by side, soldiers and workers ate better, and in some places, workers had work sheds, electric lights, and laundry service and were paid subsidies, but peasants had none of these. Some workers thus looked down on peasants. Although the official discourse lauded the "worker-peasant alliance" and the equality of workers and peasants, in reality workers were treated better than peasants and tended to discriminate against them.[165] This kind of unfair treatment and discrimination, however, was far lighter than what peasants suffered under the rural-urban dual system after 1958.[166] The Bureau of Water Conservancy of Wuchang complained in 1984 that, although they participated in high-water control on the

---

159 XX000065-WS01-1272-0016.

160 XX000065-WS01-1385-0002.

161 *Jiangling difang zhi* 1984: 96; 100-8-125: 11; *Xiaogan diqu shuili zhi* 1996: 57.

162 *Xiaogan diqu shuili zhi* 1996: 57.

163 *Qianjiang shuili zhi* 1997: 145.

164 27-2-2.

165 XX000129-WS01-0086-0001; XX000129-WS01-0005-0003.

166 Under this system, it was almost impossible for rural residents to move to cities and enjoy the benefits of urban residents. Although rural residents have been able to freely move

same dike, urban participants received four to five times greater subsidies than those for rural participants. The bureau demanded that all participants be treated equally.[167]

Whether in urban or rural areas, the PRC state had such great power that mobilizing people for high-water control was not difficult. This can be seen, for example, in the efforts in the town of Fenshui, Hanchuan County, in 1964 to protect its 12.7 kilometer dike from the rising floodwater. More than ten thousand laborers added an additional meter to the height of the dike within 24 hours.[168] In another example, in Tianmen in 1975, two thousand peasants walked 60 li overnight to the dike-site assigned to them for high-water control; some even walked 80 li.[169] And in 1983, the neighboring county of Mianyang mobilized a hundred thousand laborers overnight to reinforce its dikes.[170]

Although laborers were generally easy to recruit in the early PRC, some cadres clearly harbored old bad habits and abused laborers in old fashioned ways. For example, during the 1954 high-water control season, a cadre in Zhijiang berated a laborer who came late and punched him in the face, and also humiliated five other laborers by ordering them to kneel down.[171] In another instance, a sub-county leader in Shishou beat five laborers while dealing with an emergency, and argued that he would be killed if he did not beat them, implying that he would be executed if the dike was breached, but he would only be jailed if he beat them.[172] In still another case, in 1955 some cadres in Hanchuan were denounced for using violence to dragoon people—including a pregnant woman, a lame person, and a convalescent—to go to a high-water control site.[173] And, in a final example, some cadres in Hanyang even tied up villagers to force them to go to a high-water control site.[174] All these abuses were reminiscent of the treatment meted out by brutal foremen in the Republican era.

As we have seen, in addition to providing virtually all the labor for high-water control in the Qing, the local people also had to shoulder almost

---

to cities since the 1979 rural reform, they are still denied some benefits available to urban residents, particularly in mega cities.

167 Wuchang 49-2-78.

168 *Fenshui zhen zhi* 1998: 177.

169 42-1-172.

170 Zhou Xiangsheng 2008: 11.

171 45-1-8.

172 39-1-16-4.

173 50-1-28a.

174 SZ18-2-92. In 1954, more than ten thousand "reform-through-labor" prisoners were used for emergency repairs during the high-water control repairs of the Han River dikes. It was reported that these prisoners worked incredibly hard to atone for their "crimes" (40-1-57c).

all the financial burden. Although in the Republic this began to change as the state started to provide materials for high-water control of some important river dikes, which reduced the burden on the shoulders of the locals, they still were responsible for almost all materials and expenditures on high-water control of branch river dikes and polder dikes.[175] In the PRC, more drastic changes followed, with the state gradually providing most of the required materials, which largely lifted the financial pressure off the peasants. As for the funding of high-water control, the central and provincial governments usually have been responsible for major river dikes, the county governments for important polder dikes, and local residents for all other polder dikes.[176] From 1950 to 1995 the state spent 2.41 million RMB for high-water control on dikes along the Han and the Dongjing Rivers in Qianjiang alone, and from 1951 to 1990 about 7 million RMB on the Dongjing River dikes.[177] Since all polder dikes in Zhijiang in the early 1980s were people's polders, everyone—include government employees and urbanites (with some exceptions, such as students), as long as they were protected by the polder dikes—had to pay one RMB annually for their maintenance, including high-water control; in the countryside, farmland was assessed 2.5 to 3 yuan per mu per year.[178]

At the same time, local governments also started to impose high-water control fees on urbanites who usually had no obligation to provide labor for high-water control. In Jianli, for example, each non-farming working-age resident was required to pay 4 RMB in 1994 and 25 RMB in 1996 as a high-water control fee.[179] Similarly, in Dangyang County in 1995 each working-age non-farming resident had to pay a 25 RMB high-water control fee.[180] With the growth of the state's financial power, in 2013 Hubei stopped the collection of high-water control fees,[181] a major change in the history of high-water control.

6 Sentry Shacks

During the high-water control period, places were needed to store materials and equipment and to house laborers, technicians, and overseers. These places were called *shaopeng*, a term that can be translated as "sentry shacks,"

175 Jiayan Zhang 2014: 54–58.
176 *Jianli difang zhi* 1991: 359.
177 *Qianjiang shuili zhi* 1997: 146; *Dongjinghe difang zhi* 1994: 310–11.
178 45-1-216; 45-1-233.
179 *Jianli shuili zhi* 2005: 430.
180 *Dangyang shi zhi* 2008: 181.
181 *Hanjiang difang zhi* 2015: 333.

although from the Qing to the present, many other names have been used from place to place. Initially, the shacks were rude portable structures, but gradually they were replaced by temporary grass shacks and then permanent structures.

In Jiangling, after the 1788 flood, fifty-two garrison houses (*kafang*) were built on the Wancheng Dike, one every 500 zhang (about five thousand feet).[182] In the reign of Daoguang, because many dikes in Hubei were in remote places with no nearby residents, a top official suggested that the locals build one reed shack for every three li of dikes, with each shack manned by two people (one leader and one laborer) with a gong. The two were to patrol the dike day and night and beat the gong to rally laborers if they spotted a threat.[183] It is unclear whether this suggestion was carried out, or, if it was, how widely it was practiced. But it is certain that it had been implemented in Tianmen during the reign of Daoguang, where during the high-water control period, each polder was required to build one reed shack, each manned by three laborers who were to undertake minor repairs in daytime and to patrol and inspect the dike at night.[184]

In his Regulations on High-Water Control, Lin Zexu explained why such places were needed. During the high-water control period many people and materials were gathered in some important locations, but if there were no facilities to house them, "those people would have no place to take a rest, and there would be no place to store the materials." But such sites were limited to important locations; much simpler shacks were the rule along other sections. This was precisely the case in Jiangling, where in the Qing most shaopeng were portable shacks, consisting of a bamboo mat as a roof, wooden boards in the interior as beds, with wooden poles on both ends for carrying the shack. This kind of shaopeng, which could be transported by two people, was called a "carrying shack" (*taipeng*). There were more than two hundred of these shacks on the Wancheng Dike. They were torn down when the high-water control period came to an end.[185] In Qianjiang in the Qing, the local people built a thatched cottage on every li of dikes for the same purpose.[186]

The situation did not change much in the Republic. Shashi, for example, had only one permanent storehouse on its dikes. During the high-water control period, people put up makeshift portable shacks to supplement the single

182 *Jiangling difang zhi* 1984: 95.
183 *Zaixu xingshui jinjian* 2004: 163.
184 *Tianmen xian zhi* 1989: 143.
185 Ni Wenwei 2002 [1885]: 151; *Jiangling difang zhi* 1984: 96; *Shashi shuili difang zhi* 1994: 105.
186 *Qianjiang shuili zhi* 1997: 142.

permanent storehouse.[187] In another example, in 1947 in Yongfu polder in Songzi, each team was required to set up two shaopeng (each of which housed one dike maintenance worker) for high-water control.[188] In Shishou, all the shaopeng on the major river dikes were thatched shacks. To make things worse, the social chaos at that time likely caused their collapse due to lack of maintenance; thus, even these shacks had almost entirely disappeared by around 1949.[189]

More changes occurred in the PRC. There was a general trend to replace temporary shacks with permanent structures. This was very clear in Shishou where, in the early PRC, one straw shack (*caopeng*) was built on every kilometer or two of dikes, but these were gradually replaced by one-story brick-walled and tile-roofed barracks, each of which accommodated around thirty laborers during the high-water control period. At the end of 1985, there were 309 such structures in Shishou: one on every 1.3 kilometers of major river dikes (72 in total), one on every 1.28 kilometers of branch river dikes (134 in total), one on every 1.65 kilometers of people's dikes, and one on every 2.1 kilometers of polder dikes. Since the early twentieth-first century, all shaopeng in this county are one story brick-walled and tile-roofed structures and multistory barracks. In years when there is a threat of catastrophic floods, makeshift shaopeng will be built between existing buildings.[190]

This trend can be seen in other places as well. In the 1950s there were more than six hundred straw shacks on the dikes of Qianjiang, all of which were built by peasants who not only provided the labor but also the materials. In the 1960s, 111 brick-walled and tile-roofed shaopeng were built, and materials were provided by the state and laborers by the production brigades.[191] Similarly, in Jiangling in the 1950s peasant laborers had to bring materials themselves to put up makeshift shaopeng. Since 1964, however, 93 permanent brick and wooden barracks, called "sentry posts" (*shaowu*), have been built on the Great Jingjiang Dike, and 127 have been built on the other, less important, dikes.[192] The shaowu in Jianli were makeshift structures in the early PRC, but on the major river dikes they were later replaced with permanent buildings, usually built on a platform on the inner-side dikes. At the end of the twentieth century there were 96 such

187 *Shashi shuili difang zhi* 1994: 105.

188 100-10-423: 61.

189 *Shishou xian zhi* 1990: 141.

190 *Shishou xian zhi* 1990: 141; *Shishou shi zhi* 2014: 174. Most of these barracks, as I observed, are two-story structures made of reinforced concrete and prefabricated panels.

191 *Qianjiang shuili zhi* 1997: 146, 268.

192 *Jiangling difang zhi* 1984: 100.

buildings.[193] By 2010, there were 299 on the major river dikes along the Yangzi River in Jingzhou alone.[194]

One source indicates that in 1953 in Songzi the shaopeng on major dikes were 6 meters long by 4 meters wide by 4 meters high, and on minor dikes 4 meters long by 3.5 meters wide by 3 meters high.[195] In Jianli in 1957, newly built permanent shaowu were usually 35 square meters (7 meters × 5 meters).[196] But in Hanyang in 1964, the size of a brick and wood shaopeng was 50 square meters.[197] Where dikes were especially important and the local economic conditions were relatively advanced, bigger and stronger shaopeng were built. Thus, in Shashi in the 1950s, one makeshift one-story shaopeng was built on every kilometer of dike before the high-water control period. Each shaopeng had three rooms and could accommodate twenty to thirty people. If there were more people than these shaopeng could house, nearby residents' houses could be borrowed, or makeshift portable shacks could be erected. In 1964 seven brick-and-wood bungalows were built as the headquarters for high-water control, and in 1986 four two-story structures were built. The latter could accommodate a total of five hundred to seven hundred people.[198]

7 Breakwater Forests and Riverside Parks

Breakwater forests have at least four functions: to break waves, to provide materials for high-water control, to provide firewood and income for the local people, and to improve the environment.

Using breakwater forests to protect river dikes has long history in China, particularly along the Yellow River,[199] but also in the Jianghan area. At least as early as the Southern Dynasties (420–589) there was a breakwater forest along the Wancheng Dike, and the forest (a belt about 150 meters deep) still existed in the Qing but was destroyed in the Republic.[200] Other sources, however, indicate that this forest might have declined, or been at least partly destroyed, in the Qing. According to Ni Wenwei, when a segment of the Wancheng Dike broke in 1733, and was rebuilt, there were huge willow trees (*Salix matsudana*

193 *Jianli difang zhi* 1991: 248.
194 *Jingjiang difang zhi* 2012: 487.
195 33-1-16.
196 *Jianli difang zhi* 1991: 248.
197 30-1-37.
198 *Shashi shuili difang zhi* 1994: 105.
199 *Hanchuan xian zhi* 1873, vol. 9: 2ab.
200 *Jingjiang dadi zhi* 1989: 82.

Koidz) along the dike. But when Ni was appointed the prefect in 1872, he found that there were no longer any trees. He asked his clerks why people did not plant trees along the dike and was told that they were ordered by a previous prefect to chop down all the trees near the dike. He then checked and found that that prefect only asked the locals to remove the trees on the body of the dike, since those trees damaged the dike, but he did not object to planting willow trees on the nearby river beaches.[201]

One source mentions that willows were planted along the Han River dikes as early as the Tang dynasty (618–907).[202] In 1830 about ten thousand willows were planted along the Dongjing River dikes in Jianli alone. In 1930 the central government issued regulations to encourage the planting of trees along dikes, and in 1936 the Bureau of Engineering Work on the Yangzi and the Han Rivers also tried to plant breakwater forests along the dikes in Jianli and Jiangling. But it seems this was far from successful, since, by the end of the Republic, there were few trees along the dikes in Jianli.[203]

Again, things changed substantially in the PRC. Although trees were grown along the Jianghan dikes for long time, it is likely that most dikes in the region were not protected by breakwater forests before 1949, possibly because of the social chaos or mismanagement of dikes in the Republic. Thus, one expert viewed the planting of breakwater forests as a major change in dike management in the early PRC. It was said that people in some places had already started to plant trees along dikes in 1952. In the 1954 high-water control, it was found that wherever there was a breakwater forest the damage to the dike was minor. Thus, a breakwater forest of willows was planted on the river side of Jianghan's dikes.[204] The trees were usually planted in five to ten rows, sometimes more than twenty rows—or five to eight rows for important dikes and three to five rows for less important ones.[205] An expert explained in detail why willows were selected: they are the easiest to plant, they can grow almost anywhere and on almost any type of soil; they are resistant to waterlogging and can normally survive after being submerged in two meters of water for two months; they have a large canopy; they have a long leafy stage; and they can break waves from early spring to late autumn. He claimed that if there were no satisfactory methods to break waves, it would be very difficult to ensure the

201 Ni Wenwei 2002 [1885]: 87, 140–42.

202 *Tianmen shuili zhi* 1999: 105.

203 LS1-5-598; *Jianli shuili zhi* 2005: 390–91.

204 Liu Yun 1982.

205 SZ113-3-542-009; 54-1-58.

integrity of the dikes during the high-water control period. In short, breakwater forests were extremely important.[206]

Breakwater forests can also provide materials for high-water control. As discussed earlier, among the materials needed are plant matter such as straw, bamboo, and firewood. Willow trees are a plentiful source of twigs that can be used in high-water control in multiple ways. As a result, in Honghu in the 1960s, the locals no longer needed to provide firewood, wooden stakes, and rice straw for high-water control on dikes along the Yangzi and the Dongjing Rivers since breakwater forests had been planted.[207] Because a breakwater forest had been established earlier along the Han River dike, the state saved 6.7 million RMB on breakwater funds from 1964 to 1985 alone.[208]

The income from breakwater forests had been regarded as a way of "supporting dikes with dikes" in Mianyang.[209] This income mostly went to the nearby villages. Commonly, the income was split, with 20 percent going to the state, 10 percent to the commune or township government, and 70 percent to nearby production teams.[210]

By the same token, breakwater forests can also provide firewood for local communities. Up to the early 1980s, crop residue was still the main fuel consumed in rural China,[211] including the Jianghan area. The locals traditionally relied on plant matter for fuel, and it was not unusual that they ran short of fuel, since the straw they harvested was insufficient to meet their needs.[212] This was still the case in the PRC up to the turn of the twentieth-first century. In Tianmen, for example, twigs pruned from its breakwater forests along the Han River provided nearly 1.9 million jin of firewood for 261 production teams in 1960, which covered more than one month of consumption.[213] In 1963, the breakwater forests in Tianmen again provided 1.6 million jin of firewood, as well as materials to make 18,154 farm implements, and 2,184 boards for building houses, cattle pens, and pigsties.[214] In 1962, because of the breakwater forests, the anti-wave materials and labor spent on the Great Jingjiang Dike in Jiangling were reduced by 95 percent,[215] an enormous drop. In the 1970s, the breakwater

206 Wang Zhihua 1986a.
207 *Honghu xian zhi* 1992: 123.
208 *Hanjiang difang zhi* 2015: 169.
209 SZ113-2-311-006.
210 00046-01-0124.
211 Smil 1984: 149.
212 Jiayan Zhang 2014: 194–95.
213 SZ113-2-195-004.
214 SZ113-2-274-009.
215 48-1-183.

forests in Hanchuan could provide 30 million jin of firewood for local communities annually, a huge benefit for the local people.[216] In some places trees grown outside the dikes (facing the river) have been used as breakwater forests, while trees planted inside the dikes have provided firewood and high-water control materials.[217] These trees can also provide timber.[218] Thus, dawn redwoods (*Metasequoia glyptostroboides*), a deciduous conifer that tolerates wet soil and provides building timber, were widely planted inside the dikes.[219] Since the 1990s, in some places willows have been replaced by dawn redwoods and Italian poplars (*Populus euramevicana cv.'I-214'*).[220] Italian poplars are very commonly used in making fiberboard and other wood-based composites.

Sources from the early PRC that discussed the functions of breakwater forests only mentioned high-water control and a few valuable products such as firewood and timber. But more recently, particularly since the turn of the twentieth-first century, recognition of the value of breakwater forests had expanded to include environmental protection and landscaping.

Although some observers have been skeptical of the PRC's official statistics on "forests,"[221] the achievements of "greening the dikes" in the Jianghan area are readily visible. From the 1950s to the 1990s, governments at various levels in the Jianghan area made great efforts to plant breakwater forests along dikes. For instance, from 1949 to 1972, 1.88 million trees were planted along the dikes in Shishou alone, and in Qianjiang 6.81 million trees were planted along its 171 kilometers of dikes from 1952 to 1995.[222] As a result, by the 1980s a universal pattern of "one river, two dikes, and four belts of forest" emerged in the region.[223] Evidence of this is visible at the Distance Dike in Zhongxiang, which seen from the air looks like "a long green dragon."[224]

These forests have come to be regarded as "a green ecological defense for areas along the rivers."[225] Since there are several thousand kilometers of dikes on the Jianghan Plain, the acreage under breakwater forests is extensive. In

216 46-1-15c.

217 *Jiangling difang zhi* 1984: 86.

218 *Songzi shuili zhi* 2008: 89.

219 Liu Yun 1982. The locals often stealthily chopped down breakwater trees to use them as firewood or to build pigsties, or sell them, or trade them for alcohol or other things. These sorts of activities could be done by individuals or entire production teams (SZ113-2-195-004; 22-1-329-005; 40-1-20-3; XX000065-WS01-1386-0028).

220 *Honghu fenxuhong gongcheng zhi* 2016: 164; *Hanjiang difang zhi* 2015: 169.

221 Marks 2017: 319–20.

222 40-1-20-3; *Qianjiang shuili zhi* 1997: 9.

223 Liu Yun 1982; Wang Zhihua 1986a.

224 *Zhongxiang xian zhi* 1990: 374.

225 *Hanjiang difang zhi* 2015: 168.

Tianmen, for instance, there were 10,400 mu of breakwater forests along the Han River in 1985, 6,700 mu of which were outside the dike.[226] The forests are so extensive that they can also improve the local environment by holding water, purifying the air, and even modifying the weather. At a time when many Chinese, particularly in cities, are increasingly plagued by a deteriorating environment, these forests are a green treasure for the local people, who benefit from them in unique and unexpected ways. According to Vaclav Smil, in the Maoist era "deforestation would appear to be the country's most critical environmental problem."[227] This points to the immense significance of breakwater forests in the Jianghan area in the early PRC.

At the turn of the twentieth-first century, with the increasing demand for a higher quality of life, the Jianghan people found a new use for dikes and their exclusion zones: they can be an ideal place for entertainment, leisure, and tourism. Indeed, governments at various levels have invested heavily in building parks along dikes precisely for this purpose.

Dike greenification can be traced back to 1964, when the central government launched a campaign to choose "five-goodness-dikes" through public appraisal. One of the criteria by which these dikes were judged was the extent to which they reflected "good dike greenification."[228] As early as 1987, dike greenification in Hanyang had already been regarded as an important part of urban greening and was included in the urban greenification plan.[229] Four years later, a river beach park was built in Changfeng Township, Qiaokou District, Wuhan.[230] Similarly, greenification proceeded in other districts of Wuhan. In Jianghan District, for example, from the end of 1998 to 2000, the local government spent 240 million RMB to transform the Dragon King Temple (Longwang miao) section of the dike, historically a dangerous section, into a scenic spot with multiple functions—sightseeing, leisure, business, gardens, and high-water control.[231] On the other side of the Yangzi River, across from the Longwang miao section, after the 1998 flood the local government also repaired and rebuilt the dike and its surroundings by reinforcing the dike, removing houses and factories, and planting grass and trees, and turned the area into a new greenificated scenic spot.[232] From 2005 to 2012, more than 400 million RMB was invested in Hanyang District to improve its riverside

226 *Tianmen xian zhi* 1989: 139.

227 Smil 1984: 9.

228 45-1-15.

229 27-2-4.

230 *Changfeng xiang zhi* 2016: 162.

231 *Jianghan qu zhi* 2007: 8, 936–38.

232 *Qingchuan jie zhi* 2005: 20–21, 105–07.

scenic belts along the Yangzi and Han Rivers.[233] In Qingshan District, since the 1990s the local government has even turned the Yangzi River dike under its jurisdiction into a unique "dike green landscape" by linking many public parks together.[234] In 2010 alone, the government of Dongxihu District spent 7.35 million RMB on planting 180,000 trees along its dikes, and laid down 1.14 million square meters of sod, which gradually turned its dikes into a "tourist landscape belt."[235] At the same time, a mega-park—the well-known Hankou Yangzi River Beach Park—was built along the river beach. This prize-winning project receives millions of visitors every year. These changes in the Wuhan area since 1990 were collectively dubbed "the development and formation of dike parks." By the year 2000, about three hundred thousand people a day toured, exercised, and entertained themselves in these parks.[236]

Similar developments took place in small towns. After the 1998 flood, the local government not only reinforced the segment of the Great Jingjiang Dike in Haoxue, which was historically dangerous, but also added streetlights, traffic isolation blocks, and brick-paved sidewalks to the dike, and greened the surroundings, turning this dike segment into an attractive spot for leisure activities and entertainment.[237] When touring the dikes along the Yangzi River, the Han River, and their tributaries in the Jianghan area, one can find numerous riverside parks that have been built on different dike segments in recent decades. People of all ages flock to these parks.

---

233 *Hanyang qu shuiwu zhi* 2015: 111–12.

234 *Qingshan qu zhi* 2006: 114.

235 *Dongxihu qu jian zhi* 2012: 80.

236 *Wuhan shi zhi 1* 2006: 374, 674.

237 Zhang Xiantang 2004.

CHAPTER 3

Annual Repairs

During the high-water season in summer and autumn, the local people had to be very careful in carrying out high-water control to prevent their dikes from rupturing. In the dry season in winter and spring, they also faced the task of maintaining the dikes by carrying out annual repairs, which included closing of any breaches, occasionally constructing new dikes, and reinforcing weak ones. These weaknesses included dikes or dike sections being insufficiently high, or too thin, or having other defects. If these problems were overlooked and were not fixed in a timely manner, the result in the following high-water season could be disastrous. Thus, since the end of the Qing, some people began to argue that annual repairs were more important than high-water control.

1 The Dike Headman and the Dike Maintainers/Keepers System

In the preceding chapter, we noted that, during the high-water control period, shacks (shaopeng) were built on dikes to store materials and tools and to house high-water control workers. Another function of these structures, particularly the permanent buildings, was to provide shelter for dike maintainers whose main duties were to take care of dikes, such as doing some minor repairs and removing weeds, termite nests, and badger burrows.

According to the dike tithing system—a system of dike headmen (dijiafa) established by Zhao Xian, the prefect of Jingzhou, in 1560—a group of people were assigned the job maintaining the Wancheng Dike. After the 1788 flood, this group included 2,077 people divided into 67 subgroups, each with 31 members—including 1 dike administrator (dizhang), 5 dike heads (dijia), and 25 laborers. Each of the subgroups was responsible for maintaining 500 zhang of the dike. They all lived near the dike and were called up through a yearly rotation. After the fifteenth day of the eighth lunar month, both the dizhang and dijia were required to go to the dike to set up a shack. They were exempted from all miscellaneous corvée and were given some farmland. In winter, each laborer had to build five earth-oxen to be used in the following high-water control season.[1]

1 *Jingjiang dadi zhi* 1989: 263–64; Jiayan Zhang 2014: 52.

 | DOI:10.1163/9789004739482_005

This system, as noted, was designed for the Wancheng Dike. For other river dikes and polder dikes, the organizations and the number of their members responsible for annual repairs varied greatly across time and place, but all of them also included one leader, a few assistants, and several laborers. Both the leaders and laborers who served in such organizations came from villages that were protected by the dike in question.[2] According to Yu Changlie, maintenance or annual repairs were crucial to ensuring the integrity of dikes, because if problems were not discovered in winter, and annual repairs were not completed in time, it would be difficult to maintain the integrity of a dike when the annual high water came. Even if the dike was saved, the cost would be far higher than that of annual repairs. Thus, government officials in charge of dike affairs were advised to take ten to fifteen days to carefully examine the condition of dikes under their jurisdiction and to carry out annual repairs accordingly.[3] A prefect of Jingzhou in the mid-nineteenth century claimed that annual repairs were his first priority.[4] In 1887, the prefect of Jingzhou also argued that the annual repairs of river dikes should be the top priority, since emergency repairs during high-water control could be no more than a temporary remedy. Annual repairs, on the other hand, could forestall an emergency.[5]

The Qing organizations responsible for annual repairs continued into the early Republic, but changes soon appeared. For some important dikes, the state established specific institutions to maintain them. One such institution was the Bureau of Engineering Work on the Yangzi and the Han Rivers, founded in 1932, with a full-time staff, some of whom were hydraulic technicians armed with modern technology, not laborers recruited from nearby villages. Other river dikes and all polder dikes, however, were still maintained by villagers as was the case in the Qing.

In the PRC, a hierarchical system of organization was established to manage all dikes. The lowest level in this system consisted of the maintenance stations/points. In Jiangling, for instance, there were fifteen maintenance points for the Yangzi River dikes in 1961, each with a director who received a regular monthly salary from the county. In addition, depending on the length of dike it maintained, each point had one or two informal dike keepers who earned workpoints from their production teams but also got a small monthly living allowance. In 1982, there were twenty-one maintenance points for all the major river dikes in Jiangling, staffed with a total of 208 dike keepers who were

2 Jiayan Zhang 2014: 52–53.
3 Yu Changlie 1999 [1840]: 124.
4 Ni Wenwei 2002 [1885]: 166.
5 Shu Hui 2002 [1896]: 447.

sent by nearby communes and townships and received a small monthly living allowance. Many had worked on dikes for so many years that it was said "they consider the dikes as their home."[6] At the end of the 1980s, there were thirty-six maintenance points along the Great Jingjiang Dike, each with one director and two to seven dike keepers.[7]

In Jianli, there was one dike keeper for every five kilometers of the Yangzi River dike in 1957; in the 1970s, there were a total of 173 dike keepers responsible for the maintenance of the county's dikes. These dike keepers worked and lived on the dikes (in shaowu barracks) all year round. For this, they earned workpoints from their production teams but also received a small monthly living allowance. After the 1979 rural reform and the abolition of workpoints, however, some dike keepers left since they could not cash in their workpoints. To retain these workers, their monthly living allowance was increased, and they were given priority in contracting for the platform at the foot of the dike where they could raise crops to supplement their income. At this time, there is one dike keeper for every kilometer of river dike, and one for every two kilometers of polder dikes.[8] The dikes along the Dongjing River had 141 maintenance points and 481 dike keepers in 1990.[9]

Dike keepers did more than take care of dikes. They also watched over breakwater forests, and thus, in Xiaogan, they were called "dike-and-forest-keepers" (*dilin guanliyuan*). Many of them were retired party secretaries or leaders of peasant associations or people with social standing from nearby brigades, similar to the heads of well-to-do families and members of the gentry who served as dike managers/directors in the Qing and the Republic. After the 1979 rural reform, remuneration according to workpoints was abandoned, and the dike-and-forest-keepers lost their income. Most of them returned home to farm. In 1985, funds were raised to resolve their remuneration. There were 412 dike-and-forest-keepers in Xiaogan in 1990. Their pay, however, became a burden in the 1990s. This led to them being gradually laid off in Hanchuan, and by 2007 the county no longer had dike-and-forest-keepers of this kind. All of them since then have been replaced by formal, full-time government employees.[10] By 2010, all maintenance teams/points on the Han River dikes were disbanded, and maintenance was taken over by government employees.[11]

---

6 *Jiangling difang zhi* 1984: 78.

7 *Jingjiang dadi zhi* 1989: 264.

8 *Jianli shuili zhi* 2005: 387–88; *Jianli xian zhi* 2013: 194.

9 *Dongjinghe difang zhi* 1994: 269.

10 *Xiaogan diqu shuili zhi* 1996: 140; *Xiaogan shi shuili zhi* 2011: 133–35.

11 *Hanjiang difang zhi* 2015: 178.

A similar change occurred much earlier in the maintenance of the Yangzi River dikes. In Gong'an, the dike keepers started to receive a regular salary as early as 1979.[12] The 2005 reform in which the agricultural tax was abolished required all dikes to be managed by formal government employees, thus by that year all informal dike keepers of the Great Jingjiang Dike were dismissed and were replaced by government employees.[13] Similarly, in Jiangling, all rural dike keepers were dismissed by 2006.[14] This switch, an outgrowth of the reform of rural governance after the abolishment of the agricultural tax, is a milestone in the management of the Jianghan dikes.

2 Termite Nests, Badger Burrows, and Embankment Streets

A routine task for dike maintainers/keepers was to carefully watch over dikes to uncover and fix any hidden dangers. The Jianghan dikes are subject to hidden threats that could cause ruptures, such as termite nests, badger burrows, house foundations, graves, sewers, tree stumps, stakes, slug bricks, kilns, and even military engineering projects. In this section, the discussion turns to the two most common dangers, termite nests and badger burrows, plus embankment streets.

2.1 *Termite Nests*

In China there is a popular saying that "one termite nest can cause the collapse of a thousand li dike." And in fact, termite nests are a major threat to Jianghan's dikes. If they are not removed, they might cause a dike to fail. The dike maintainers in the Qing who were assigned to live on the dikes and watch over them were expected to remove the nests and fill in the resulting holes, thus preventing leaks and possible collapse.[15] According to one source, the Wancheng Dike had more termite nests than any other. These could be spotted and marked in the high-water season and then treated during the following annual repairs. The best solution according to this source was to dig them out and then fill the hole with dirt mixed with lime.[16]

But in the late Qing that solution was expensive and impractical. It was difficult to track down all the termite nests, and there was the danger of digging

12 *Jingjiang fenhong gongcheng zhi* 2000: 273.
13 *Jingjiang difang zhi* 2012: 487.
14 *Jiangling xian zhi* 2015: 131.
15 *Jiangling xian zhi* 1876, vol. 8: 23a; *Songzi xian zhi* 1869, vol. 4: 10ab.
16 Yu Changlie 1999 [1840]: 130.

too deep into the inner-core of the dike, which could severely weaken the whole dike.[17] Termite nests could be huge. In Songzi in 1947, a nest inside a dike was found to be as large as a small house. It took 3,500 jin of lime to fill it.[18] However, not all termite nests could be discovered by dike keepers and properly dug out and filled. Thus, in the Republic, termite nests seriously threatened the Great Jingjiang Dike. In addition, there were other threats, such as graves and tree stumps. All of these would cause the dike to burst.[19] From 1644 to 1935, the Great Jingjiang Dike suffered frequent ruptures, many of them were caused by termite nests.[20] Early in the PRC, it was found that about half of this dike had been damaged by termite nests. From 1949 to 1953, when the locals were mobilized to search for hidden threats to the dike, they found a total of 52,151, including many termite nests. In the 1954 high-water control, 94 percent of the identified threats were termite nests; from 1950 to 1990, the dike suffered 6,150 various threats, 41.5 percent of which were caused by termite nests.[21] For this reason, a great deal of work had been done in the PRC to find a solution to this problem. In 1954, a state-funded termite control engineering team for the Great Jingjiang Dike was established. The work of this team greatly contributed to the security of the Great Jingjiang Dike in the following decades. The team's termite control technology even earned it a national science and technology award in 1978. By 1982, 35,930 termite nests were eliminated from the Great Jingjiang Dike.[22] Similar action has been taken in Zhijiang since 1955.[23] A termite control team of the Yangzi River dike was also established in Jianli in 1972 to deal with termite nests on its major river dikes.[24]

Termite nests have also been the major hidden threat to the Distance Dike along the Han River. From 1949 to 1974, 14,607 hidden threats to the Distance Dike—of which 1,358 were termite nests—were eliminated. The largest termite nest reportedly could hold several people.[25] From 1950 to 1975, 8,227 termite nests on the Han River dikes were destroyed.[26] In the first five decades of

---

17 Shu Hui 2002 [1896]: 434–35.
18 100-8-108: 58.
19 *Jingjiang difang zhi* 2012: 219.
20 XX000065-WS01-1589-0001.
21 *Hubei shuili zhi* 2000: 536; *Jingjiang difang zhi* 2012: 224; SZ113-2-274-003.
22 *Jiangling difang zhi* 1984: 79, 148.
23 *Zhijiang xian zhi* 1990: 170.
24 *Jianli xian zhi* 1994: 159.
25 54-1-129; Wang Zhihua 1986b.
26 *Hanjiang difang zhi* 2015: 89.

the PRC, a total of 100,000 or more termite nests were eliminated from all the dikes in Hubei.[27]

2.2 *Badger Burrows*

Badger burrows are generally not as serious a threat as termite nests, but nonetheless they can be very damaging. Thus, badger burrows and termite nests are usually mentioned together in the documents. According to Yu Changlie, badger and rat burrows were the most harmful threats to long dikes and should be completely eradicated. Eradication methods included smoking, netting, and noosing, all with the help of hunting dogs.[28] In one case, a prefect of Jingzhou in the late Qing spotted a badger burrow in the winter and dug it out the next spring—thirty or so badgers were captured in the process. But badgers are clever, and they might escape, but later return after their burrows were filled. Hence, completely eliminating them is challenging.[29]

This problem continued in the Republic. From the mid-nineteenth century to 1955, badger burrows caused five ruptures of the Han River dikes alone, four in the Qing (in Tianmen) and one in the Republic (in Qianjiang).[30] The collapse of the dike of a polder in Songzi in 1946 was also attributed to a badger burrow.[31] Therefore, eradicating badgers was an important part of dike work. In 1946, residents who lived along the dikes of the Yangzi and Han Rivers were given a reward of 10 kilograms of flour for each badger they captured.[32]

After the PRC was established, the new state immediately took action. According to one source, the government of Zhongxiang County offered a bounty of 15 jin of husked rice for each badger killed. In 1950, it paid a total of 17,000 jin[33]—in other words, more than a thousand badgers were exterminated. According to another source, at the end of 1949, the Zhongxiang County government issued orders that badgers, as well as pigs and other animals that were harmful to dikes, be captured and killed and paid a bounty of 4 kilograms of husked rice for each animal eradicated. By the end of 1951, 6,100 badgers had been destroyed, and badger burrows on the dikes of Zhongxiang were basically eliminated.[34] Badger burrows, however, were not that easy to eliminate.

27 *Hubei shuili zhi* 2000: 493.
28 Yu Changlie 1999 [1840]: 130.
29 Shu Hui 2002 [1896]: 434.
30 SZ113-2-077-001.
31 100-8-103: 110.
32 LS30-1-0889-004.
33 54-1-5a.
34 *Zhongxiang shuili zhi* 1998: 66. The source did not make clear whether 6,100 included badgers, pigs, and other animals, or only badgers; based on the context, it seems the number referred to badgers alone.

In 1953, for example, 194 badger burrows were found and filled up on Zhijiang County's dikes, and sixteen badgers were killed; in the following year, 64 badger burrows were again on these dikes, and fifteen badgers were killed.[35] The locals had to be constantly vigilant.

2.3 *Embankment Streets*

Although the damage caused by both termite nests and badger burrows involves mostly non-human factors, the problems caused by embankment streets (*dijie*) are entirely man-made. The factors behind the emergence and formation of embankment streets are simple. First, the relatively high elevation of dikes naturally attracted nearby people during and after floods, and some of these people put up huts as a refuge, and a few also stayed on after the floodwater had retreated.[36] This gradually led to the construction of embankment streets and markets. Second, since dikes are usually located where the land is naturally elevated, and some are also close to river courses, they provide a safe and convenient area for local people to gather to trade goods, and, in due course, to build shops and houses on the dike, thus forming embankment streets, such as those in some small towns along the Han River.[37]

The largest and most famous embankment streets, however, are in Shashi. As one of the most important water transportation entrepôts in the mid-Yangzi River valley, the city of Shashi had embankment streets as early as the Tang dynasty. The streets then slowly spread in the following dynasties, and eventually flourished in the Qing and the Republic. In the Republic, there were at least four embankment streets in Shashi, stretching for several miles.[38] Since many of the main roads in Shashi in the early PRC were built on dikes dating back to the Tang-Song era, the city has been regarded as "a city on an embankment."[39] Another place with long embankment streets was Haoxue, where streets on the Great Jingjiang Dike stretched for several miles, and even some government offices were built on the dike. The embankment streets of Haoxue reached their heyday before the Anti-Japanese War.[40]

It is unclear how many market towns were located on Jianghan's dikes before 1949, but even sporadic data indicate that their number was not small. A 1949 survey found that twenty-five streets and about a hundred villages were built on and/or along the Yangzi River dike in Jianli County,[41] and at least

35 45-1-3; 45-1-8.
36 Hu Zuhe 1999 [1838]: 182.
37 *Fenshui zhen zhi* 1998: 1; *Xiliuhe qu zhi* 1986: 80–81; *Miaotou zhen zhi* 2010: 104–05.
38 *Shashi shuili difang zhi* 1994: 78–79.
39 *Shashi shi zhi 1* 1992: 278.
40 *Jingjiang difang zhi* 2012: 742; *Haoxue zhen zhi* 1986: 6.
41 *Jianli xian zhi* 1994: 151.

seventeen towns were built on dikes along the Yangzi and Dongjing Rivers.[42] In 1950 it was reported that twenty-five towns (counting towns and streets as one), with thousands of homes, were built on and/or alongside the Dongjing River dikes.[43]

The local economy and people have surely benefited from these embankment streets and market towns, but they also present a threat to the dikes. First, the construction of houses itself is harmful to dikes since it is liable to lead to leaks, and makes maintenance and repair more difficult, and also creates other problems. Second, householders have been reluctant to proactively report any potential dangers found in their houses out of fear they would be torn down. This in fact caused a dike rupture in Xiaogan in 1950 when an owner did not report a leakage in his house but tried—and failed—to fix it himself.[44] Third, householders also built outhouses, pigsties, and the like on dikes, which increased the threat of leakage and collapse, and some planted bamboo and trees, which attracted termites and badgers. And, finally, some business owners even opened oil press shops on a nearby river dike, which not only weakened the dike due to vibrations, but also lured turtles, which dug chambers that could cause a collapse.[45]

Thus, ensuring the integrity of dikes requires the removal of residential and commercial buildings and streets. In fact, in the Laws and Regulations of the Qing Dynasty, completed in the reign of Qianlong (1736–1795), only preexisting buildings on dikes that did not block dike work were allowed to remain, and no new ones could be built.[46] That there were nonetheless countless embankment streets/houses showed that the local people did not take these regulations seriously. After the 1788 flood, the Qianlong emperor immediately issued an edict to remove all the houses—about ten thousand—on the Wancheng Dike. But this edict had limited effect. Some houses were removed, but many still stood. About a hundred years later, Ni Wenwei found it was very challenging to remove densely packed houses on the Wancheng Dike. In 1832, for example, to reinforce the dike section in Haoxue, all the houses on the embankment streets were demolished, but the locals immediately rebuilt them and restored the embankment streets to boot. In the gazetteer he compiled, Ni added a note explaining that "people hate to leave a place where they have lived a long

42 *Jianlin shuili zhi* 2005: 390.

43 *Dongjinghe difang zhi* 1994: 78–79.

44 *Jingjiang difang zhi* 2012: 742; *Xiaogan diqu shuili zhi* 1996: 111.

45 Ni Wenwei 2002 [1885]: 138, 143.

46 Zhang Youyu and Gao Chao 1991: 140.

time, making it difficult to ban [embankment streets]."[47] Out of sympathy, the official in charge of a project to reinforce the dike section with embankment streets in Shashi in 1891 did his best to remove as few houses as possible.[48] The Republican government continued to issue regulations banning or ordering the removal of houses on dikes, but it did not have much success either,[49] except in one township in Songzi in 1946, when the locals were willing to rebuild their houses off the embankment street after the Japanese destroyed it.[50]

The situation only began to dramatically change in the PRC. From 1951 to 1954, about a thousand houses were removed from the Yangzi River dike in Honghu, and from 1951 to 1956, all embankment streets and 4,876 households (4,700 on the Yangzi River dike, and 176 on the Dongjing River dike) in Jianli were removed.[51] In 1952, to reinforce the Great Jingjiang Dike, all 1,316 houses on the embankment streets of Shashi—1,993 households and 7,446 people—were removed within ten days; from 1952 to 1955, all the embankment streets in Haoxue were removed; the houses on the embankment streets in other towns in Jiangling were also removed. The embankment streets on the Great Jingjiang Dike, which had existed for centuries, eventually disappeared in the early PRC. Thus the maintenance, repair, and reinforcement of the dike was no longer hindered by houses built on it. Since 1980, houses within fifty meters of the dike were also gradually removed.[52] The same occurred along the Han River dikes. In Zhongxiang, all 509 houses were removed from its dikes in 1952.[53] In Qianjiang, fifteen small towns and 4,948 houses were removed from its dikes from 1952 to 1958.[54] In the first ten years of the PRC, 6,754 houses in total were removed from atop the Han River dikes.[55] Market towns and thousands of houses were also gradually removed from the dikes along the Dongjing River in the early PRC.[56] To be sure, it was very difficult to rid the dikes of all those houses and market towns, as the archives in Honghu and Gong'an for the

47 Ni Wenwei 2002 [1885]: 26, 95, 137.

48 Shu Hui 2002 [1896]: 427.

49 *Dongjinghe difang zhi* 1994: 271; *Jianghe xiufang jiyao* 1971 [1936]: 194; 100-8-104: 3, 49, 60, 87.

50 100-2-371: 20–23.

51 *Honghu xian zhi* 1992: 119; *Jianli shuili zhi* 2005: 390.

52 *Shashi shuili difang zhi* 1994: 79–80; *Haoxue zhen zhi* 1986: 6; *Jiangling xian zhi* 1990: 327; *Jingjiang difang zhi* 2012: 283, 742.

53 *Zhongxiang shuili zhi* 1998: 67.

54 *Qianjiang xian zhi* 1990: 317.

55 40-1-180.

56 *Dongjinghe difang zhi* 1994: 81.

year 1961 reveal.[57] In fact, if one drives on the dikes on the Jianghan Plain today, one will still find houses on them in many places.

It is clear, nevertheless, that the PRC state has done a much better job of removing termite nests, badger burrows, and streets from dikes than its predecessors. But it would be unfair to say that the previous governments did not recognize the seriousness of these problems nor that they failed to take any corrective actions. On the contrary, dike maintainers in both the Qing and the Republic were aware of the problems and tried hard to solve them, but, for various reasons, largely failed. If all one has are shovels, hoes, and other simple implements, digging down into the core of a dike runs the risk of undermining it. In the PRC, however, workers have been armed with new technology and machines, alleviating that concern. In the case of embankment streets, the problem the state prior to 1949 faced was that it did not deeply penetrate, nor have firm control of, local society. Furthermore, private ownership in the Qing and the Republic made it difficult for dike maintainers to tear down houses and shops on embankment streets.

3 Recruitment and Assignment of Laborers

Other than dike maintainers/keepers who lived on dikes all year round, or at least part of the year, and were responsible for daily maintenance, from time to time, a great number (and sometimes a huge number) of laborers were also recruited for such annual repairs as rebuilding damaged dikes, reinforcing weak ones, and constructing new ones.

In the Qing and the Republic, laborers who worked in high-water control were usually fit and strong, although for annual repairs both elderly men and youth sometimes were called upon to do some light jobs, though this was usually not recommended. Besides, working as high-water control laborers was usually compulsory, whereas those who participated in annual repairs could get some payment, particularly for work-relief projects.

The laborers who worked on annual repairs were usually recruited from areas protected by the dikes they repaired. For example, in Jingshan in the Qing, recruits for annual repairs were apportioned based on the amount of land they worked, but traditionally owners of lake land had to contribute three times more laborers than owners of mountain land. Furthermore, men who lived close to dikes would themselves go to the dikes to fulfill their duties, but

57 SZ113-2-195C.

those who lived far away might avoid doing any work by making a cash payment to the government, and in turn the government then used this money to hire dike laborers.[58] The Jingshan County gazetteer of 1882 pointed out that this was actually a good thing. It argued that if the people doing the annual repairs lived close by, they would not be careless since sloppy work could endanger the dike, putting their own lives in jeopardy.[59]

For polder dikes, residents of some neighboring polders cooperated well, while others did not. In some cases, cooperation even among residents within the same polder was wanting. This, however, does not necessarily mean that the people involved simply refused to cooperate. For instance, a Qing source mentioned that there were so many dikes in Mianyang that it was difficult to endlessly deploy people and expect them to cooperate. Instead, it was suggested that the responsible officials should avoid rigidly sticking to the law and instead be pragmatic and act in line with local conditions.[60]

The basic principle behind recruiting and assigning dike laborers generally still applied in the Republic, although the newly established dike bureaus might directly contract a dike project to a contractor who then hired laborers to complete the work, thus eliminating the reliance on laborers assigned by their villages. Or in some instances the two could be combined: the bureau could assign projects to the local government, which then chose a contractor within its jurisdiction, and the contractor was responsible for recruiting laborers and completing the project.[61] In Songzi, one of the major responsibilities of the annual repairs and high-water control agencies of people's polders was to organize the polders' residents to carry out annual repairs (see Chapter 4). In 1942, the government of Hubei, to reduce the burden on the people, ordered that dike fees should not be collected by each annual repairs and high-water control agency of people's polders, but rather by the township offices, on behalf of the county government. After discussion with county bureaus and annual repairs and high-water control agencies of people's polders, the Songzi County head reported back to the governor of Hubei that this method was very difficult to implement mainly because the township offices were already burdened with a heavy workload.[62] Thus, until the end of the Republic, in Songzi dike fees were still collected by the annual repairs and high-water control agencies of each people's polder.

---

58 *Hubei An Xiang Yun dao shuili ji'an* n.d., 11: 2ab.

59 *Jingshan xian zhi* 1882, vol. 4: 3b.

60 *Hubei An Xiang Yun dao shuili ji'an* n.d., 11: 2b–3b.

61 *Jianghe xiufang jiyao* 1971 [1936]: 152–53.

62 100-6-1060: 40, 46.

As in the Qing, in the Republic landlords were only responsible for land taxes, while their tenants had to answer all calls for dike work, at least in Jianli.[63] In Songzi, on the other hand, it was reported in 1948 that landlords in Datong polder were responsible for dike fees, while tenants were responsible for dike labor, although the landlords paid their tenants some money for this work.[64] In Xiaogan it was reported that the burden of dike fees was so heavy that the local GMD government even used force to drive people, including children from families that could not pay their dike fees, to work on dike projects. An old peasant in Tianmen also said that in the Republic he was forced to labor on the construction of the Distance Dike even though he was only thirteen years old. No matter what the circumstances, working on dikes could be exhausting, and many people were injured, and some were even disabled.[65]

Sometimes people might voluntarily work on dikes. In 1911, the Han River dike was breached in the town of Shayang in Jingmen, and in the following winter and spring, a great number of people were hired to rebuild it. In the final days as the breach was being closed, thirty to forty thousand people went to the worksite every day. Some went voluntarily and worked for free.[66] These volunteers likely benefited from this dike; they might well have been unwilling to provide free labor to repair a dike on the opposite bank of the Han River.

As one might expect, there were also people who refused to go to dike worksites to undertake the tasks assigned to them. In Songzi, for example, the local government, at the request of the annual repairs and high-water control agencies of people's polders, frequently sent armed police to pressure these people to complete the work assigned to them.[67] In the CCP's discourse, this is a perfect example of the GMD using violence, or the threat of violence, to force people to work, but the truth is that these shirkers did not fulfill their duty and were not in any sense victims.

In the PRC, although most laborers working on annual repairs have still come from the countryside—that is, they have been mostly peasants—the methods of recruiting them and the ways of apportioning the tasks have been subject to constant change. In the early 1950s, work was apportioned to individual households according to the size of their land and number of working-age people in the family. During the collective era, work was first apportioned to entire counties (still based on area of land and size of the workforce), and then later it was

63 Hubei sheng zhengfu mishuchu tongjishi 1938: 11.
64 *Songzi shuili zhi* 2008: 178.
65 *Xiaogan shi zhi* 1992: 840; 42-1-158a.
66 *Jingmen shi shuili zhi* 1989: 166; Yang Jingqu 1982.
67 100-2-198: 50–52.

assigned to communes, brigades, and production teams. Finally, since 1981, it has again been apportioned to individual households but based on the area of the land, population, and workforce of the entire village.[68]

This is just a very general summary for the Jingjiang dikes—the situation varied greatly across time and place. At the very beginning of the PRC, considering the importance of dikes and their poor condition, the new state immediately took action to reinforce the existing dikes and to build new ones. This needed a huge number of laborers. The state's solution was to launch a mass campaign to mobilize newly "liberated" peasants to work on numerous dike projects. The locals thus had to shoulder a very heavy burden. Mainly based on a system of administrative allocation, peasants were mobilized and assigned to build dikes near and far. In Jiangling, for example, from winter 1949 to spring 1954, every year twenty to thirty thousand laborers were sent to repair and reinforce the county's major dikes.[69] According to a 1951 report, 3.9 million laborers participated in that year's annual repairs in Jingzhou. These laborers fell into three categories: those who were responsible for the work assigned to them, and had to provide their own meals and received no pay (26 percent of the total number of laborers); those who worked on assigned jobs but also received a state subsidy (15 percent); and those who worked entirely for pay, including laborers on work relief (59 percent).[70] But gradually, the first category became dominant in the following decades until the early 2000s when compulsory labor was abolished.

Although in the Qing and the Republic it was very difficult to ensure that the local people would cooperate in dike maintenance, and, in fact, there conflict was more common than cooperation in dike management on the Jianghan Plain,[71] in the early 1950s, under the slogan "all peasants under heaven belong to one family," the state had remarkable success in getting peasants to help each other. Thus, in Songzi, residents from Changshou polder and Sanhe polder helped residents from Shangxing polder with their annual repairs, and laborers from Jingshan were deployed to build dikes in Qianjiang.[72] The stability of the new state and its powerful party-state system made it possible to

---

68 *Jingjiang difang zhi* 2012: 508.

69 *Jiangling xian zhi* 1990: 326.

70 40-1-13. The subsidies varied. In Jingzhou in 1961 they were 0.8 yuan per standard laborer per day on annual repairs of major river dikes, and 0.4 yuan for secondary river dikes (D35-1-255), or 0.5 catties of grain per standard laborer per day on annual repairs of the Han River dikes in Tianmen in 1974 (42-1-149). In general, only laborers who worked on major river dikes and some important branch dikes and people's dikes were given subsidies.

71 Jiayan Zhang 2014: 58–64.

72 33-1-6; 40-1-73.

transfer people from place to place to work on dikes, obviating the need to rely on nearby people alone. After the 1954 flood, for example, the state mobilized laborers from four counties to reinforce the Great Jingjiang Dike: 62,000 laborers from Jiangling, and 70,000 from the neighboring counties of Qianjiang, Tianmen, and Jingmen.[73] In another example, in winter 1973, 480,000 laborers from six counties participated in the construction of the Honghu flood diversion project in Honghu.[74] And in December 1974, the central government began to fund a project to reinforce the Great Jingjiang Dike, bringing 89,588 laborers from Jiangling and three neighboring counties in the following spring. These laborers moved about 5.6 million fang of dirt, the greatest amount in one year in history.[75] Cooperation of this sort was rare in the Qing and the Republic in the Jianghan area.[76]

With the coming of collectivization, peasants' income relied entirely on workpoints. Hence local governments paid close attention to the workpoints earned from dike projects, otherwise, peasants would not be motivated. Thus, a regulation adopted in Jingzhou in 1957 provided that peasants who participated in annual repairs should earn the same or more workpoints than those who remained home and farmed. Once collectivization was achieved, sideline production and annual repairs could be pursued at the same time, which contributed to the success of many projects.[77]

Witnessing people coming from other places to work on their dikes naturally pleased the locals and buttressed their support for the new government, but those who had to shoulder the heavy burden of dike projects were not so happy, and, to evade dike work, some of them even refused to accept land allocated to them during Land Reform (which began after the defeat of Japan in World War II and ended, with a few exceptions, in 1953). Many peasants had to spend a quarter of their time on annual repairs. In Hanyang in 1953, for example, one household with 29 mu of land but only two laborers, was required to work 96 days on the dikes to complete their quota, which, of course, greatly reduced the time available for farming. In another example, in Gong'an, on average a laborer had to spend 164 days—111 working days plus 53 rain days—on dike work per year. One unlucky household had seven members but only one laborer, and this person was responsible for providing 221 days of

73 *Jiangling xian zhi* 1990: 327.

74 Jingzhou shi shuiliju et al. 2008: 175.

75 *Jiangling xian zhi* 1990: 327.

76 It was not unusual for the imperial state to draft "tens of thousands of laborers" to reinforce the Yellow River dikes (Ling Zhang 2016: 183).

77 40-1-125; 40-1-129.

labor a year! Some complained that the situation was even worse than in the Republic, since people could no longer escape the dike work assigned to them. In Qianjiang, the heavy demand for dike work even drove some local women to marry out or move out. To express their unhappiness, some even exaggeratedly complained that although there were only 365 days a year, they spent 390 days on dike work. And there were many heartbreaking stories. In one extreme case the wife of a landlord in Qianjiang committed suicide because no one in the family could go to the dike worksite to fulfill the assigned workload, and some who carried dirt to the dike worksite purportedly died of exhaustion.[78]

Again and again, both cadres and the public realized that a balanced allocation of workload was the key to the success of any dike project. In 1953, the state recommended that laborers in Jingzhou should be expected to work ten to fifteen days a month in winter, and no more than 60 percent of all laborers should be engaged in dike work, which was still a very high percentage. After 1953, the burden of dike work was reduced, and the policy gradually changed so that those who benefited from a particular dike were responsible for working on it.[79] Keep in mind that this burden was mainly physical, not financial, which was very different from the previous eras. Another difference was that in the early PRC landlords were also required to work on the dikes, unlike in the Qing and the Republic, when they usually shifted that burden onto their tenants.

In 1950 in Xiaogan, when 122 polders were flooded, the new county government desperately needed money to rebuild the dikes and to provide relief for the flood victims. The situation was so urgent that the county government resorted to bringing in laborers from unaffected areas to help rebuild the dikes, but some landlords still refused to pay their dike fees. A certain Sun, who once served as the head of his polder for thirteen years and peculated dike fees, not only openly refused to pay his dike fee, but also secretly held two meetings of landlords and rich peasants and bribed members of the peasant association. He was arrested and sentenced to death. After that, the number of people from his polder who went to dike worksite increased from 1,700 to 2,300.[80] These six hundred additional people were likely intimidated by the punishment meted out to landlords like Sun.

The burden grew and reached a peak in the GLF. In Tianmen, each working-age laborer had to complete 125 units of "standard labor" (*biaogong*)

78 SZ34-2-412ABCD.

79 SZ1-2-155; SZ1-5-59A; SZ34-2-412AB.

80 81-1-3b.

in water conservancy annually and work for five months.[81] The work time was reduced to forty-two days in 1961, and the workload was reduced to thirty to forty units of standard labor in the 1970s, and twenty-five to thirty after 1988.[82] In Shishou in 1964, on average each rural laborer had to spend five months on water conservancy, including annual repairs and the construction of new dikes, and high-water control.[83] Although not all this work time and "standard labor" were used for dike projects, five months a year is nonetheless astonishing. One can also imagine how heavy the burden of water conservancy was in the 1950s–1960s. But without the solid foundation of water conservancy that was laid at that time, the agricultural growth in the following decades might well have been impossible.

From its very beginning, the PRC government encouraged everyone to work on the annual repair sites. In 1949 in Mianyang, for example, all men aged from seventeen to fifty, if they were not disabled, were required to put in time on dike work. Although men younger than seventeen and older than fifty and all women had no obligation to work on dikes, they were praised if they voluntary worked.[84] In the following years, however, these people would not be allowed to work on dikes, a subject discussed below.

In the early 1950s, the state still took traditional customs into consideration. In Songzi in 1953, for example, during the annual repairs period, laborers were given twenty days off for the Spring Festival,[85] the biggest and most important traditional holiday in China, and a time when people usually make every effort to return home and celebrate with their family. But in the collective era, with the growing demand for labor on ever more water conservancy projects, and also because of the new revolutionary ideology of that era, laborers were encouraged or ordered to stay at their worksites without a single day off even during the Spring Festival. On the Dongxihu Farm worksite, about seventy thousand laborers continued to work without a break the Spring Festival of 1957.[86] And at the worksite of the Zhang River Reservoir project (from 1958 to 1966), some laborers from Tanglin Village spent three Spring Festivals laboring away without visiting home.[87] Rao Mintai, the head of this project, spent five

81 One unit of standard labor in Tianmen consisted of picking up 60 jin of dirt and carrying it 250 meters and repeating this 60 times (SZ68-2-226); in Qianjiang the standard was carrying 1 cubic meter of dirt over 354 meters (*Qianjiang shuili zhi* 1997: 266).

82 *Tianmen shuili zhi* 1999: 273–74.

83 39-1-68-12.

84 Yingcheng 49-1-2a.

85 D35-1-37.

86 Wang Pingsheng 2004: 61.

87 *Tanglin cun zhi* 2008: 66.

Spring Festivals on the worksite with the laborers.[88] On a 1974 diking project in Honghu, most laborers had to stay on their worksites during the Spring Festival, which was regarded as "a revolution in the field of thought"; one militia laborer did not return home even after his mother died, as he considered the dike work was more important than his personal affairs.[89] Something like this would be unimaginable and impossible today.

4 Women, Laborer Management, and Machinery

As noted earlier, not all dike laborers were men. Women, however, usually did not participate in dike work in the Qing due to the pressure of social customs, and in any case their bound feet would have made some physical tasks very difficult. In the early twentieth century, some women began to unbind their feet and work in fields, especially in dry land areas but generally not in rice paddies.[90] Some also started to work on dike worksites. In a case in Qianjiang in 1946, a pregnant woman was forced to go to a dike worksite, where she was beaten and suffered a miscarriage. And two other women, a mother-in-law and daughter-in-law, were beaten to death.[91] In another case, a Communist leaflet in the early 1949 excoriated the Nationalists for forcing pregnant women to work on dikes.[92] Since these cases were reported in PRC sources, they might have been exaggerated, but they at least demonstrate that some women indeed worked on dike worksites in the Republic, which can be verified by contemporary materials. For example, in Songzi in 1944, laborers, including some women, were hired to rebuild a dike section breached in the preceding year.[93] From a cultural perspective, it is understandable that women usually did not go to dike worksites to work alongside men. Furthermore, dike work is physically demanding. According to a 1951 report on annual repairs in Songzi, before 1949 women "very rarely" went to work on dikes.[94] It is safe to say that, before 1949, the main workforce in rural China consisted of men, which was particularly the case on dike worksites.

In the early PRC, especially during Land Reform and after, women were treated more or less the same as men, at least when it came to distributing

88 Jingzhou shi shuiliju et al. 2008: 22.

89 42-1-158a.

90 Jiayan Zhang 2014: 192–94.

91 40-1-195.

92 *Jingjiang difang zhi* 2012: 823.

93 100-10-204: 50–58.

94 33-1-6.

the land equally among all villagers. This dramatically changed women's social status in Chinese society and doubled China's labor force almost overnight. Although the revolutionary ideology of gender equality glossed over the differences in the physical strength between men and women, it stimulated women to engage in work that previously was done by men alone, such as farming in the fields and laboring in water conservancy projects, include carrying dirt on dike worksites, traditionally a man's job.

Even so, women were generally encouraged to work on the farm, freeing up men to go to dike worksites. Some women, however, also participating in diking. In 1951, a village in Mianyang was asked to send 200 laborers to a dike worksite, but more than 280 laborers, including 18 women, "voluntary" went.[95] In Hanyang, it was reported that about 20 percent of laborers working on dikes were women, some of whom even engaged in wielding a *wo* (a large stone, typically with ropes attached to the side, used to compact soil),[96] the toughest job on a dike worksite. And in both Shishou and Yingcheng, women accounted for a third of laborers on dike worksites, and some also operated a wo, as did some women in Zhongxiang.[97] Women, however, generally still only earned 70 percent of what men did.[98] At the end of 1951, the government of Jiangling ordered all men aged from seventeen to fifty years and women from eighteen to forty-five, as long they were not physically disabled, to go to the annual repairs sites in rotation.[99] In 1955 in a ward of Hanyang, more than a third of the laborers working on the ward's annual repairs worksite were women.[100] In 1956, in a township in Xiaogan, female laborers accounted for 35 percent of the total labor force on that year's annual repairs, and in Shishou, they comprised 60 percent of Changshou Township's labor force on its water conservancy sites.[101] In Ouchi, women comprised up to 70 percent of dike laborers.[102]

Especially during the GLF, there were instances where the demand for women's labor reached indefensible extremes. In Yingcheng, for example, even some lactating women were sent to work on water conservancy sites, leaving their babies at home to be fed by their grandparents with thick soup.[103] In Honghu at the same time, women who traditionally did not engage in

95 SZ1-5-65.
96 30-1-4a.
97 39-1-6-2; Yingcheng 49–1-9; 54-1-3a.
98 81-1-3a.
99 48-1-5a.
100 50-1-35a.
101 81-1-45; D35-1-89.
102 00046-001-0088.
103 Liang Yin'guo 1995: 103.

agriculture also participated in the construction of water conservancy projects that involved enclosing lakes and building dikes.[104] In 1970 in Pengxing Township, Xiaogan County, women also participated in a dike project during its peak time, and some production teams went so far as to bring cradles to the worksite.[105] This was impossible before the PRC, and it is also unthinkable now.

According to Kay Ann Johnson, these kinds of changes do not really demonstrate the victory of "gender equality," but rather the imposition of a "double burden" on rural women, since they were required to do what was formerly considered to be men's jobs and yet they still had to do almost all the household chores after they returned home from their daily work.[106] Not surprisingly, this generation of rural women strongly supported the one-child policy launched around 1979.[107]

But women did receive some special consideration on worksites. One source indicates that women's "special problems" should be taken into account, and another notes that, if a project lasted more than one month, women would be granted three days off with a food allowance per month for their menstrual period and a monthly "hygiene payment" as well.[108] In 1959 in Wuchang, for couples who worked on the same dike worksite, there was a "husband-wife meeting room" particularly reserved for them to live in rotation.[109] Even judged by today's workplace ethics, this is a rather humanized arrangement. These female laborers were by and large young and capable women, not pregnant or middle-aged women. In urban areas, female workers were reportedly also eager to participate in diking. In 1974 in Wuhan, women claimed they could do whatever men could do, and many went to work at the annual repairs site and were lauded for "holding up half the sky."[110]

Not all women who worked on dikes, however, were inspired by the quest for gender equality. On an annual repairs worksite in 1955 in Hanchuan, a woman with a three-month old baby was asked to go home, but she refused, saying that if she left she would die of starvation.[111] That could have been merely an excuse, but it also could have been true. In the same year, even pregnant women, old women, and elementary school children were also found on another annual

---

104 *Honghu xian zhi* 1963: 295.

105 *Pengxing xiang zhi* 2011: 62.

106 Johnson 1983: 157–77.

107 Hershatter 2011: 180.

108 D35-1-207b; D35-1-21.

109 Wuchang 49-2-18.

110 XX000065-WS01-1829-0008.

111 50-1-35a.

repairs worksite in Hanchuan. They were hoping to earn some husked rice.[112] All of them were not supposed to be at dike worksites in the first place.

Even so, many sources indicate that men were still the main workforce on dike worksites, and women were commonly encouraged to stay home to work in the fields. For example, in 1952, no female laborers went to dike projects in the third team of a township in Jiangling.[113] In 1954 in Jingzhou, thirty-two new latrines were dug on a dike worksite, but only four of them were for women,[114] suggesting there were far fewer female than male laborers. On Qianjiang's worksite to rebuild the Great Jingjiang Dike in the same year, more than 93 percent of the laborers were male. They had 212 latrines, while their female peers had only 20.[115] Other than carrying two baskets of dirt on a shoulder pole as male laborers did, women also served as cooks on dike worksites, or washed clothes for other laborers, or provide other logistics service by delivering firewood and grain to dike worksites.[116]

Some of those women, such as Xin Zhiying from Songzi, even became model laborers. In spring 1952, the 19-year-old Xin and her one hundred-plus fellow laborers walked one and a half days to the construction site of the Jingjiang flood diversion project, toting their baggage and rations with them on shoulder poles. Her job was to break up large stones, which, using the old method, was tough and slow. She invented a new and more effective way of breaking stones. For this she was selected as one of China's few above-rank model workers and was received by Chairman Mao in Beijing. Mao told her "both men and women are the same in the new society" and encouraged her to improve water conservancy in her hometown. After that, she dedicated her entire life to this mission. Although she once nominally served in the government at different levels, from township to province, she remained in her village as a peasant who earned workpoints, although she was the party secretary of her village. She had worked so hard on so many water conservancy projects that she suffered from many maladies, including some serious gynecological disorders. Over the years, she earned numerous political accolades, something beyond the reach of most of the people of China. She is also highly respected by her fellow

112 50-1-35b.

113 SZ34-2-412C.

114 D35-1-41a. These latrines also collected night soil for farming (48-1-32). At that time, night soil was still important in agriculture—chemical fertilizers only became widely applied in the 1980s or even the 1990s.

115 D35-1-50a.

116 38-1-7a; Liu Qiansheng n.d.: 121; 40-1-35a; 40-1-57a.

peasants and has been mentioned in many gazetteers of her home county and home township.[117]

Laborers often had to walk long distances to get to dike projects. In Xin's case, her hometown was not far from the worksite, but due to the limited means of transportation back then, she had to walk a day and a half. Long journeys on foot were common in the early PRC. In 1959, twelve thousand laborers from Mianyang walked about three hundred miles in about fifteen days to work on the Danjiangkou Reservoir project.[118] In Tanglin Village in Jingmen, many villagers had been frequently sent out to participate in several dike projects, such as the construction of the Jingjiang flood diversion project, the reinforcement of the Great Jingjiang Dike, the construction of the Zhang River Reservoir, the maintenance of the Han River dikes in Zhongxiang, and the reclamation of Dachaihu—a project to reclaim land in Zhongxiang along the Han River abandoned after the 1935 flood. Many of these dike projects are distant from Tanglin Village. Working conditions were poor, working hours were long (from six in the morning to six in the evening), and the work was heavy—each male laborer had to transport two to three cubic meters of dirt, and female laborers no less than 1.5 cubic meters per day.[119] Therefore, it was important to improve the living condition of laborers on the worksites, such as providing fresh green leafy vegetables and ample hot water.[120] Many worksites brought together veritable armies of laborers—numbering in the hundreds of thousands—and thus hygiene on the worksites was crucially important as well. In 1954, 193 laborers died due to illness on Jingzhou's worksites alone, and in Tianmen in 1955, fifty-three people were paralyzed because their work sheds were too damp. Thousands of medical staff were sent to worksites, and in some sites, clinics or even hospitals were established.[121]

Medical care made a very important contribution to the success of dike work. Bubonic plague frequently broke out in China from the Yuan dynasty (1279–1368) to the early twentieth century and caused disastrous socioeconomic consequences.[122] In the 1970s, when I was a child, more than once we heard horrific stories about death caused by rats on dike project worksites—not bubonic plague, but hemorrhagic fever (transmitted by the striped field mouse [*Apodemus agrarius*]). It is difficult to verify those stories, but deaths from

117 *Shadaoguan zhen zhi* 2009: 295; *Songzi shuili zhi* 2008: 215–16; Xin Zhiying 2005.

118 Fu Zhongjin 2006.

119 *Tanglin cun zhi* 2008: 66–67, 71.

120 Wuchang 49-1-6.

121 *Jingzhou weisheng zhi* 1990: 14, 194–95; SZ18-2-92.

122 Benedict 1996; Cao Shuji and Li Yushang 2006.

hemorrhagic fever did occur in several cases, such as in winter 1965 and spring 1966, on the site of a river project where 1,221 laborers were infected, and 69 (or, according to some sources, 83) died.[123] In another case, in Shayang in 1983, after floodwater was discharged, hemorrhagic fever broke out and caused the death of 220 people.[124]

Even laborers who did not live far from worksites faced the difficulty and inconvenience of commuting on foot. In 1816 in Qianjiang, because their homes were distant from the dike worksite, laborers arrived at worksite late but left early to go home. Thus, they did not work the whole day, and consequently the dike project stretched over several years but still remained uncompleted.[125] This sort of situation was still common in the PRC. In 1951 in Jianli, some laborers who lived 10 li away from their dike worksites had to spend a long time on the road.[126] Thus, it was suggested that laborers try to board in peasants' houses 5 to 8 li away from the dike. If there were not enough houses, the laborers had to build work sheds, and bring their own rice straw.[127] Because of a shortage of housing, some laborers in Jianli in 1951 were even encouraged to sleep on boats.[128] Since many laborers had to stay in other people's homes, it was important to maintain a good relationship with their hosts.[129]

It is easy to understand why a project would need hundreds of thousands of laborers, since traditionally humans were the main source of power for any dike project. On dike worksites in Jingzhou until the 1950s, only simple tools and equipment were available: dirt was moved by laborers carrying a pair of baskets on a shoulder pole, and other operations were carried out using shovels and wo. Later wheelbarrows, walking tractors, tractors, tipcarts, and trucks were introduced.[130] In the early PRC the wo was still the chief device for compacting dirt, but caterpillar tractors began to be used in the mid-1960s.[131] In 1978 in Jianli, some machines were first used in annual repairs to replace humans in transferring dirt.[132] In recent decades, the use of machinery in dike projects has become increasingly common, and on some projects in Jingzhou District work

123 *Hubei shuili zhi* 2000: 692; Liu Zhenggang 2008.
124 *Shayang xian zhi* 2013: 311.
125 Shuili dianlibu 1991: 615.
126 22-1-021-012.
127 D35-1-41b.
128 1-1-029-022.
129 48-1-21.
130 *Jingjiang difang zhi* 2012: 232–33.
131 *Hubei shuili zhi* 2000: 1231.
132 *Jianli xian zhi* 1994: 158.

has been entirely done by machinery.[133] In Hanchuan, all water conservancy construction in the twenty-first century has also been done with excavators.[134]

Coincidentally, the upsurge in the building of water conservancy projects, including dikes, gradually cooled down in the 1980s. At the same time, the development of China's non-farming economy drew tens of millions of villagers from all across China to work in towns, cities, and coastal areas. Looking at a few typical cases in Hubei, in 2005, 60 percent of the households in Tanglin Village in Jingmen and 70 percent of the labor force in Zhugengtan Township, Qianjiang, worked outside.[135] Some peasants from Lengshui Township, Hanchuan, even worked abroad.[136] In Xiaxindian Township, Yunmeng, in the old days peasants usually went out to fish in the slack season, but after 1979 they took up non-farming jobs, mostly construction work, all over the country.[137] In Jianli, almost all peasants farmed before 1949, but by 2008, a third of the rural workforce worked or engaged in business in cities.[138] At the same time, the system of compulsory labor for water conservancy projects gradually ended by the turn of the century. In Xiaogan the system was finally abolished in 2005, and since then it has become increasingly difficult for the local government to mobilize and recruit villagers to work on dike projects. Thus, machines began to be used.[139] The same thing happened in Hanchuan. When the agricultural tax was abolished in 2005, the system of compulsory labor for water conservancy projects was also abolished, leading to a shift to the use of machinery and hired laborers.[140]

This has been the general trend in Hubei. Before the 1970s, most water conservancy projects relied on the manpower of the local people. In the 1980s and afterward, such projects, especially large and state-funded projects, began to rely on construction companies selected via public bidding. Since 1979, the state could no longer deploy rural laborers for water conservancy projects at will, a change that paralleled the rural reform. Rural laborers who work on these projects are very different from those who worked on projects in earlier years: they are now hired by construction companies, not sent by their production teams; they receive wages, not workpoints or subsidies; they need to have some familiarity with the technology involved, and they need to do more than

133 *Jingzhou qu zhi* 2013: 82.
134 *Hanchuan shi zhi* 2010: 826.
135 *Tanglin cun zhi* 2008: 2; *Qianjiang shi zhi* 2011: 747.
136 *Lengshui zhen zhi* 2012: 166.
137 *Xiaxindian zhen zhi* 2012: 99.
138 *Jianli xian nongye zhi* 2011: 150.
139 *Xiaogan shi shuili zhi* 2011: 233–34.
140 *Hanchuan shi zhi* 2010: 206.

just provide physical power.[141] This is another of the profound changes in the history of the PRC.

5 Funding

In general, the funds for annual repairs in the Qing came from two sources: the state and the local people. Funding was drawn from the earth fee or dike fee (apportioned according to farm acreage), donations, fines, interest, loans, a great variety of taxes, and state appropriations.[142] In theory, anyone with land protected by a dike had to pay dike fees; even members of the gentry and degree-holders (men who had passed the imperial civil service exams), by law exempt from corveé labor, also had to pay dike fees.[143] But in reality, the burden of dike fees was uneven: the poor paid more and the rich paid less.[144] Sometimes the fees were particularly onerous, so much so that in some extreme cases people had to sell their wives and children to come up with the money.[145]

Suffering from frequent water calamities in the mid-nineteenth century, the Jianghan people became even more impoverished and craftier. If the dikes were to be maintained separately, the rich would secretly transfer the dangerous sections to the poor; if the dikes were to be repaired cooperatively, small households might refuse to pay the dike fee citing the excuse of poverty.[146] In this case, it seems they were using "weapons of the weak" against the wealthy, which implies that small or poor households were not always the victims or the exploited party in the struggle for power within China's rural communities.[147]

The earth fee continued in the Republic, as did the problems associated with it. In 1926, the state started to levy a special dike tax in thirty-six counties with dikes in Hubei, and used the revenue for high-water control and annual repairs of the major dikes along the Yangzi and the Han Rivers.[148] In theory, this tax was levied based on the extent of land owned, but in reality—at least in Honghu—landlords used many excuses to shift the burden to their tenants, and some poor peasants even lost their land to landlords because they could

141 *Hubei shuili zhi* 2000: 1185–88, 1208.

142 Jiayan Zhang 2014: 54–58.

143 *Gong'an xian zhi* 1874, vol. 3: 80a; Chen Zhenhan et al. 1989, vol. 3a: 111.

144 *Jingshan xian zhi* 1882, vol. 4: 17ab.

145 Chen Zhenhan et al. 1989, vol. 2: 408; *Xiangdi cheng'an* 1969, vol. 1: 135b.

146 Hu Zuhe 1999 [1838]: 157.

147 On "weapons of the weak," see Scott 1987.

148 LS31-4-58.

not pay the heavy dike fees.[149] It was not easy even for the common people to pay their dike fees. Some of them complained in 1931 that they had to walk long distances and spent two to three days just to pay the fees.[150] And some had to sell their land to pay the fees.[151] It is no wonder some even went so far as to advocate abandoning dikes because the dike fees were oppressive.[152]

This has long been viewed by the Communists as a perfect example of rural exploitation before 1949. Some newly available archives, however, provide a different picture. For instance, in Desheng polder in Songzi in 1947, landlords were responsible for dike fees, but many of them were absentees, making it very difficult for the dike managers to collect their dues, and thus their tenants were asked to pay the fees, in the expectation that their landlords would repay them.[153] A similar custom can be traced back to the Qing. In the town of Yuekou in Tianmen, in the early nineteenth century all land leases stipulated that the tenants had to "help" their landlords in annual repairs. Usually the leases stipulated that one laborer had to be provided for each mu of rented dry land. Some even made it clear that if the tenant refused to "help" the landlord with dike work, the latter had the right to find a new tenant.[154]

In the early PRC, people no longer had to pay the earth fee or dike tax,[155] and instead, most major dike projects were covered by state appropriations, with the local people providing the labor. Small or local dike projects were the responsibility of local government, the basic principle being that those who benefit from a particular dike must pay for its annual repairs.[156] Starting in 1972 Jingshan County, for example, launched a water fee to cover expenditures on the county's dikes.[157] And in 1994, the government of Hubei began to collect a dike maintenance fee from all industrial and commercial enterprises and individuals (2 percent of their regular tax), agricultural enterprises and individuals (currently at the rate of 5,000 grams of rice per mu per year), and

149 *Honghu xian zhi* 1963: 120–21; *Honghu xian zhi* 1992: 101.

150 LS1-5-849.

151 LS20-11-00001162-001.

152 *Zhongxiang xian zhi* 1937, vol. 3: 30a.

153 100-1-99: 49; 100-8-108: 47.

154 Zhang Jianmin and Tang Gangmao 2014: 509–88.

155 In 1951 in Jiangling, Land Reform had not yet been carried out, and landlords were required to take full responsibility for water conservancy of the land they rented out; and clan land, temple land, and charitable land—all of which were exempt from water-control fees before 1949—were required to shoulder the burden of water conservancy (48-1-5a).

156 To balance the burden of dike fees, there was a proposal in 1954 in Hanchuan County to collect an interim house property tax in some of the county's towns (50-1-15a). It is unclear whether this proposal was approved and implemented or not.

157 *Jingshan xian zhi* 1990: 214.

enterprises and individuals engaged in forestry, animal husbandry, and fishery (currently 6,000 grams of rice per mu per year). The revenue from this fee is used for annual repairs of dikes along the Yangzi River, the Han River, and some of their important tributaries.[158] From 1990 to 2003, the local government collected tolls on the Great Jingjiang Dike, 40 percent of which were used for its maintenance.[159]

Since the turn of the twentieth-first century, loans, even international loans, have been widely used for dike projects. As early as the 1980s–1990s, loans from the World Bank helped fund enlarging some river dikes in Qianjiang.[160] In the 2000s, World Bank loans were also used in Shishou to reinforce its major dike along the Yangzi River.[161] In fact, World Bank loans and national debt were the main sources of funds for the reinforcement of the major dikes along the southern bank of the Jing River.[162] After 1998, almost all funding for dikes came directly from the central government. From 1987 to 2007, total expenditure on dikes in Honghu was 2,290.3 million RMB (or 277 million USD), of which 2,266.8 million came from the central government, and only 235.5 million from local sources.[163] In other words, about 99 percent of the investment in dikes in this county was funded by the state. In Hanyang, the years 2000–2005 were regarded as "the golden age of dike building," with 200 million RMB (or 24.2 million USD) invested to reinforce its dikes.[164] After the 1998 flood, the major funding for the reinforcement of dikes along the four southward branch rivers of the Jing River was also provided by the central government.[165]

In imperial China, another source of funds for dike work consisted of donations, which were also encouraged by the state. It was common for Qing officials to donate their *yanglian yin* ("silver for honesty," or "honesty-nurturing stipend") to dike work, such as in 1825, when a dike project needing 131,700 or more taels of silver was funded by a donation of "silver for honesty" from officials in Hubei, as well as donations by local gentry and salt merchants in Hankou. The prefects of Jingzhou also frequently donated their "silver for honesty" for dike work. Naerjing'e, the governor-general of Hunan and Hubei, donated 20,000 taels of silver to rebuild a breached dike in Zhongxiang in 1836. And a county magistrate was even promoted in 1830 because he donated

158 *Jingjiang difang zhi* 2012: 508–09.
159 *Shashi qu zhi* 2007: 115.
160 *Qianjiang shuili zhi* 1997: 12.
161 *Shishou shi zhi* 2014: 19.
162 *Jingjiang difang zhi* 2012: 314.
163 *Honghu shi zhi* 2014: 219–20.
164 *Hanyang qu shuiwu zhi* 2015: 111.
165 *Jingzhou shuili zhi 1* 2016: 762.

to dike work in Qianjiang. In the late Qing, it was common for merchants to donate money for the annual repairs of the Wancheng Dike.[166] Personal donations to dike projects continued in the Republic. In an example from the year 1925, the funds for a stone dike project in Xiantao were mainly donated by a mathematician of local origins.[167]

This long tradition in the Jianghan area, and elsewhere throughout China as well, however, disappeared in the PRC. Under the private ownership system in the Qing and the Republic, some people became rich, and China's tradition of charity encouraged them to donate money for building communal facilities, such as roads, bridges, and dikes. The state also encouraged donations to dike projects by honoring donors with a commemorative arch or plaque. Under the public ownership system in the early PRC, however, personal wealth was no longer glorified, and no one was super rich either. Furthermore, the government and the collective were financially responsible for almost all dikes. Even the smallest polder dike was taken care of by a production team, not by individuals, and hence there was no need for contributions except after disasters, when the public was called upon to donate to relief and reconstruction efforts. China's rapid economic development in the past several decades has given birth to a great number of upstarts—millionaires and even billionaires—who might donate money to build rural roads and houses for their fellow villagers or help the senior citizens in their hometowns,[168] but no one has donated to dike projects.

Fines for dike projects also disappeared in the PRC. In the Qing and the Republic, the formal salary of governmental officials—from provincial governors to county magistrates—was small, but many of them actually were very rich (through inheritance or corruption or by investing) and they were able to pay huge fines. For example, after the 1788 flood, Governor-General Shu Chang was fined the vast sum of 40,000 taels of silver,[169] and in 1827 the former magistrate of Jingshan was fined 20,061 taels.[170] Public ownership in the PRC, however, has made it impossible to hold cadres financially responsible for dike ruptures. While it has not been unusual in recent decades to find bureaucrats who have embezzled a huge amount of money, no bureaucrat would publicly admit that he or she is very wealthy. Instead, the punishment for incompetent bureaucrats in the PRC is not financial, but administrative or political. To be

166 Ni Wenwei 2002 [1885]: 189; *Zaixu xingshui jinjian* 2004: 39, 109, 200, 223, 586–87.

167 *Mianyang xian zhi* 1989: 129.

168 *Pengxing xiang zhi* 2011: 412.

169 Ni Wenwei 2002 [1885]: 210.

170 *Zaixu xingshui jinjian* 2004: 81.

sure, this kind of punishment might have a serious negative impact on, or even end, one's political career.

Looking at the sources of dike fees we can also see that the management of dike affairs differed over the years. In the Qing and the Republic, both the local state and the local people played a role, with bureaucrats, gentry, and villagers all having duties and responsibilities; but in the PRC almost all dikes came to be managed by the government at various levels, except for some very small polder dikes that were managed by villages, but still not by villagers.

6 Annual Repairs and Mass Campaigns in the PRC

In its early years the PRC took on annual repairs of dikes as one of its major tasks. At the same time, it launched countless campaigns across the nation, and annual repairs were unavoidably tied up with those campaigns. In the early 1950s, the most important tasks or national campaigns were Land Reform, the Movement to Suppress Counterrevolutionaries (a campaign from 1950 to 1953 to crush opposition from remnants of the GMD), and the War to Resist U.S. Aggression and Aid Korea (the Korean War). The relationship between diking and these campaigns can be described as follows: Land Reform inspired peasants to work on dikes to protect their own land,[171] the Movement to Suppress Counterrevolutionaries intimidated those who were reluctant to go to work on dikes (particularly landlords and rich peasants, who might never really have shoveled dirt before), and the Korean War stimulated everyone to go to work on dikes through unrelenting political propaganda.[172]

Let us turn to several cases that illustrate these relationships. First, Land Reform. In 1949 in Mianyang, major leaders were required to stay on dike worksites and cadres had to work side by side with laborers. This was justified by the argument that "in terms of leadership, diking should be taken as a mass campaign. This is a great revolution in the leadership of dike works; [we] should be serious [and recognize] that this is the best opportunity and occasion to mobilize the masses."[173] In Jianli in 1950, diking was regarded as "a great mass campaign and a sharp class struggle," and landlords were forced to go to dike

171 48-1-5a.

172 See DeMare (2012) for a more detailed discussion of the relationship between political propaganda and mass campaigns—from the angle of rural amateur drama troupes—in the early 1950s in Hubei.

173 Yingcheng 49-1-2b.

worksites to work or to pay dike fees.[174] In Songzi, several former directors of annual repairs and high-water control agencies of people's polders (including Xu Songqiao, who was "unanimously" reelected by 648 people as the director of Changshou polder in 1947) as well as some "local tyrants" were executed.[175] In 1951, on a dike worksite in Zhijiang, five "local tyrants" were executed and three were sentenced to imprisonment; laborers claimed that they were happy to shoulder ten extra loads of dirt per person after the execution of each local tyrant.[176] And in Zhongxiang, "eight counterrevolutionaries were suppressed" on a dike worksite.[177] It is unclear why these people were sentenced to imprisonment and/or executed/suppressed, but landlords were certainly frightened and had little choice but to go to the dikes. Thus, in Xiaogan in 1951, more laborers went to dike worksites after some local tyrants were "struggled."[178] In 1952 in Mianyang, one landlord, unable to endure the exhausting work, committed suicide after he returned home.[179]

During the Korean War, propaganda to encourage people to work hard on dike projects was particularly emphasized on diking sites. The key messages were, first, that the government built dikes for the benefit of the common people, whereas it was the powerful and the rich who benefited from dike projects under the GMD; second, only if the major river dikes were well managed could the minor branch river dikes be well managed; third, only a stable dike system could guarantee agricultural production; and fourth, diking was patriotic, no less than struggling against China's external enemies.[180] Laborers working on high-water control and annual repairs were called on to learn from Huang Jiguang—a Chinese combat hero who used his body to block a machine-gun slit on an enemy blockhouse at the Battle of Triangle Hill during the Korean War—and sacrifice themselves to ensure dikes were secure.[181]

Propaganda, which was regarded as the "soul" of dike worksites,[182] was also applied in the mobilization of peasants to go annual repair worksites. The Publicity Department of Mianyang County, for example, suggested that stories of the dark side of diking in the past (before 1949) be collected and used

174 1-1-010-004.
175 100-8-172: 9; 33-1-5 (1)b.
176 45-1-1.
177 54-1-3c.
178 81-1-3b.
179 D35-1-10.
180 38-1-7b.
181 40-1-180.
182 D35-1-127a.

as propaganda materials.[183] A 1949 report notes that propaganda on annual repairs sites included the use of blackboards, radio stations (to help illiterate laborers), the mass publication of popular propaganda materials, and leaders lecturing laborers.[184] In Jianli in 1951, people were told that in the past they built dikes for landlords, but now they built dikes for themselves; in the past the Nationalists seized villagers and used them as cannon fodder, forced villagers to pay dike fees and then pocketed the proceeds, and under the guise of building dikes actually dug out dikes and flooded crops. Reportedly, such stories inspired people to go to annual repairs worksites.[185] After the 1954 flood, hundreds of thousands of laborers participated in the rebuilding projects in the following years. In September 1955, Zhang Tixue, the Hubei party secretary, emphasized that it was propaganda, not force, that would mobilize the laborers to participate in the following annual repairs, thus underscoring the crucial role of political propaganda in dike affairs.[186]

In the same year in Qianjiang, to cajole people from a mountainous area (who thought their land would not be flooded by river water) to build dikes in a lake area, the government proclaimed that "all peasants under heaven belong to one family."[187] It cannot be denied that there were peasants who really bought this discourse and were happy to help others, but most villagers likely felt they had little choice but to obey the government's orders. In 1956, for example, to complete the annual repairs of the Distance Dike on time, the Jingshan County government brought in three thousand laborers, none of whom would benefit from the dike. Most of these laborers were selected by casting lots; only a few signed up voluntarily.[188]

That there were so many people on dike worksites during annual repairs provided an excellent opportunity for the new government to promote many other mass campaigns as well as recruit soldiers and political activists. In 1955 in Hanyang alone, 326 laborers from that year's dike projects were made members of the CCP and 939 were inducted into the Communist Youth League.[189]

---

183 SZ1-5-59B.

184 Yingcheng 49-1-2b.

185 1-1-033-026.

186 Wuchang 49-1-4a. Some sources considered propaganda to have been one of the main reasons for the success of high-water control in Hanchuan in 1954. In this case, propaganda was circulated via high-water control newspapers, blackboards, posters, slide shows, and entertainment performances (50-1-16a).

187 40-1-73.

188 00046-001-0032.

189 Wuchang 49-1-2a. The same occurred on high-water control sites. In Wuhan during the 1954 high-water control, 7,349 people were recruited as members of the CCP and 17,789 as members of the Communist Youth League (*Wuchang qu zhi* 2008: 172).

Although the interaction between annual repairs and rural society has been multifaceted, here I will discuss three representative mass campaigns: the campaign to eliminate illiteracy, the mutual-aid and cooperation movement, and the public health (or patriotic health) campaign.

6.1 *The Campaign to Eliminate Illiteracy*

When the PRC was established, up to 80 percent of the people of China could not read and write, the vast majority of whom were peasants. To wipe out illiteracy, the PRC state launched a campaign to eliminate illiteracy in 1950, and literacy classes soon sprung up across the nation. Annual repairs worksites provided an excellent opportunity to teach laborers to read and write. For example, in Hanchuan in 1956, on the worksite of a flood diversion project, cadres overcame many difficulties in carrying out the campaign to eliminate illiteracy among laborers. They turned work sheds into classrooms and beds into school tables, connected teaching with learning, and had those who learned earlier teach those who learned later.[190] In the same year, on the worksite of annual repairs of a township of Qianjiang, 6,756 laborers (74.8 percent of all laborers) participated in literacy classes. One laborer said that working on annual repairs was like going to school; he wished it would continue all year round. Six once-illiterate laborers even became teachers.[191] Up until 1963, literacy classes were still very common on water conservancy worksites in Jingzhou.[192]

6.2 *The Mutual-Aid and Cooperation Movement*

It is well-known that the mutual-aid and cooperation movement in agriculture had a close relationship with Land Reform, particularly on the North China Plain.[193] On the Jianghan Plain, the two were connected to dike affairs. In fact, mutual-aid and cooperation were critical to carrying out annual repairs in the region. In Jianli in 1954, the county government recognized that many laborers on the annual repairs worksite were worried about their farms back home. It determined that the only way to overcome their concern was to set up many temporary mutual-aid teams.[194] The Jianli County government even viewed mutual-aid and cooperation as the key to the success of its annual repairs.[195] To complete the huge task of annual repairs, a township in Zhijiang in 1955

190 50-1-45.

191 40-1-111.

192 48-1-95.

193 The formation of the Geng Changsuo co-op, a well-known model organization in the early PRC, could be traced back to a land-pooling group formed in 1944 by four labor-poor households (Friedman, Pickowicz, and Selden 1991: 61).

194 1-1-092-009.

195 22-1-085-016.

organized numerous mutual-aid teams to go to dike worksites in rotation.[196] These teams also could be used to help peasants who were torn between working on annual repairs and engaging in sideline production back in village. Both annual repairs and sidelines were important to peasants, but working at an annual repairs site meant one would miss out on the income that would be generated by sideline production. Here, mutual-aid teams also came into play. In a township of Mianyang, thirty-nine mutual-aid teams were set up within two months in 1953, and two-thirds of the township's residents participated in various mutual-aid teams.[197]

6.3 *The Patriotic Health Campaign*[198]

The so-called patriotic health campaign, launched in the early 1950s, was of particular importance on annual repairs worksites on the Jianghan Plain, since the successful completion of any project largely relied on the health of the assigned laborers. But in the 1950s laborers still died on annual repairs worksites, particularly from infectious diseases. In 1951, more than thirty laborers died of meningitis on annual repairs worksites in two townships in Zhongxiang, and thirty-four died of poor sanitation and unhealthy living conditions (related to dampness in particular) on a worksite involving Jiangling and Jianli.[199] In 1952, five laborers died of illness on the annual repair worksite in Zhongxiang.[200] In the same winter, in Mianyang more than a thousand laborers got sick every day—most commonly from the flu, bronchitis, and gastroenteritis—and five died of illness on a water conservancy worksite.[201] And in Qianjiang, 9 percent of the laborers on the annual repairs worksite got sick and thirty-four died.[202]

The impact of incidents like these is clear: they reduced the available labor force, increased medical costs, lowered the effectiveness of laborers, and sometimes caused panic. All these also entailed a serious political cost to the party and the government.[203] The local governments thus were urged to take measures to deal with these problems. First, since some laborers died because

196 45-1-11c.

197 D35-1-38a.

198 The term *aiguo weisheng yundong* (patriotic health campaign) has been translated by Ruth Rogaski (2004: 2) as "Patriotic Hygiene Campaigns," in the belief that "weisheng" is best rendered as "hygienic modernity."

199 54-1-3a; D35-1-4b.

200 54-1-5c.

201 D35-1-10.

202 40-1-35a.

203 D35-1-50b.

they were sick even before they went to the worksites,[204] the elderly, the ill, and pregnant women were not to be allowed to go to the annual repairs worksites in the first place, and if they were already on the worksites, they were to be sent back.[205] Second, medical personnel were to be dispatched to the worksites and set up clinics or even hospitals there. Third, and most importantly, laborers were to be educated to take good care of both public health and their own personal hygiene. They were asked to eat hot meals, eat cooked foods, drink boiled water (it was widely believed that boiled water could prevent stomach ailments), set up beds over, not on, the ground (to avoid damp), lay down a thick layer of rice straw underneath their beds (for warmth), use public latrines (and not relieve themselves anywhere else), and so on.[206] For the sake of convenience, one dike worksite set up many mobile toilets.[207] The local governments made great efforts to promote the idea of public health and personal hygiene by issuing detailed health conventions.[208] According to a report from Qianjiang, after years of annual repairs and high-water control, by the end of 1953 the local people "had gradually formed healthy habits."[209]

A report on the completion of annual repairs of the Distance Dike in 1956 recognized that the safety and health of laborers was the key to whether annual repairs could be completed on time. Because the organizers took this issue seriously, there were no deaths in that year's annual repairs.[210] Almost all speeches, reports, and summaries in the archives emphasize the importance of public health on dike-related worksites; that is, not just on the annual repairs worksites, but also on the high-water control worksites.[211] How all of this influenced the local people may be subject to conjecture, but it is certain that in Gao Village, in neighboring Jiangxi Province, many contagious diseases had been wiped out in the Mao era.[212]

However, inculcating good habits of public health and personal hygiene is certainly not easy. In 1960, early in the construction of the important Zhang River Reservoir, it was found that the laborers on the worksite slept on the ground, had no boiled water, and relieved themselves everywhere.

204 For example, two-thirds of those who died of illness on the worksite of annual repairs in Qianjiang in 1952 were patients before they went to the worksite (40-1-35b).

205 D35-1-127b.

206 40-1-35c.

207 Wuchang 49-2-18.

208 40-1-35d.

209 40-1-52.

210 00046-001-0032.

211 1-1-271-012; 40-1-35a; 40-1-35d.

212 Mobo Gao 2018: 148.

Unsurprisingly, nearly 9 percent of them got sick. Some of them, it was reported, were ill because of food shortages at the time.[213]

Although annual repairs in the PRC have encountered many problems, the achievements have been remarkable. Take the Great Jingjiang Dike for example. Around 1949, not only was the dike thin, but it also had three major weakness: it contained hidden dangers, such as termite nests, badger burrows, and house foundations; it was susceptible to leakage along its base; and some segments were in danger of collapse, caused by torrents that scoured the foot of the dike. All these were threats that should have been solved by annual repairs. From the year 345 to the founding of the PRC in 1949, the dirt used on the Great Jingjiang Dike amount to only 29 million cubic meters, and stone but 0.23 million cubic meters. In comparison, from 1949 to 1984, 70.97 million cubic meters of dirt and 4.64 million cubic meters of stone were used to reinforce the dike. In other words, the volume of dirt used on this dike within the first thirty-five years of the early PRC (1949–1984) was about two and a half times greater than that used in the preceding 1,604 years, and the volume of the stone was more than twenty times greater. This dike thus became higher, wider, and stronger.[214] More reinforcement projects were carried out after 1985, particularly after 1998. For all dikes in Jingzhou, the volume of dirt used from 1949 to 2010 (980 million cubic meters), was three and half times greater than the volume used in the preceding 1,604 years.[215] At the same time, hundreds of thousands of hidden threats were removed. On the dikes in all of Hubei, more than 240,000 hidden dangers were removed from the 1950s to the 1990s.[216] These achievements largely explain why dikes frequently broke in earlier centuries but suffered far fewer ruptures in the PRC.

---

213 *Zhanghe shuiku zhi* 1991: 158–59.

214 *Jiangling xian zhi* 1990: 326.

215 *Jingjiang difang zhi* 2012: 4. The 980 million cubic meters of dirt was enough to build a wall 4 meters wide and 6 meters tall running all the way around the equator.

216 *Hubei shuili zhi* 2000: 493.

CHAPTER 4

People's Polders: Annual Repairs and High-Water Control Agencies

There is little concrete information on the management of polder dikes in the Qing era. Sources usually emphasize that polder chiefs (*yuanzhang*) should be carefully selected from among men who were fair (thus unlikely to practice favoritism), astute and capable, and well-off (and hence might eschew embezzlement), otherwise it will be difficult for them to carry out their duty.[1] According to the Jiangling gazetteer of 1794, powerful clans controlled the position of polder chief for generations.[2] But how these men were selected and the extent of their power and authority remain unclear. It was not until the Republic, especially in the 1930s and 1940s, when many polders established annual repairs and high-water control agencies, that more information on polder management became available. These agencies and their directors quickly disappeared in the early PRC as the state took over the management of polder dikes. Thus, these agencies and their directors turned out to be a phenomenon specific to the Republic. Past scholarship has largely ignored the management of polder dikes in the Republic—or more precisely, it has rarely explored the subject in detail. No doubt this is because of a dearth of information. But newly available archival materials, particularly in the Songzi City Archives (formerly the Songzi County Archives), provide a rich trove of information.

A pro-Communist secretary of the Songzi County government under the GMD cooperated with underground CCP agents, managed to preserve some nine thousand or so volumes of archives in caves during the chaotic years of 1948–1949, and handed them over to the Communist government after it took power.[3] Documents related to the management of people's polders used here are only a tiny portion of this rich archival collection. These voluminous archives remain largely ignored by scholars. In terms of the subject of this chapter, the articles relying on these archives that have been published so far treat the relationship between water control and rural social structure and rural administration, dike disputes, the connections between polders and

1 Hu Zuhe 1999 [1838]: 199.
2 *Jiangling xian zhi* 1794, vol. 8: 10b.
3 Yang Yongkui 1999.

 | DOI:10.1163/9789004739482_006

rural market towns, and the formation of polders respectively.[4] None discuss the people's polders' annual repairs and high-water control agencies and their directors, the core of the management of people's polders.

Given their ideological stance, most PRC publications have portrayed the management of people's polders in the 1930s and 1940s very negatively. Almost unanimously they describe the directors of annual repairs and high-water control agencies of people's polders as "evil" gentry who seized their positions by force and cared only about bullying and exploiting villagers and peculating dike fees. Their image in the materials in the archives, however, is far more complicated. Based on the newly available archival materials, particularly in the Songzi City Archives,[5] this chapter provides a more rounded discussion of this organization, its structure, operation, and leadership, its relationship with rural administration, and how all of these can help us rethink dike management in the 1930s–1940s in particular and in the late Qing, the Republic, and the PRC in general.

1 Establishment and Structure of the Agencies

Because of the paucity of information, one can only conjecture about how many polders established an annual repairs and high-water control agency in the 1920s. However, many people's polders began to set up such organizations in the 1930s, particularly after the 1931 flood. After that flood, the government called for the establishment of a bureau of annual repairs and high-water control which would be responsible for the major river dikes in Hubei.[6] Likely influenced by this order, many people's polders in the Jianghan area also established or were required to establish such an agency at roughly the same time. The sources show that such agencies emerged in 1932 in Shishou and Mianyang and in 1934 in Qianjiang.[7] In Songzi, the county government officially ordered the establishment of such agencies in January 1934; many polders there established an agency that year.[8] Also in 1934, there were 36 annual repairs and high-water control agencies of people's polders in Zhijiang. These organizations were also widely established in other counties, such as Hanyang,

4 Lü Xingbang 2011, 2015, 2019a, 2019b, 2020.

5 For this reason, in this chapter, all polders outside Songzi are indicated by the name of the county to which they belong.

6 LS31-4-1170.

7 Xiao Zijun 1991; *Dongjinghe difang zhi* 1994: 253; LS1-5-0790-001.

8 LS1-5-0607-0018.

TABLE 4.1 Number of personnel in the annual repairs and high-water control agencies in selected people's polders in Songzi, 1944–1947

Name of polder	Area (*mu*)	Number of personnel	Year
Changshou polder	88,000	50*	1947
Peixing polder	25,000	37	1945
Datong polder	107,350	31	1944
Sanhe polder	28,382	25	1945
Jiangsong Taibao polder	14,025	17	1944
Zhisong Zhongcheng polder	3,800	14	1946
Futai polder	7,100	13	1946
Caixue polder	925	6	1946

* *Including personnel of three sub-agencies or branch offices.*

SOURCES: 100-1-182: 33; 100-2-256: 43; 100-2-364: 15; 100-8-30: 1; 100-8-49: 7; 100-8-87: 83; 100-8-97: 15–17. THE AREA OF THE POLDERS IS FROM 100-2-242: 114 (1946).

Xiaogan, Jiangling, and Gong'an.[9] However, to "save money and trouble," many small polders did not have an agency.[10]

In most polders that had such agencies, there was only one, although some large polders had several sub-agencies or branch offices. The agency personnel usually included a director, one or two deputy directors, one or more dike managers (*didong*) and dike maintainers (*dibao*), and a support staff, which typically included such positions as secretary, accountant, supervisor, office clerk, cook, dike security guard, fee expediter, sluice gate keeper, and occasionally an engineer or technician, as well as others. Not all agencies had all these positions; the types and number could vary greatly depending on the size of a polder. The personnel in the agencies could be as few as six or as many as fifty (Table 4.1).

As Table 4.1 shows, the number of an agency's personnel was not necessarily commensurate with the size of the polder. Datong polder, at 107,350 mu, was the largest in Songzi, but its agency had only thirty-one personnel in 1944, whereas Caixue polder contained only 925 mu, but its agency had a staff of six in 1946.

There were no universal standards for the salaries of agency personnel nor for their term in office. In fact, standards differed even within a single polder

9 *Hubei xianzheng gaikuang* 1934: 67, 643, 905, 1001, 1038.

10 LS7-2-2707[1]-001.

and for a single polder in different years. The budget of Yongfu polder in 1944 and of Peixing polder in 1945 and 1946 are good examples of these differences (Tables 4.2–4.4).

TABLE 4.2 Budget for regular expenses, Yongfu polder, 1944

Personnel		**Monthly pay (cash, yuan)**		**Term in office (months)**
Title	*Number*	*Wage*	*Subsidy*	
Director	1	480		12
Dike manager	1	400		7
Dike maintainer	3		213.3	7
Clerk	1	400		12
Dike security guard	1	240		12
Cook	1	240		12

SOURCE: 100-8-154: 7–8

TABLE 4.3 Budget for regular expenses, Peixing polder, 1945

Personnel		**Monthly pay**		**Term in office (months)**
Title	*Number*	*Husked rice (dou)*	*Cash (yuan)*	
Director	1	3		5
Dike manager	3	3		5
Dike maintainer	12		600	5
Office clerk	2	3		5
Dike security guard	2	3		5
Fee expediter	12		600	5
Cook	1	3		5
Sluice gate keeper	4		133.3	12

SOURCE: 100-2-256: 43

TABLE 4.4 Budget for regular expenses, Peixing polder, 1946

Personnel		**Monthly pay (cash, yuan)**			**Term in office (months)**
Title	*Number*	*Wage*	*Living expenses*	*Subsidy*	
Director	1		9,600	20,000	12
Deputy director	1		9,600	16,000	12
Dike manager	3		9,600	12,000	12
Dike maintainer	12			8,000	6
Secretary	1	25,000			12
Clerk	1	20,000			12
Accountant	1	20,000			12
Dike security guard	3	12,000			12
Cook	1	12,000			12

SOURCE: 100-2-364: 36

The income of the directors in these three cases is even more complicated. In Yongfu polder in 1944, the pay of the various personnel of its annual repairs and high-water control agency belonged to different categories: "wages" for the director, dike manager, office clerk, dike security guard, and cook; and "subsidies" for dike maintainer. In Peixing polder, some personnel received pay in cash while others received pay in kind ("husked rice") in 1945. In 1946, the payment included wages, subsidies, and living expenses. Since the posts of director, deputy director, and dike manager were all compulsory positions or "voluntary posts," the incumbents received no "salary." Instead, they were given some living expenses and subsidies since they planned and prepared for high-water control in the agency's office all year round. But these explanations provided by the sources are likely excuses. If their living expenses and subsidies were totaled, the income of both the director and deputy director was higher than that of the secretary, and 3.2–3.7 times the income of a dike maintainer; while in 1945, the payment to different personnel was largely equal—the director and the cook got the same monthly income. Because of inflation, the monthly payment to dike maintainers in 1946 was 13.33 times the amount in 1945. In Yongfu polder in 1944, the payment also varied greatly, and there was a clear hierarchy among different personnel: the director's pay was about 2.3 times that of a dike maintainer. Also, in this polder some staff worked all year round, while others only worked for seven months. In Peixing polder in 1945 most personnel only worked for five months, but the sluice gate keepers worked for twelve; in 1946, they all worked year round except for dike maintainers, who only worked for half a year.[11] It is unclear why there were so many changes in this polder within such a short time.

In most cases, payment was made in cash, although in some instances it could also be in kind, including husked rice, such as in Datong polder in 1944 and in Upper Bailizhou polder in Zhijiang in 1950, ginned cotton in Shangxing polder in 1947, and soybeans in Yonghe polder in 1949.[12]

2 Operations

The most striking feature of the governance of people's polders was that it was to some extent democratic. In Songzi in the 1930s–1940s, both the director and deputy director of the annual repairs and high-water control agencies of people's polders were publicly elected by representatives or a general meeting of all residents of their polders. These representatives were usually household heads who lived in the polder in question, and they convened several meetings a year—Sanhe polder, for example, had a regulation that required a meeting be held every three months.[13] Aside from elections, all other important issues were discussed and decided upon in these meetings—issues related to annual repairs and high-water control, such as budgets and expenditures, the period and materials for high-water control, and other business. These meetings could be held in the office of the agency, in a temple or a nunnery, in an inn or a teahouse, or in someone's residence. Depending on the size of a polder and the meeting time, the attendees could be as few as fifteen (in Yongfeng polder in 1946), or as many as 235 (in Xihu polder of Hanyang in 1947), when a meeting dealt with a dispute over the appointment of the director.[14] But the record was in Songzi in 1948, when 648 residents of Changshou polder attended their year-end general meeting to elect a new director. There were so many attendees that an election via ballots was impractical, so they simply voted by raising their hands, twice. Understandably, under such circumstances, it was hard to show disagreement, thus the immediate past director, who convened the

11 100-2-256: 43; 100-2-364: 36-37; 100-8-154: 7-8.

12 100-1-182: 33; 45-1-6: 2-3; 100-8-43: 80; 100-1-97: 92.

13 100-8-125: 120.

14 100-8-120: 56; LS31-4-1407.

meeting and announced his resignation at the beginning of the meeting, was "unanimously persuaded to stay on."[15]

To discuss and make decisions on some special issues, many agencies in Songzi also had special or ad hoc committees such as an audit committee (these were very common) and a levy committee, a fundraising committee, a high-water control materials preparation committee, a high-water control committee, a dike work verification committee, an examination committee (which was assembled during the transition of the directorship), a land measurement committee, and an assessment committee (to determine the market value of dike fees paid in kind).[16] The sources note that since the task of the director of annual repairs and high-water control was onerous, residents in Sanhe polder in 1948 even proposed and set up a property preservation committee, which had its own bylaws and detailed rules and regulations.[17] All these various committees are a testament to the highly autonomous nature of community organizations at the time. Based on the records of neighboring Hunan Province, these organizations were designed that way: in 1935, the government of Hunan ordered each polder along Lake Dongting to establish a dike affairs agency, which was to be "an organization for the people to handle dike affairs themselves; [its] characteristics are similar to a people's autonomous organization."[18]

In some polders dike managers were also elected by representatives. Dike managers were assistants to directors, below directors but above dike maintainers. It was expected that they would have a certain level of education and qualifications, such as knowing when and how to build dikes, to strengthen important dike sections, to control high-water, and to mobilize the polder's residents.[19] In some cases, both dike managers and dike maintainers were "publicly recommended" because of their extensive dike experience.[20] Polder residents, however, frequently contested the appointment of a director or a dike manager without going through an election. In 1947 in Datong polder, a dike manager clashed with the director and secretly transferred his position to a certain Lü, who was denounced as "ignorant and incompetent and corrupt by nature." Lü automatically renewed his tenure the next year without an election. Thirty-two representatives of Datong polder signed a petition in which

15 100-8-172: 11-17.
16 100-8-2: 12, 38; 100-8-49: 34; 100-8-69: 31; 100-8-97: 174; 100-8-108: 8, 47; 100-8-149: 174.
17 100-8-125: 115-17, 129.
18 *Hunan sheng binhu zhoutu shichatuan baogaoshu* 1947: 106.
19 100-2-469: 18.
20 100-8-38: 62.

they claimed that since even governmental officials were now elected, Lü must be recalled.[21] A similar event occurred in 1948 in Hanyang, where the director of a polder wanted to resign, and asked several times for an election to select someone to succeed him.[22] It seems that elections had been widely accepted by residents of people's polders as the sole legitimate way to choose directors and dike managers.

General meetings of the polder residents were conducted according to agreed procedures. The following is an abridged translation of the Provisional Rules on Procedures of the Polder People's Representative Association of Sanhe Polder, Songzi, 1948:

> Article 6: Attendees cannot leave without the chair's permission.
>
> Article 7: Under any of the following circumstances the chair must order the attendee to leave: carrying weapons, being drunk, making noise, and disturbing the meeting and refusing to stop.
>
> Article 21: Decisions are valid if more than half of the attending representatives agree to a motion … The forms of voting include voting by a show of hands, voting by sitting and standing, and voting by secret ballot.
>
> Article 25: Minutes of each meeting must be kept.
>
> Article 26: Minutes must include the following: number of the meeting; meeting time; meeting location; names and numbers of representatives and non-voting attendees; names and number of absentees; name of the chair; name of the recorder; reported items; discussed items; extempore motions or examined items; the form of voting and the numbers of yes and no votes.[23]

I have particularly translated some articles of these provisional rules on procedures because when I first used the archives, I wondered why they all recorded items in the same way. Once I read these rules, I realized that is the pattern which had to be followed. The people who managed the general meetings were required not only to record the activities of the meeting in detail, but also to keep minutes. Thus, there is a large batch of minutes in the Songzi City Archives, a rare phenomenon in the Jianghan Plain (and likely in the

21 100-2-469: 18-20.

22 LS31-4-1405.

23 100-8-125: 123–26. The head of Songzi County rejected the organization of this representative association and Sanhe polder decided to "temporarily suspend" it in May 1948 (100-8-125: 131; 100-8-2: 30), but the minutes of almost all the meetings of Sanhe polder followed these rules.

nation). These minutes reveal a slice of the operations of the annual repairs and high-water control agencies of people's polders in Songzi: it is certainly not a case of some local despots controlling the agency and using it to savagely bully and oppress the local people, as is widely claimed in PRC publications, particularly gazetteers. Instead, although these agencies were nongovernmental organizations, their operations actually were very formal, at least in terms of procedures.

Some may argue that it would be naïve to take these records at face value since the recordkeepers would want to make the accounts look perfect. There are sources that in fact mention that other elections at that time were just for show, and bribery was common.[24] In a primary rural election in 1947, for instance, of 900-plus ballots in a bao (each bao had a hundred households) in Songzi, only one was authentic—all the rest were fake.[25] In 1948 in nearby Zhijiang, there was also widespread bribery in the election of national assemblymen involving offering money and banquets to voters.[26] A survey of neighboring Jiangling in the 1930s also reveals that it was common for the rich to exploit the poor in dike affairs (particularly in a polder along the Hudu River), by, for example, the manager and wealthy households paying fees ahead of time, and thus ensuring that only the rich could serve as managers.[27] But it is still hard to imagine anyone in Songzi would offer money or banquets to bribe people to get elected the director of a polder's annual repairs and high-water control agency, since this was basically a service post with little power and few financial benefits, at least according to the archives, as discussed below.

Sanhe polder also proposed some "provisional rules for elections of the polder people's representatives" in 1948. According to these rules, any man or woman over the age of twenty who had lived in the polder for more than six months had the right to vote for representatives. And among them, anyone over the age of twenty-five who had certain qualifications such as having been a schoolteacher for more than one year, or having served in a professional organization for more than one year, and who knew dike affairs well would have the right to be elected as a representative, but those who had political problems such as having betrayed the government or having been deprived of their civil rights, or financial problems such as engaging in corruption, or personal

24 *Songzi xian minzheng zhi* 1985: 16–17.

25 100-10-421: 17. The baojia system was an administrative system based on households, with 10 households making one jia, and 10 jia one bao. The bao head was roughly the same as the village head.

26 *Zhijiang minzheng zhi* 1990: 43.

27 *Hubei xianzheng gaikuang* 1934: 905, 1004.

problems (e.g., bad habits or addictions) had no right to vote or be elected. For unknown reasons, members of the armed forces (soldiers and policemen) were not allowed to be elected as representatives either.[28] In fact, according to the above-mentioned procedures, a person who carried a weapon was not even allowed to attend the meeting.

Although these rules were proposed by Sanhe polder, there is evidence that similar rules might have been applied in other polders as well. In 1948, a certain Wang was "elected" the new director of annual repairs and high-water control agency of Zhongcheng polder in Zhijiang and Songzi counties. Wang informed the county head of Songzi that he wished to resign because he lived in Peixing polder, he was illiterate, 90 percent of the residents in Zhongcheng polder had more land than he did, and most importantly, the election was secretly manipulated and was illegal. In this case, the county head ordered another election since Wang did not live in Zhongcheng polder.[29]

Both autonomy and free elections impacted the political character of the agencies in Songzi in the 1940s. Considering the variety of committees, the frequent elections, and the provisional rules for elections of polder residents' representatives, it is apparent that these organizations operated in a fashion very close to that of a democracy. In several cases, polder residents indeed used "democracy" as a reason to reject unparliamentary appointments of agency personnel. The same phenomenon also can be found in other places on the Jianghan Plain. In 1947 in Hexiang polder in Hanchuan, for example, a group of residents charged that the director had been appointed by the county head and many dike managers were appointed secretly, and complained that this did not conform to the principles of democratic elections.[30] Mark Elvin once referred to the similar management of water control in late imperial Shanghai and Guangdong as "a sort of proto-democracy."[31] But it is clear that the people in the Jianghan area believed what they practiced or wanted was true democracy.

3 Directors

Unlike the directors of the bureaus of annual repairs and high-water control of river dikes, who were selected and appointed by the county head, the directors of the annual repairs and high-water control agencies of people's polders

28 100-8-125: 112–13.
29 100-8-149: 204.
30 LS31-4-1404.
31 Elvin 2004: 117–18.

were usually elected by the people. One of the most important tasks of the above-mentioned meetings was to elect the director. To guarantee the legitimacy of these elections, a government official was to be present at the meeting as a monitor, otherwise the election would be regarded as illegitimate and invalid. To avoid cheating, elections were mostly conducted via secret ballots: one person first distributed the ballots, then he or another person collected the ballots; after the ballots were collected, they were counted by a teller, a recorder, and a scrutineer. The candidate who received a simple majority of votes won the election. In case of a tie, the scrutineer decided the winner by drawing lots. In some cases, candidates who came in second were elected as the deputy director.[32] In theory, their term in office was one year; but in reality, the term could be three years or even up to ten years in some polders, or even eighteen years in a polder in Hanyang.[33]

It is unclear how candidates were chosen. In one case, each *zong* (an administrative unit, consisting of one or more villages, within a polder) selected one person as a candidate.[34] But it is still unclear how each candidate in a zong was selected. According to revised regulations issued in 1947, the director and deputy director of the annual repairs and high-water control agency should be well-known local elites with extensive experience of dike affairs.[35] In Shishou in 1948, the meeting convener (the current director) and election monitor (the local township head) were required to nominate six to twelve such people as candidates before the meeting.[36] Simply put, in theory these candidates should have been well-known and familiar with dike affairs as well, but in reality that was not always true. In a case in point in 1945, a certain Cheng reported that his brother was elected the new director of the annual repairs and high-water control agency of Xida polder, but his brother knew nothing of dike affairs and was doing business outside and rarely came home.[37] Eventually, the current director had to continue to serve another year. It is worth mentioning that education was not a requirement. In another case in Songzi, a certain Shen wanted to resign from the directorship because he was a peasant without any

32 100-8-38: 63; 100-8-97: 77–78, 158–60; 100-8-149: 205.

33 100-8-49: 11; LS31-4-1405.

34 100-8-149: 207. In Songzi at that time, each polder was divided into several zong, but it is unclear how large a zong was. According to one source, a certain zong in Shangxing polder had only about 500 mu of land, but another zong in the same polder had five bao and 2,400 mu (100-8-43: 11, 83). It is likely a zong included one or more villages.

35 LS1-2-0484-007-001.

36 LS31-4-371.

37 100-8-160: 90.

formal education.[38] If education had been required, as an uneducated person, he could not have been a candidate for the directorship in the first place.

In some cases, directors did not seek reelection, but the representatives of polder residents, believing that some of these directors were incapable of fulfilling their duty, demanded an election. In Datong polder in 1946, twelve representatives called for the removal of the director since he did not examine the dike in winter and failed to cooperate with dike managers and dike maintainers. The polder's residents elected a more capable person to replace him.[39]

Directors did not have to be natives of the polder where they served. In 1947, Zhou Zhongshu, the director of the annual repairs and high-water control agency of Tianwangtang polder, wanted to resign since he was a native of Huanggang, a county downstream the Yangzi River. Because of the Japanese invasion, he had withdrawn to Songzi with troops and had been there for nine years, and now wanted to return to his hometown to visit relatives. His request was approved, and a new director was elected.[40]

PRC publications widely portray directors as aggressively pursuing personal gain. In fact, however, not all of them received compensation. Although, as noted earlier, in some polders, the directors usually received monthly wages or some subsidies, in other polders they received no compensation at all since serving as a director was regarded as an obligation to the community. For example, in 1948 in the annual repairs and high-water control agency of Tongxin polder in Songzi, all the personnel—one director, two deputy directors, six dike managers, and thirteen dike maintainers—were "positions without payment."[41] Thus, in Songzi, many people who were elected as directors were unwilling to serve and frequently wanted to resign, giving such excuses—including self-depreciating ones—as they were incapable, ill, or illiterate, or that they had served as director too long, they were too old, they had family-related hardships, they were attending school, and so on.[42] While some may have offered one excuse, others offered several. In 1945, a certain Wang, the former director of Yixing polder's annual repairs and high-water control agency, who resigned due to overwork and illness. One year later, he was surprised that he again was elected as the director of Desheng polder, even though he did not attend the election, which was not a general meeting of the polder's residents. In 1946 he was seventy years old and even sicker than before. His

38 100-8-93: 59.

39 100-1-182: 119–20.

40 100-8-38: 59.

41 100-8-69: 13.

42 100-1-182: 1; 100-8-103: 10; 100-8-106: 1, 11.

request that he be relieved of his post was approved, and the original director was ordered to organize another election and remain in charge until a new director was elected.[43]

Resignations required, first, the approval of the general meeting of the polder's residents, and, if that was forthcoming, then the approval of the county government. In most cases, resignations were accepted, but directors could be required to stay in their position if an emergency made that desirable. In 1946 in Desheng polder, for instance, the director wanted to resign, and the general meeting of the polder's residents approved the resignation and elected a successor. But the county government rejected the resignation because dike repairs were unfinished and there was an urgent need for high-water control.[44]

Since directors in Songzi frequently sought to resign, being a director was apparently not an attractive job, or at least not a pleasant or profitable one. In 1948 in Peixing polder, several meetings had to be held just to elect a new director, since both the outgoing director and the newly elected director refused to accept the position.[45] In Yongfeng polder, a certain Li was so angry that he accused the previous director of manipulating the election and "deliberately shifting the misfortune" to him by electing him as the new director. Li considered the directorship a bad job, and he demanded a new election.[46] This was nothing new. For example, in Jianli in the late Qing, some dike managers (*dongshi*) went bankrupt because they had to pay dike fees in advance with their own money.[47] In other locations on the Jianghan Plain, however, some people vigorously competed for the job. In Hanyang in 1946, the director of the annual repairs and high-water control agency of Xinhe polder was accused of bullying and exploiting residents in the name of diking. After he became director, according to his accusers, he became the richest person in the entire polder. People complained that he treated the polder as his "fiefdom." However, after an investigation, the county government explained that he was actually a public-spirited person and was elected by the public.[48] In another case, a director in Yao polder in Mianyang was accused of getting the position through bribery when the area was occupied by the Japanese. People complained that he knew nothing about dike affairs—he only knew how to extort money from the residents, which he continued to do even after the war ended. For example,

43 100-1-99: 115–16.
44 100-1-99: 45.
45 100-8-108: 122–31.
46 100-8-149: 107.
47 *Jianli xian zhi* 1872, vol. 3: 11a.
48 LS31-4-1406.

he and his team arrested, tortured, and fined a peasant to force him to pay more and work longer. His fellow residents were outraged that they were still at his mercy in 1947, two years after the surrender of the Japanese.[49]

In a polder located between Hanyang and Hanchuan, a certain Lin in 1937 was accused of getting the directorship via the power of his clan. Residents of the polder accepted that they had to contribute both money and labor, but they could not tolerate the fact that Lin knew nothing about dike affairs. He even took dirt from the bottom of the dike to block an opening—an extremely dangerous move—and did not prepare enough dirt for high-water control ahead of time, although he collected a great deal of money from residents for this very purpose.[50] In another polder in Hanyang, some residents in 1947 also complained that the director was appointed by the county head, not publicly elected.[51]

A polder is also a micro society, a factor that complicated a director's task. In an acquaintance-based society like rural China, where everyone was connected with everyone else, a director, who usually was a member of the polder community, would have no chance of successfully discharging his duty unless he could navigate the treacherous waters of interpersonal connections and act impartially and vigorously. That was exactly the reason the director of the annual repairs and high-water control agency of Jiangsong Taibao polder (a polder shared by Jiangling and Songzi) wrote in a 1944 report that the agency's personnel must "not spare anybody's feelings" in demanding that those who did not complete their work pay a certain amount in fees to make up for the work they failed to do.[52] In another case, many of the residents in a polder in Shishou had migrated from Hunan and frequent clashed with the natives. But the director of the annual repairs and high-water control agency, by the name of Deng, worked well with both parties, and was also relentless in pursuing people who did not pay their dike fees.[53] The source on Deng in the Hubei Provincial Archives particularly highlights his unremitting determination to collect dike fees.

Being a director could sometimes involve not merely dealing with residents' personal feelings but also their violent acts. In 1923 in Mianyang, for example, a member of the powerful Yan clan who served as the director of annual repairs and high-water control of Yaojiu polder pocketed the fees he collected from

49 LS31-4-1408.
50 LS36-1-69.
51 LS31-4-1404.
52 100-10-186: 18.
53 LS31-4-1724.

residents for high-water control, and then sneaked away when the dike was in danger of collapse. A stalwart youth from the Tang family in the polder led outraged residents in tearing down Yan's house. In search of revenge, the Yans proposed young Tang's father as the new director. In the following high-water season, Mr. Tang and his two sons stood at the most dangerous spot on the dike to protect it, while the Yan clan organized about a hundred members, armed with weapons, surrounded the Tangs, and planned to throw them into floodwater once the dike ruptured. But the Tangs were not deterred. They used their own money to incentivize the high-water control laborers, and the dike eventually held. After the autumn harvest, the residents gladly repaid the Tangs double.[54] It is clear being a director was definitely a challenging job in this polder. In another case, in 1935, the dike of Baili Islet polder in Zhijiang collapsed in July, but the directors of annual repairs and high-water control agency pocketed the funds for repairs. The dikes broke again in the autumn, and the floodwater inundated millet that had been sown after the previous flooding. Some residents of this polder were so angered that, with knives and spears in hand, they stormed the houses of the two directors. One was captured and killed, while the other was able to escape, but his 8-room house was burned down. Intimidated by the power of the people, the county government, although knowing the truth, did not dare take any steps to charge the people involved.[55]

It was also in Mianyang that the annual repairs and high-water control agencies of people's polders did not do much in the way of annual repairs. The directors collected fees but did not actually reinforce the dikes. Worse still, they countenanced jerry-rigging. Furthermore, people tended to protect only their own polders, hindering the discharge of logged water.[56] It is little wonder that Mianyang, located in the core area of the Jianghan Plain, suffered from frequent water calamities before 1949.

Unlike Songzi, where directors were elected by their fellow polder residents and then appointed by the county government, in Hanyang both the director and deputy-director were appointed by the county government,[57] but it is unclear if they were also elected by their fellow polder residents first. Guo Dehai, the director of the annual repairs and high-water control agency of Xihu polder in Hanyang, claimed in 1934 that he was "publicly recommended" as the director. Since he was a merchant, he was reluctant at first to take up the position, but he overcame his fear of difficulty out of the sense that he had a

54 *Mianyang Chenchangqu xiangtu zhi* 1987: 299.

55 Guo Yuneng 1989.

56 *Mianyang Chenchangqu xiangtu zhi* 1987: 69.

57 *Hubei xianzheng gaikuang* 1934: 67.

duty to serve the community. He spent a great deal of his own money on dike projects in this polder. But at the same time, some "dike sticks" (those who were in charge of dike projects but pocketed dike fees) were only interested in enriching themselves by misappropriating dike fees.[58]

In some cases, directors were very powerful. Zhao Nan'gai, resident of Shishou, is a good example. According to a PRC gazetteer, Zhao was a typical bad landlord. In a year with water calamities, he made loans to poor peasants who usually were unable to repay and thus lost their land to him. After water calamities in which silt formed more beaches or islets, Zhao enclosed the new land and rented it out. In doing so he increased his landholdings from 100 or so mu in 1902 to more than 4,600 mu by 1945.[59] According to other sources, however, Zhao was a more complicated figure. His father was a rural private school teacher who also acted as a lender and invested in rice mills, and then purchased land. Zhao Nan'gai, the oldest son, herded cattle as a child, but as an adult continued to purchase land. The Zhao family became the largest landlord in Shishou, and by 1949 owned more than 4,000 mu in three counties of two provinces. Zhao Nan'gai became the director of the annual repairs and high-water control agency of Jiuhe polder in 1932. He led residents in reinforcing polder dikes and dredging channels, but he also collected heavy dike fees since he also wanted to profit from his position. To win support, he also reduced or forgave rent in poor years. His power peaked in the years of 1945 to 1947, when he was concurrently the director of two polders' annual repairs and high-water control agencies, and at the same time served in the Shishou County government. In 1946 some residents accused Zhao of an array of offenses such as seizing cattle and quilts of recalcitrant villagers to force them to pay their dike fees. He was also accused of embezzling dike fees or loaning the fees out and pocketing the interest. Investigators, however, reported that he was honest and there was no evidence he had engaged in embezzlement. Moreover, he had reluctantly accepted the position of director, and he worked hard to block breaches in two polders. In 1948, when a wealthy landlord refused to pay his share of high-water control fees, Zhao sent troops to arrest him. After the landlord paid the fee, he was released.[60]

Both ideologized new gazetteers and relatively less partisan sources from the Republic make it clear that Zhao was no doubt a powerful figure. In Hanchuan in 1937, two residents of a polder sued the director of the annual repairs and high-water control agency for changing dike lines without authorization and

58 LS1-5-0764-001.

59 *Shishou xian zhi* 1990: 156–57.

60 LS31-4-60B; LS35-3-007-002; Xiao Zijun 1991.

bullying the weak.[61] This director was unquestionably a powerful figure as well. This points to a dilemma noted by Hu Zuhe, a local expert on water control, one-and-a-half centuries earlier: if a person was forced to take the position of the leader of annual repairs in a polder, he must be a coward and hence could not be effective at collecting fees; if, on the other hand, the person got the position corruptly, he could not be relied on to do a good job.[62]

In fact, cases reported in archival sources show that throughout the Jianghan Plain there were many directors who were guilty of wrongdoing. For example, in 1934, the head of Jiangling County asked the governor of Hubei to arrest a former director from Jiangling who had absconded to Wuchang with public funds he embezzled.[63] Some directors in Zhijiang in 1934 were certainly not "experienced in dike affairs" and got their position through bribery.[64] This sort of wrongdoing was also found in the management of river dikes as well. In a case in Jiangling in 1946, a director of the bureau of annual repairs and high-water control of the Great Jingjiang Dike got the position through bribery and knew nothing of dike affairs. Instead, it was said he was good at smoking opium, gambling, frequenting prostitutes, embezzling relief funds, and beating a laborer who dared to stop his jerry-rigging dike works.[65] In another case, on August 1, 1946, a director of annual repairs and high-water control of a township in Qianjiang escaped from custody after he had been accused of opium-smoking and stealing and selling relief flour.[66] It is unclear whether he was elected or appointed, but according to Songzi's regulations, he was not qualified to be the director since opium smokers (as persons with bad habits) were unqualified for the position.[67]

The experience of Songzi shows that not all polder leaders were rich or powerful; more commonly they were simply average people. But the sources also reveal that if a leader was not "astute" (capable), it was hard for him to carry out his duty. Documents in the Songzi archives portray almost all directors positively, but in the provincial archives the picture is less flattering. For example, the provincial archives contain a report on a case in Peixing polder where the director of the annual repairs and high-water control agency was sued in 1937 for overcharging dike fees as well as illegally arresting people or fining them for not paying the fees on time. Some of these people were bankrupted,

61 LS36-2-081-002.

62 Hu Zuhe 1999 [1838]: 184–85.

63 LS1-3-0297-004.

64 LS1-5-0854-001.

65 LS30-1-0200-001; LS30-1-0203-003.

66 LS7-6-0006269-003.

67 100-8-125: 112.

or according to the sources, even died.[68] The director in this case behaved like a typical "local tyrant." There were almost no cases of "dike vermin" in the Songzi archives, but in the neighboring county of Zhijiang in 1944, there were many "dike vermin" who embezzled dike fees and were arrested or punished by the newly arrived county head.[69] Simply put, the image of these directors is a rather mixed picture.

As late as 1950 directors of annual repairs and high-water control agencies of people's polders still existed in such places as Zhijiang and Hanchuan.[70] It is unclear when they were disbanded. According to a 1950 report, "all counties abolished their old system of annual repairs and high-water control agencies."[71] If this is so, then the position of director would have disappeared in that year.

The PRC archives usually emphasize that during the Republic, directors were landlords, and dike managers were rich peasants.[72] As the above discussion reveals, however, this is not true in every instance. In many cases, directors did have some economic power or social status in their community—without such standing, it would have been very difficult for them to do their job.

4 Income and Expenditures

The very name "annual repairs and high-water control agency" indicates the major tasks of these organizations: annual repairs and high-water control, both of which needed money. Thus, income and expenditures were the major items discussed in an agency's meetings.

In most cases, an agency's main source of income was the dike fees collected from the polder's residents. The fees could be assessed according to the acreage of land a household owned, no matter the quality or grade of the land.[73] More commonly, however, different grades of land were assessed different amounts of dike fees. For example, in Sanhe polder in 1945 fees were 50 yuan per mu for upper-grade land, 40 yuan per mu for middle-grade land, and 30 yuan per mu for lower-grade land.[74] Fees also could be in the form of goods. In Datong polder in 1948, fees were paid in unginned cotton: 4 jin per mu for upper-grade land, 3 jin per mu for middle-grade land, and 2 jin per mu for

68 LS63-2-005-006.
69 LS1-5-855A.
70 45-1-6; 50-1-1a.
71 SZ113-2-11.
72 33-1-2; SZ113-2-11.
73 E.g., 100-8-104: 36; 100-8-160: 33.
74 100-8-87: 64.

TABLE 4.5 Fees assessed according to grades of land in Songzi, 1940s

Location	Fees assessed per mu			Type of expenses	Year	Sources
	Upper-grade land	Middle-grade land	Lower-grade land			
All polders	0.5 yuan	0.45 yuan	0.4 yuan	Office expense	1942	100-6-1060: 37
Desheng polder	2,000 yuan	1,500 yuan	1,000 yuan	Dike fee	1947	100-1-99: 49
	5 liters of soybeans	4 liters of soybeans	3 liters of soybeans	Earth fee	1948	100-1-99: 80–81
Datong polder	15 yuan	10 yuan	5 yuan	Office expense	1944	100-1-182: 43
	80 yuan	60 yuan	40 yuan	Office expense	1945	100-1-182: 73
	4 jin of unginned cotton	3 jin of unginned cotton	2 jin of unginned cotton	Office expense	1948	100-8-99: 118
	15 liang of ginned cotton	10 liang of ginned cotton	4 liang of ginned cotton	Office expense	1949	100-8-99: 155
Sanhe polder	27 yuan	22 yuan	17 yuan	Earth fee	1944	100-8-87: 41
	50 yuan	40 yuan	30 yuan	Dike fee	1945	100-8-87: 64
Quanmei polder	50,000 yuan	40,000 yuan	30,000 yuan	Earth fee	1948	100-8-103: 146

lower-grade land. In the following year, fees were 15, 10, and 4 liang of ginned cotton, respectively.[75] In Desheng polder they were 5 liters of soybeans per mu for upper-grade land, 4 liters per mu for middle-grade land, and 3 liters per mu for lower-grade land (Table 4.5).[76] In a rare case, even broad beans were collected to pay off dike fees.[77]

Before the collection of these fees, notices would be publicly posted to inform all the residents. To encourage timely payment, those who paid early would be given a discount, and those who paid late were fined. Although some

75 100-8-99: 118, 155.
76 100-1-99: 80–81.
77 100-8-120: 25.

residents paid on time, more commonly they did not. Frequently, dike security guards or even local police were sent to collect from residents in arrears.

In many cases, directors reported that residents intentionally delayed payment,[78] but often people simply could not scrape together enough to pay dike fees on time. For example, in Sanhe polder in 1945, some did not pay their fees on time because they were away on business or were working as laborers far from home.[79] Also in Sanhe polder, in 1947 the director asked the county head to bring charges against those who did not pay their dike fees on time. Ten households were charged; two were headed by women who were most likely widows and were probably experiencing real hardship.[80] In Xida polder in April 1946, of 55 households that did not complete their dike work on time, 27 rented their land out to tenants. A family shrine and a charity hall, which also likely rented out their land, were included on the list of delinquents. After the county head ordered the local police chief to urge these people to complete their dike work, a team of four armed policemen visited those households in June.[81]

Since both annual repairs and high-water control were time-sensitive, directors often had to divert funds or borrow money or pay the fees themselves in advance, or ask those who had more land to pay in advance and then be reimbursed later. In a case in point, in 1946 in Peixing polder, the agency lacked funds for high-water control, so, after discussion, it was decided to borrow money from wealthy households in the polder and pay them back after the completion of that year's high-water control.[82] For this reason, in land transactions both the buyer and the seller were required to go to the polder's annual repairs and high-water control agency to register their deed to make clear who was responsible for dike fees; otherwise, the original owner would still be required to pay the fees even after the transaction, or the agency could ask the new owner to cover the seller's arrears.[83] Indeed, in the Republic, the payment of dike fees could serve as legal evidence of landownership in disputes over the title to land.[84]

Many of those who did not pay their dike fees were absentee landlords. In 1947, this became such an important issue that it was brought up at that year's second general meeting of the residents of Peixing polder. After

78 100-2-47: 163; 100-8-155: 136.
79 100-8-87: 60.
80 100-8-125: 67.
81 100-8-104: 10-12, 45, 47.
82 100-2-364: 42.
83 100-6-1060: 36; 100-1-99: 104.
84 LS19-4-6552.

deliberation, they decided to ask tenants to pay for their landlords.[85] In other words, the tenants were made responsible for the dike fees. PRC sources have condemned landlords' shifting the responsibility for dike work onto their tenants, a typical example of "class oppression" or "feudal exploitation." On the surface, it is clear the burden was transferred from the landlords to their tenants, but was it also true that these tenants were "oppressed" or "exploited"? That depends. In 1945 in a high-water control meeting in Datong polder, a motion made at the last moment proposed that if tenants built dikes for their landlords, those landlords should compensate their tenants 300 yuan per fang of dirt—the motion was passed.[86] In 1947 in Desheng polder, since absentee landlords paid so little of their dike fees and it was difficult to extract payment from them, the director and dike managers of the polder's annual repairs and high-water control agency proposed requesting their tenants to pay their rent to the agency as dike fees—that is, these tenants paid dike fees on behalf of their landlords, and they did not have to pay rent later, which was regarded as good for the landlords, the tenants, and the agency, or "killing three birds with one stone." The agency submitted this proposal to the county government, and it was approved.[87] In all three cases, the decision was made by the polder's residents in their general meetings or the agency's personnel, and was not imposed on tenants by their landlords. In a rare case, in 1946 in Yixing polder, fifteen petty merchant families who rented houses in a small rural town sued their landlords, contending that they had already paid their rent and thus bore no obligation for dike fees and dike work. They eventually won their case.[88]

Again, we should be cautious about generalizing what happened in Songzi, since it is also true that in other places this burden was normally shifted onto tenants, such as in the eighteenth century in Baiju polder in Jiangling, where landlords were only responsible for half of the dike work and costs of annual repairs, with their tenants shouldering the remaining half.[89] In Tianmen in the mid-nineteenth century, tenancy contracts specified that tenants were responsible for dike work, and in Jianli in the 1930s, landlords were responsible for taxes, but their tenants bore the burden of dike work.[90]

Residents were also responsible for earthwork. According to most records, residents were responsible for a certain amount of *tufei*, which was generally

85 100-8-108: 47.

86 100-1-182: 60.

87 100-1-99: 49.

88 100-8-120: 104–22.

89 *Jiangling xian shuili zhi* 1984: 62.

90 Zhang Jianmin and Tang Gangmao 2014; Hubei sheng zhengfu mishuchu tongjishi, comp. 1938: 11.

understood as an "earth fee."[91] But in people's polders in Songzi in the Republic, the obligation was actually listed as two separate things: *tu* (dirt) and *fei* (fee); that is, residents had to pay a certain amount in fees, and were also responsible for a certain amount of earthwork. The latter obligation required working on annual repairs and/or new dike construction sites. The fee was paid to the annual repairs and high-water control agency to cover the costs of its operations, as well as the purchase of materials for both high-water control and annual repairs. Finally, earthwork was assigned to individual households.[92] We can see this, for example, in Sanhe polder in 1944, where each mu of upper-grade land was assessed a dike fee of 27 yuan and 1.5 fang of dirt.[93] Similarly, in Changshou polder in the same year, each mu of land was assessed 1.2 fang of dirt and 10 yuan in dike fees.[94]

Land owned by charities was exempted from dike fees in the Qing, but apparently not in the Republic. In Xida polder in 1946, of the fifty-five households that did not pay their share on time, one rented land from a charitable organization, which meant charities were responsible for earthwork as well.[95] In 1948 in the same polder, a resident named Sun frequently asked the annual repairs and high-water control agency to exempt some of his land from dike fees because he had devoted that land to charitable purposes, such as supporting a free ferry. But the agency rejected his request since the land was still under his name, and the county government had no policy on exemptions. Besides, it was felt that a decision on such a matter should be made by all the polder's residents.[96]

Aside from annual repairs, the most significant outlay of these agencies consisted of what could be described as regular expenditures (e.g., salaries, subsidies, and administrative expenses, and high-water control expenses), and irregular or temporarily expenditures (for such things as construction, emergency repairs, rebuilding, transportation, communications or "social expenditures"), and costs associated with deity worship. Not surprisingly, a large portion of expenditures was devoted to salaries and/or subsidies of the personnel of the agencies. In addition, there were expenditures for meals, including for members of the annual repairs and high-water control agency, for laborers during the period of high-water control, for craftmakers, and for guests.

91 Jiayan Zhang 2014: 55–57.
92 100-8-93: 14.
93 100-8-87: 41.
94 100-8-97: 18.
95 100-8-104: 12.
96 100-8-170: 24–25.

In Xida polder, expenses even included pig feed and haircuts.[97] In Jiangsong Taibao polder in 1945, revenue from the high-water control fee was used to cover expenditures for such things as lanterns, candles, husked rice, vegetables/toufu, cottonseed oil, and salt. Although most of these items were listed under "high-water control expenditures," they were actually not directly used for high-water control, but for meals.[98]

In some polders, expenditures categorized under "construction" actually included high-water control materials. In Peixing polder in 1946, for example, expenditures on construction were used to purchase lumber, nails, lime, ramie, rice straw, reeds, and cotton fibers.[99] Since these materials would be purchased by the agency, individual residents would be relieved of the responsibility for providing them. This was also true in Datong polder in 1945, although the list there was longer and more diverse: "lanterns, candles, torches, bamboo hats, straw shoes, bamboo strips, coir ropes, reeds, reed mats, and soybeans."[100] Since the annual repairs and high-water control agencies of people's polders were public organizations, not part of the formal government, they had to buy everything from the market and could expect no discount.[101]

Among these expenses, social expenditures warrant a closer look. Social expenditures were used to build connections with local society and officials. They were not directly related to dike repairs and high-water control, but were nonetheless officially included in expenditures, which meant that they were accepted by the general meeting of the polder's residents, presumably because they understood that the agency had to frequently ask local officials for help, a subject discussed in detail below.

Expenditures for worshipping deities is also a subject worthy of some discussion. Although the GMD in general regarded deity worship as superstitious, it was still widely practiced in the Republic (see Chapter 5). This was precisely the case with the people's polders' annual repairs and high-water control agencies in Songzi, where deity worship was routinely practiced each year before the beginning of annual repairs. However, the ceremonies were not always taken seriously. In one case, it is said that because the so-called sacrifice fee was hard to collect, the agency did not want to worship dike deities. The issue was discussed at a general meeting, and it was decided that in the future the

97 100-8-69: 7; 100-8-99: 189; 100-8-108: 170–71; 100-8-104: 283.

98 100-8-49: 24.

99 100-2-364: 36–37.

100 100-1-182: 59.

101 100-8-97: 8.

agency would perform the ceremony if there were surplus funds.[102] In the minutes of a meeting of dike affairs in Futai polder, the agency's attitude toward deity worship was explained in detail: the worship of deities was purely superstitious and had nothing to do with the integrity of dikes, and thus, to save money, should not have been practiced. The agency, however, engaged in worship simply because, in response to public demand, it had done so every year in the past and the practice had become a "must" routine.[103] The same held true for forbidding butchering. After the 1931 flood, the Hubei government took a step it clearly understood was useless for relief: it ordered all government departments to forbid butchering for five days to show the people that the authorities were sympathetic to their suffering.[104]

In the first case, the worship of deities was cancelled due to a shortage of funds. In fact, it was common for polder agencies to have a deficit. In Peixing polder in 1945, for instance, the agency's annual revenue was 370,328.60 yuan, but its annual expenditures were 497,311 yuan, leaving a deficit of 126,982.40 yuan.[105] In Datong polder in 1946, the deficit was more than a million yuan, and even then the budget was so small that it barely covered that year's annual repairs.[106] Running deficits could threaten the very existence of an agency. At the end of 1947, Zhou Shuwu, the director of the annual repairs and high-water control agency of Peixin polder, reported to the county head that because he was too poor to fund the agency, he had taken a drastic step to prevent the mushrooming of deficits: he dismissed all the agency's personnel and closed the agency. However, his report was rejected by the county head, who told Zhou that, no matter what, he had to organize people to conduct annual repairs.[107] The county head was right to point out that winter was a good time for annual repairs, but Zhou clearly was at an impasse since he had to pay for the repairs himself, but did not have the wherewithal to continue to do so. Many a time, one of the key items discussed in various meetings of the agencies was how to solve the problem of deficits.

Both past scholarship and PRC gazetteers have contended that corruption was rife among directors. Archival materials, however, reveal a different picture. In Xida polder, for example, Xiao Litian, the director of the annual repairs and high-water control agency, was accused of corruption by resident Sun Yitai

102 100-1-97: 139.

103 100-8-42: 109–10.

104 Hubei sheng shuiliting and Hubei sheng dang'anju 1999: 37.

105 100-2-256: 51–53.

106 100-1-182: 96.

107 100-8-108: 118.

on September 2, 1945. Sun also mentioned in his accusation that a dike security guard was said to be extorting fees from residents. The county head immediately sent clerks to investigate, but after spot-checking accounts, found no evidence of corruption. Moreover, the dike security guard who was involved in extortion had already been fired by the director. To get to the truth, fifty or so elite members of the polder were called to a general meeting, and four people were recommended by general acclaim to form an ad hoc committee to check all the agency's accounts over the preceding two years. On November 11, Xiao submitted a long petition to the county head to refute Sun's accusations. Xiao's petition listed twelve points on which Sun was wrong. First, since all income and expenses were recorded in the account books, how could anyone get away with embezzlement? Second, Xiao often provided money in advance to cover shortages. Third, over the five years that Xiao had been director, Sun was the only person to accuse him of wrongdoing. Fourth, Sun was in arrears in paying his dike fees and had once destroyed a dike to build a kiln; Xiao had sent staff to press Sun to pay his share and rebuild the destroyed dike. Sun thus resented Xiao and invented stories about him. And so on. On November 30, Xiao submitted another report explaining why Sun had accused him. On December 28, 1945, the ad hoc committee concluded that all income and expenses matched the figures in the agency's accounts, and no evidence of embezzlement was found. Sun agreed to pay his arrears for dike work, and someone had already repaired the dike Sun destroyed. Xiao was in the clear and the case was closed.[108] Partly because the agency kept detailed accounts, Xiao was able to prove his innocence. In another case, in 1947 the former director of annual repairs and high-water control agency of Peixing polder was also accused of embezzling dike fees based on a report of an interrogation and examination committee, but the case was withdrawn through mediation because the committee misunderstood many items listed in the accounts.[109]

Evidence in account books of people's polders in Songzi indicates that directors rarely embezzled dike fees. There was an exception in Deshen polder in 1946, in which the director was accused of embezzling dike fees, mainly by falsifying the accounts, and each charge had supporting evidence. There was no response from the accused director, but the county government apparently verified the charges.[110] However, in another document, the director countered the accusation by claiming that the accuser did not complete his dike work but

108 100-8-93: 2–4, 10, 14, 18, 22–24, 27.

109 LS36-2-066-003.

110 100-8-95: 1–7.

nonetheless demanded he be paid.[111] Based on the Songzi archives, the annual repairs and high-water control agency of every polder prepared an annual budget and issued expense reports, which were open to the public. That is to say, financial operations were apparently transparent and that, at least in theory, should have reduced the chances of embezzlement. In 1951, the new government discovered that 24,738 jin of husked rice had been embezzled by the staff of the former annual repairs and high-water control agencies of two polders in Songzi.[112] This is a huge amount of rice. Since the source offers no detailed explanation of this statement, it is hard to tell whether the rice was really "embezzled" or just retained by the members of the staff. In other places, however, the sources claim there were many cases of agency directors engaging in corruption. In Zhijiang, for example, a document in the Hubei Provincial Archives asserts that the dike fees of most polders in the county were controlled by "dike vermin."[113]

Outside Songzi, the archives contain detailed records of income and expenditures of the dike engineering bureau of a polder in Yingcheng for the years 1922 to 1936. A former dike manager of this agency submitted accounting records to prove he was innocent of corruption. According to him, he actually sold his property to cover the agency's expenditures.[114] But it is unclear if it was common for polders in other places in the Jianghan region to keep detailed income and expenditures records. Because of the frequency of accusations of corruption elsewhere in the Jianghan area, even though there were few or no transparent reports in those places, we should be leery about believing everything recorded in reports from Songzi. We should bear in mind that "accounting" is an art, and that no one would be so foolish as to expose his embezzlement of fees in records or accounts. A document in the Xiaonan District Archives claims that every old polder head was corrupt and created very complicated accounting records to bamboozle the peasants.[115]

According to regulations on annual repairs and high-water control agencies of people's polders issued by the government of Hubei in 1947, all the positions in these organizations were unpaid. The only compensation consisted of some meals during construction and high-water control, and these had to be approved by the county government.[116] It is clear that members of

---

111 100-10-307: 32, 42–43.
112 33-1-5 (1)c.
113 LS1-5-855A.
114 Yingcheng 49-1-3.
115 81-1-3b.
116 LS1-2-0484-007-001.

the agencies in Songzi did not follow this regulation at all, inasmuch as they received income via multiple conduits. What then was exactly the relationship between people's polders and local governments?

5 People's Polders and Local Governments

In the Qing and the early Republic, government personnel usually only acted as supervisors in organizations responsible for annual repairs and high-water control in people's polders. That was still the case in the 1930s and 1940s, but by then the state had deepened its penetration into the governance of polders, a basic unit of rural society in the Jianghan area in the Qing and the Republic.

First, the state ordered many polders to establish annual repairs and high-water control agencies after the 1931 flood. It turned out that these agencies could not function well without the involvement of the local government. The county government was required to send a representative to monitor agency elections, and the elected director or other members of the agency needed a letter of appointment from the county government to take up their duties. In the revised regulations on annual repairs and high-water control agencies of people's polders issued in 1938 by the Hubei provincial government, both dike managers and dike maintainers were to be chosen from among current rural leaders.[117] Past scholarship thus contended that polder management organization roughly overlapped with or were parallel to the rural administrative systems in the Republic,[118] but the actual situation was much more complicated.

In some places, a large polder included several small polders, and a small polder included one or more villages. And in places a large village had more than one polder. In multi-village polders, the director of annual repairs and high-water control clearly enjoyed a higher social status than a village head. For example, in Qianjiang in 1947, a director served concurrently as the township head,[119] and in Xihu polder in Hanyang in 1945 the director of the annual repairs and high-water control agency was also concurrently the township head, but two other township heads refused to serve as his deputy directors. In this case, it seems the director was more powerful than the township head.[120]

The same was true in Sanhe polder, a large polder in Songzi, where the township head ranked below the director in the organization of high-water control.

117 LS36-2-15.

118 Jiayan Zhang 2014: 204; Lü Xingbang 2020.

119 LS30-1-0207-017.

120 LS31-4-1407.

During the high-water control period, the heads of sub-county or township offices were required to serve on the local high-water control committee of people's polders, which understandably could only be applied to large people's polders that had jurisdiction over one or more townships. The organization of high-water control measures in Sanhe polder in 1944 made it clear that high-water control should be carried out in cooperation with the local administrative apparatus: "The director of the annual repairs and high-water control agency shall be the general team leader; the township head, the deputy general team leader; the dike manager, the sub-general team leader; the dike maintainer, the branch team leader; the bao head, the deputy branch team leader; the dike head, a squad leader; and the jia head shall be a deputy squad leader."[121] On the surface, at each level, it appears that the local rural leaders were under the leadership of the annual repairs and high-water control agency. But the reality, as we will see, was different.

In these polders, the director of the annual repairs and high-water control agency was more powerful than the township head, and hence could issue orders to the latter. Even so, in several cases if a township head was elected the director of the annual repairs and high-water control agency of a people's polder, he kept the former post and gave up the latter. For example, in 1944 in Sanhe polder, when the director of the annual repairs and high-water control agency was appointed as the head of a township, he told the county head that he could not serve in the two positions simultaneously. He was allowed to resign from his polder position.[122] This case suggests that the position of township head apparently was more attractive than that of the director of an annual repairs and high-water control agency. At the same time, some townships had jurisdiction over two or more people's polders. In such cases, the director was far less powerful than the township head. Furthermore, as noted earlier, some small polders had no annual repairs and high-water control agency at all.

In February 1947, the director of the annual repairs and high-water control agency of Peixing polder planned to mobilize all the laborers for annual repairs since the dike had been devastated in the war, but the local township office also needed laborers to transfer tax grain, giving villagers a good excuse to avoid dike work. The director then asked for help from the county head, who responded that since both the transportation of grain and the annual repairs of dikes were important, they should go hand in hand. He kicked the ball back to the director, which suggests that the county head did not give priority to dike affairs over the transportation of grain. In fact, a crossed-out original comment

121 100-8-87: 35.

122 100-8-87: 3.

on the director's request in the archive reads "do not approve."[123] Frequently, to assemble enough laborers for annual repairs, some directors might have asked the local administration to urge or even arrest those who refused to go to dike worksites to fulfill their duty. Of course, having to ask the township office to assist sometimes gave the appearance of begging for help,[124] a clear sign that the township office was more powerful than the annual repairs and high-water control agency.

The state was also indirectly involved in the governance of people's polders. The staff of annual repairs and high-water control agencies included dike security guards, whose numbers varied from agency to agency. One of their tasks was to dun people who had not paid their dike fees. But the security guards usually failed. In most cases, people who had not paid refused to heed dike security guards, and in some cases, they even beat them, but the guards could not arrest them. Hence, the directors had to seek help from the local police or county government, which had the authority to subpoena and arrest malefactors. In 1945 in Zhongcheng polder, for instance, the director of the annual repairs and high-water control agency accused resident Du Qifa of "stubbornly resisting dike policy, beating and humiliating a dike security guard, and seizing public funds by force." The county government responded by arresting Du. Then a village leader and nine village representatives sent a petition letter to the county government in which they claimed that Du had already finished his arrears of dike work, he did not beat the dike security guard, and he did not seize public funds. Soon Du was released.[125] The core issue in this case should have been Du's failure to complete his dike work. Since Du refused to heed the dike security guard, the director asked the county government for help. The purpose of the county government was to intimidate, not punish, Du; thus, it immediately released him after it received the petition. This case served as a warning to the public that anyone who refused to complete his share of dike work might be arrested. Although in this case Du perhaps did not beat the dike security guard, in another case, some residents who owed dike fees in Zhongcheng polder indeed beat the fee collector and even brawled with the local police. After investigation, the township office verified that the incident had occurred but dared not arrest the perpetrators, using the excuse that they were residents of a different township.[126] The PRC new gazetteers have portrayed the government in the Republic as a violent oppressor of the

123 100-8-108: 33.

124 100-1-99: 31; 100-8-160: 100, 128.

125 100-4-1651: 3–5.

126 100-8-149: 31–32, 36–37.

people, but in this case, it was the people who were violent, and the government the victim.

In times of disaster, the directors of polder agencies for annual repairs and high-water control often asked the county government, on behalf of the polder residents, for a tax exemption or sought relief funds from the county government and/or the Bureau of Dike Work on the Jing River. That is, directors were responsible not only for annual repairs and high-water control, but also for finding funding from outside sources or asking for tax relief, particularly after a flood.[127] However, it seems people's polders in the 1930s rarely received government relief.[128]

In dealing with people who refused to heed the personnel of the annual repairs and high-water control agencies of people's polders and continued to farm on dikes and build houses on them, despite these acts having been repeatedly banned, the agencies did not have the authority to compel compliance but instead had to report wrongdoings to the county government and ask it to take action. In other words, the regular operation of these agencies very much relied on the support of the local county government.

Disputes over dike work in polders with dikes that extended across county lines, often prompted the director of the annual repairs and high-water control agency to seek help from the government of the county in which he resided.[129] Furthermore, in the Republic, cooperation between neighboring polders on high-water control could be wanting.[130] And sometimes even the involvement of the county government failed to resolve conflicts, a result of the GMD's limited penetration of rural society.

6 Conclusion

Most PRC gazetteers portray the directors of annual repairs and high-water control agencies of people's polders in the Republic as "evil" people who grasped their position through force, bullied the residents, and peculated dike fees. The archives from the Republic, however, paint a more complicated and nuanced picture. Undoubtedly, there were some directors who were thoroughly corrupt, but more often than not, particularly in Songzi, directors were common villagers elected by their fellow polder residents to serve the community, and, the

127 LS24-3-00002802-009; LS30-1-1445-002.

128 LS1-3-0237-001.

129 100-2-466: 4, 9, 20; Lü Xingbang 2015.

130 LS31-4-1531; *Xiaogan shi zhi* 1992: 211.

records show, most of them did their best, even if their best was not always very good. Some directors found their job unreasonably difficult and unprofitable; for others, it was satisfying and rewarding. But no matter how a person became a director, and even if a director was not entirely free of the taint of corruption, he certainly should have been capable of fulfilling his duty.

The operation of the annual repairs and high-water control agencies of people's polders on the Jianghan Plain in the Republic (particularly in the 1930s and 1940s) demonstrates that, influenced by nation-building efforts and new discourses, these agencies were fairly democratic organizations. Detailed records of agencies' income and expenditures also demonstrate that some directors received no payment (except for some subsidies), and financial arrangements also left little room for them to embezzle dike fees (at least in theory). Instead, it was always difficult for the agencies to collect dike fees on time, and some directors had to cover expenditures out of their own pocket in advance. Polder agencies therefore frequently asked the local government or the police to press those who refused to fulfill their duty to complete their assigned dike work, or to urge or force payment from people who failed to pay their dike fees. Past scholarship and PRC gazetteers have treated these scenarios as perfect examples of how the GMD government "oppressed" or "exploited" the people. The reality is, however, that defaulters refused to fulfill their duty or to pay their dues—in other words, they were violators of the regulations and of their civic duty, not the victims of class oppression or "feudal exploitation."

It is also clear that annual repairs and high-water control agencies of people's polders could not function well without the involvement, whether direct or indirect, of local government. This kind of relationship reflected the increasing penetration of state power into rural society in the Republic (which was also seen in village governance in North China[131]) and was a harbinger of the transition from the indirect involvement of the state in dike management in the Republic to its full and direct involvement in the PRC. Thus, although the annual repairs and high-water control agencies of people's polders did not continue in the PRC, similar organizations emerged, particularly a hierarchical system of annual repairs and high-water control which, in the PRC, has entailed more direct involvement from the local rural administration (see Chapter 1 for more details). Many things related to dike affairs in the PRC did not arise out of thin air nor simply emerge overnight after the CCP took power; they were often inherited from the Republic.

131 Huaiyin Li 2005.

CHAPTER 5

Water Deity Worship and Dike Management

The Chinese understanding of nature and their tradition of water control were directly related to the local geographical setting. They constructed water-control engineering projects in numerous places and worshiped many water deities, but the gods they worshiped varied from place to place according to local circumstances. In dry areas, irrigation channels were particularly important; in flood-prone areas, dikes were crucial. In terms of water deity worship, in North China where people relied on well-water for irrigation, well gods were worshiped;[1] along the lower reaches of the Yellow River and the Grand Canal—places usually with ample and sometimes too much water—people worshiped the Great King (or the Dragon King);[2] on China's southeast coast, sea-going fisherfolk worshiped the goddess Mazu; and all over China people built temples to worship Yu the Great, the legendary ruler who supposedly introduced high-water control and thus "tamed the waters." The custom of worshiping water deities is deeply rooted in Chinese culture, as is evident in the earliest extant Chinese writing—inscriptions on the oracle bones (on the worship of the Yellow River deity), dating to ca. 1200–1050 BCE.[3] In later times, some water deities, such as Yu the Great and the Dragon King, become major gods worshiped across the country; people also worshiped the various water deities of their regions.

It is well known that in traditional China, water control was an indispensable part of governance. The economist Ch'ao-ting Chi, for instance, once argued that in different periods in China's history, key economic areas were distinguished and defined by their water-control projects.[4] Karl Wittfogel even claimed in his theory of "oriental despotism" that China's "despotism" was rooted in the country's water-control projects.[5] Although he exaggerated the degree of control the imperial state exercised over rural society, the emperors of the Qing dynasty unquestionably played a pivotal role in river management.[6] Scholars have argued that the practice of water deity worship and the control

1 Hu Yingze 2006.
2 Dodgen 1999.
3 Wei Ziren 2002.
4 Ch'ao-ting Chi 1963 [1936].
5 Wittfogel 1981 [1957].
6 Leonard 1996; Dodgen 2001; Lillian Li 2007: 38–73.

 | DOI:10.1163/9789004739482_007

of rain shrines were also a way for the traditional Chinese state to penetrate and subsequently control local society.[7] And many Chinese deities certainly resembled earthly bureaucrats, though there were some non-bureaucratic or "unruly gods" as well.[8]

At the same time, China has long been notorious for its frequent natural disasters. Traditional Chinese ideology particularly emphasized the relationship between the morality of the ruler and the incidence of natural disasters. According to the notion of "dynastic cycles," at the beginning of a new dynasty the ruler was virtuous, and water-control projects worked well, but if the ruler lost his virtue, the dynasty would decline and consequently water-control would deteriorate. Heaven would punish the ruler with natural disasters—floods, droughts, and earthquakes. In other words, such disasters were an indication that the ruler might have lost his virtue and thus his legitimacy to rule. Some even claim that there was a so-called hydraulic cycle that roughly paralleled this dynastic cycle.[9]

In addition, Chinese traditional philosophy emphasized that, when all was well, nature and humans were in a harmonious relationship (*tianren heyi*). To maintain this harmony, a virtuous ruler should build water-control projects to ward off floods and droughts. In fact, favorable weather was viewed as evidence of the legitimacy and brilliance of the ruler.[10] Thus, in late imperial times, according to Mark Elvin, "the importance of the correct psychological or spiritual attitudes" toward nature and natural disasters was particularly pronounced.[11] The Kangxi emperor (r. 1661–1722), for instance, emphasizing the relationship between the virtueless leader and natural disasters, implied that "if natural disasters were to recur they [governmental officials] would be to blame and would be held responsible by the people," and the Qianlong emperor supported Buddhists' (as well as Daoists') prayers for rain, though the Qing emperors' attitudes toward bad weather seems to have changed between the early years of the dynasty and the late years.[12] Nonetheless, as we shall see, the emperors' attitudes toward water deity worship also varied. In looking at water control on the Jianghan Plain, we will focus not on the philosophical and/or religious dimensions of water deity worship, but on the management side.

7 Pomeranz 1991; Tan Xuming 2009.

8 Shahar and Weller 1996.

9 Will 1985; Perdue 1987.

10 Lillian Li 2007: 13–15.

11 Elvin 1998: 215.

12 Elvin 1998: 215, 220.

Since the Jianghan Plain is a water-rich region, as one might expect the worship of water deities—stretching back to as early as the Zhou dynasty (ca. 11th century BCE–256 BCE)—was an integral part of local culture.[13] By the Ming-Qing era, especially the Qing dynasty, the Jianghan area became subject to frequent water calamities. Between 1739 and 1911, for example, the plain suffered more water calamities than any other area of the Yangzi River valley.[14] Scholars have found that these water calamities were caused mainly by the enclosure of polders, the spread of river dikes, and either mismanagement or ineffective management of the dike system.[15] The relationship between water deity worship and dike management, however, remains cloudy.

According to the historians Yang Guo and Chen Xi, from the Song dynasty to the Qing dynasty, people who lived along the Yangzi River on the Jianghan Plain worshiped a variety of water deities, which could be natural gods, such as river gods, or personalized gods, such as Guan Yu (also known as Guan Gong). Furthermore, water deity temples were usually located in important places.[16] Their research only occasionally mentions that some of these temples were built to protect dikes, and some local officials were posthumously apotheosized to honor their achievements in water control. However, they did not elaborate on the relationship between water deity worship and dike management, nor did they discuss the Republic and the PRC.

In his study of the temple—polder relationship, the historian Lu Xiqi found that, because of frequent floods, the Jianghan people built many temples and worshiped manifold deities to protect their polders. These deities usually were considered efficacious within only a single polder, a feature that became especially clear in the late Qing.[17] From the perspective of polder management, Gao Yan has argued that it was their psychological fear of frequent water disasters, particularly floods, that induced the Jianghan people to build many polder dikes in late imperial times as well as temples to house deities that were worshiped for protecting the dikes. Again, these deities only protected a certain polder, and this exclusivity helped the formation of a sense of community within a polder. Understandably (though not explicitly), water deity worship was an inseparable part of polder dike management.[18] Lü Xingbang has found that in Sanhe polder, in Songzi County, in the 1940s, the worship

13 Yang Hua 2007.
14 Zhang Jiayan 2011: 60.
15 Will 1985; Jiayan Zhang 2014; see also Zhang Jiayan 1997.
16 Yang Guo and Chen Xi 2008: 314–50.
17 Lu Xiqi 2011.
18 Yan Gao 2012: 69–97.

of water deities was a routine practice in polder dike management.[19] None of these scholars, however, has probed the relationship between the worship of water deities and the management of river dikes.

The management of dike works, particularly river dikes, was one of the most important official duties of prefects and magistrates in the Jianghan Plain in the Qing.[20] Some of these officials frequently visited temples to worship water deities and pray for the integrity of the dikes. This practice continued under the Republic—even though some Nationalists were atheists who opposed "superstition" such as the worship of water deities.[21] The ideological influence of the atheist Communists, however, was far more powerful. The PRC government not only deeply penetrated rural society, but it also devoted a great deal of effort to taming rivers and reinforcing dikes and dams. This resulted in a very substantial reduction in the frequency of dike ruptures and inundations. As a result, the local people evidently no longer felt a need to pray to the water deities for dike protection. All of this changed the tradition of water deity worship and its role in dike management. This chapter, from an environmental perspective, by concentrating on the government and its officials, examines the changing relationship between the worship of water deities and the management of dikes—particularly the Wancheng Dike—in the Jianghan Plain over the past three centuries, a topic that to my knowledge has rarely been touched by researchers.

1 Water Deity Temples on the Jianghan Plain in the Late Qing

In traditional China, it was said that the mighty Dragon King could control the rain that fell from the sky and water on the ground. In the flood-prone Jianghan area, the local people had a tradition of performing the dragon dance to persuade the Dragon King to prevent floods or grant rain in times of drought.[22] Worshiping in temples, however, was more popular than performing the dragon dance—the dragon king temples once found all across the Jianghan area were a testament to this.

The Dragon King, however, is only one of many water deities worshiped by the Jianghan people. In the Song-Qing period (960–1911), those who lived along the Yangzi River in the Jianghan area worshiped no less than twenty-seven

19 Lü Xingbang 2011.
20 Jiayan Zhang 2014: 48–51.
21 Nedostup 2009.
22 *Jiangdi xiang zhi* 2006: 233.

water deities: fourteen natural gods and thirteen personalized gods.[23] People who lived along the Han River valley in the Jianghan area worshiped similar gods and built similar temples.

TABLE 5.1 Name, number, and distribution of water deity temples in the Jianghan Plain at the end of the Qing dynasty

Names	Number	Distribution
Guandi miao (King Guan Temple)	75	Jiangxia (3), Hanyang (2), Xiakou, Hanchuan (5), Xiaogan, Mianyang (8), Yunmeng (2), Yingcheng (7), Jingshan, Qianjiang, Tianmen (5), Jingmen (3), Dangyang (4), Zhijiang (3), Songzi (2), Jianli (5), Jiangling (8), Gong'an (6), Shishou (7)
Shuifu miao (Water God Temple)	22	Xiakou, Hanchuan, Mianyang (4), Yingcheng, Jingshan, Tianmen, Jingmen (4), Dangyang (2), Shishou (2), Jianli (2), Zhijiang (3), Xiaogan[a]
Longshen ci (Dragon God Shrine)	13	Jiangxia, Hanyang, Xiaogan, Mianyang, Yunmeng, Jingshan, Tianmen, Jiangling (2), Jianli, Gong'an, Zhongxiang (2)[b]]
Longwang miao (Dragon King Temple)	12	Jiangxia (2), Yingcheng (4), Jingmen (2), Gong'an, Shishou, Qianjiang, Zhongxiang[c]
Yuwang gong (King Yu Palace)	7	Jiangling, Shishou, Jianli, Hanyang, Mianyang (2), Songzi[d]
Yuwang miao (King Yu Temple)	5	Hanyang, Jingmen, Songzi, Jiangling, Zhongxiang[e]
Tianhou gong (Empress of Heaven [Mazu] Palace)	5	Yunmeng, Shishou, Jingshan, Mianyang, Qianjiang[f]
Yangsi miao (Yangsi Temple)	5	Tianmen, Shishou (3), Xiaogan[g]
Yangong miao (Revered Mr. Yan [Shuzi] Temple)	4	Jiangxia, Jingmen, Jiangling, Jianli
Longshen miao (Dragon God Temple	3	Jiangxia, Jingmen, Dangyang
Jiangshen miao (River God Temple)	2	Jiangxia, Jiangling[h]

23 Yang Guo and Chen Xi 2008: 314–50.

TABLE 5.1 Name, number, and distribution of water deity temples (*cont.*)

Names	Number	Distribution
Dongting miao (Dongting Temple	2	Jiangxia, Jianli
Baodi si (Dike Protecting Monastery)	2	Zhongxiang, Qianjiang[i]
Qinglong miao (Black Dragon Temple)	2	Shishou, Hanchuan[j]
Tianfei miao (Tianfei Temple)	2	Jianli, Shishou[k]
Zhenjiang an (Containing the river Nunnery)	1	Zhongxiang[l]
Xiaogong miao (Revered Mr. Xiao Temple)	1	Jianli
Xiaogong ci (Revered Mr. Xiao Shrine	1	Jiangling
Jianghan shen ci (Jianghan God Shrine)	1	Jiangxia
Zhenjiang wang miao (Containing the River King Temple)	1	Jingmen
Jiangdu gong (River Ditch Palace)	1	Jiangling
Jiangdu miao (River Ditch Temple)	1	Gong'an
Zuzhang shen ci (Zuzhang God Shrine)	1	Jiangling
Dingjiang wang miao (Pacifying the River King Temple)	1	Zhijiang[m]
Heishen miao (Black Spirit Temple)	1	Jianli
Leigu miao (Thunder Ax Temple)	1	Jianli
Zhenjiang gong (Containing the River Palace)	1	Shishou

TABLE 5.1 Name, number, and distribution of water deity temples (*cont.*)

Names	Number	Distribution
Shenying miao (Spirit Response Temple Temple)	1	Jiangxia
Qingyuanjun miao (Qingyuanjun Temple)	1	Jiangling
Nigong ci (Revered Mr. Ni Shrine)	1	Jiangling
Xiegong ci (Revered Mr. Xie Shrine)	1	Shishou
Xinzhou si (New Islet Monastery)	1	Qianjiang
Dashi an (Guanyin Nunnery)	1	Qianjiang[n]
Ouyang jiangjun miao (General Ouyang Temple)	1	Jianli

a *Xiaogan xian zhi* 1883, vol. 2: 15b.
b *Zhongxiang xian zhi* 1867, vol. 5: 33a.
c For Qianjiang: *Qianjiang xian zhi* 1879, vol. 7: 8a; for Zhongxiang: *Zhongxiang xian zhi* 1867, vol. 5: 31b (The name of the temple is Longwang gong [Dragon King Palace]).
d For Hanyang: *Hanyang xian zhi* 1868, vol. 26: 3a; for Mianyang: *Mianyang zhou zhi* 1894, vol. 3: 26a; for Songzi: *Songzi xian zhi* 1869, vol. 2: 35a.
e For Jiangling: *Jingzhou fu zhi* 1880, vol. 27: 9a; for Zongxiang: *Zhongxiang xian zhi*, 1867, vol. 5: 26a.
f For Shishou: *Jingzhou fu zhi*, 1880, vol. 27: 24b; for Jingshan: *Jingshan xian zhi*, 1882 vol. 2: 17a; for Mianyang: *Mianyang zhou zhi* 1894, vol. 3: 27a; for Qianjiang: *Qianjiang xian zhi* 1879, vol. 7: 9b.
g For Xiaogan: *Xiaogan xian zhi* 1883, vol. 2: 17a.
h Recorded as Jiangshen ci (River God Shrine) in *Jiangling xian zhi* 1876, vol. 5: 12a.
i For Zhongxiang: *Zhongxiang xian zhi* 1867, vol. 5: 27b; for Qianjiang: *Qianjiang xian zhi* 1879, vol. 7: 7b.
j For Hanchuan: *Hanchuan xian zhi* 1873, vol. 8: 26b.
k For Shishou: *Shishou xian zhi* 1866, vol. 2: 56a.
l *Zhongxiang xian zhi* 1867 vol. 5: 30b.
m *Jingzhou fu zhi* 1880, vol. 27: 41a.
n *Qianjiang xian zhi* 1879, vol. 7: 5ab.

SOURCES: UNLESS OTHERWISE NOTED, ALL INFORMATION IN THE TABLE COMES FROM *HUBEI TONGZHI* 1921: VOL. 27–31: 892–994. THE TABLE IS ONLY A ROUGH LIST BASED ON GAZETTEERS (TEMPLES RECORDED AS "CURRENTLY COLLAPSED" ARE NOT LISTED); MANY TEMPLES MIGHT NOT HAVE BEEN RECORDED IN GAZETTEERS.

As Table 5.1 shows, by the late Qing dynasty the most numerous water deity temples in the Jianghan Plain were dedicated not to the River God or the Dragon King, but to King Guan. King Guan, the god enshrined in these temples, is Guan Yu, a famous general in the period of the Three Kingdoms (220–280). Guan Yu has special importance and influence in the Jianghan area as he had been the prefect of Jingzhou (most of the Jianghan Plain is within the jurisdiction of Jingzhou Prefecture), and almost every Chinese knows the story of how Guan Yu lost Jingzhou (to the rival state of Wu) due to his negligence. According to recent research, Guan Yu was first worshiped in Jingzhou by the locals, then became a national deity in the Tang-Song era (618–1279).[24] In the Qing dynasty, Guan Yu was mainly worshiped as a military god, parallel with Confucius as a civil god; and the King Guan temples were among the official temples in the Qing dynasty. Two official worship ceremonies were held in these temples every year.[25] Because of the mighty power of Guan Yu, the Guandi miao (King Guan temples) were usually built in prominent locations. For example, there was a Guandi miao on the Wancheng Dike named the "Zhendi Guan miao"—the Dike-Protecting Guan Yu Temple.[26] Not all King Guan temples, however, were built to contain floodwater. In Qianjiang, for instance, a King Guan temple served as the guild hall of merchants from the North China province of Shanxi.[27] On the North China Plain, Guan Yu was the major patron and protector of the local people.[28]

The second most numerous water deity temples were the Shuifu miao (Water God temples). An inscription from 1804 on a stele in Jingshan explains that the local administrators worried that water from the Han River might destroy the wall surrounding the (county-seat) town of Jingshan, so they built a temple, which they named Shuifu (Water God)—since "water is one of the Five Elements, the god is there and should be worshiped."[29] The Five Elements were traditionally metal, wood, water, fire, and earth. Inasmuch as water is abundant on the Jianghan Plain, Water God temples were built in most counties. Some temples may have had different names, such as the Ouyang Jiangjun Temple in Jianli,[30] but were also built to worship the Water God.

The third and fourth most numerous water deity temples were dedicated to the Dragon God (Longshen) and the Dragon King (Longwang), respectively.

24 Cai Dongzhou 1999.
25 *Hanyang xian zhi* 1868, vol. 11: 6a.
26 Ni Wenwei 2002 [1885]: 90.
27 *Qianjiang xian zhi* 1879, vol. 7: 9b.
28 Duara 1988: 139–48.
29 *Jingshan xian zhi* 1882, vol. 23: 35a.
30 *Hubei tongzhi* 1921: 992.

Important local officials such as prefects and county magistrates were expected to host two official worship ceremonies in the Dragon God Shrine every year, one in the spring and the other in the autumn.[31] A gazetteer recorded that in Hanyang, the Dragon King Temple was located outside the city wall and along the Yangzi River, and that "the river has an undercurrent. According to legend, a dragon hides there, [and thus the people] built a temple to worship him."[32] The residents built these temples in the hope that dragons would protect the region from disasters or bring water in times of drought.

Both the fifth and the sixth most numerous water deity temples were dedicated to the same figure, Yu the Great, the national water-control hero and the legendary founder of the Xia dynasty, or the first hydraulic engineer in Chinese history. According to the Zhongxiang County gazetteer of 1867, the King Yu Temple in Zhongxiang, located at the south end of the county seat, was built because the local people suffered from floods and thus sought Yu's protection.[33] Residents in other counties built King Yu temples for the same reason.

Some of these temples were built by merchants. In Shishou, for example, the King Yu Temple was built by merchants from the province of Hunan, the Zhenjiang Gong (Containing the River Palace) was built by merchants from the prefecture of Wuchang, and the Water God Temple was built by the local residents in collaboration with merchants from other regions.[34] Mazu temples in Shishou and Jingshan were built by people from Fujian Province,[35] where Mazu was a very popular deity.

Temples were also built by the local government. In Jiangling, for example, both the River God Shrine and the Dragon God Shrine were built with government funds.[36] According to historian Lillian M. Li, state-sponsored temples and shrines in the North China Plain were not only places of worship but also symbols that the government highly valued water control.[37]

Officials, therefore, must have believed water deity worship was important for the control of floods and the maintenance of the integrity of dikes. Shu Chenglong, the prefect of Jingmen Department from 1743 to 1755, wrote a note inscribed on a stele to document the building of a Baihe si (White Crane Monastery), which

31 *Yunmeng xian zhi* 1840, vol. 2: 15a; *Xiaogan xian zhi* 1883, vol. 6: 10a.

32 *Hubei tongzhi* 1921: 913.

33 *Zhongxiang xian zhi* 1867, vol. 5: 26a.

34 *Shishou xian zhi* 1866, vol. 2: 53a, 54a.

35 *Jingzhou fu zhi* 1880, vol. 27: 24b; *Jingshan xian zhi* 1882, vol. 2: 17a.

36 *Jiangling xian zhi* 1876, vol. 5: 12ab.

37 Lillian Li 2007: 71.

> was previously located on Xianrengudi [Immortal Ancient Dike], ten li [5,000 meters] south of the new town. ... According to legend, there was an immortal who flew on a white crane over this place when [people] began to build the dike. ... After the dike was built, [people thus] named it the Immortal [dike], and built a monastery on it, which they called the White Crane [monastery]. ... Many years later, the monastery collapsed. [One year the local people] moved the monastery about one li west of the dike. The dike at Qianjiang ruptured in that year; people all laid the blame on the movement of the monastery. ... Five counties had suffered from floods for two hundred years. In 1743 I came here to serve as the prefect and immediately issued orders to reinforce the dike; [I] lived on the construction site day in and day out, and therefore was able in every possible way to inquire into the details of the local situation and came to realize that the restoration of the Immortal Ancient Dike was a permanent solution. ... I think, after rebuilding the dike, that it is impossible not to worship the deity and put the dike under his protection. Thus I donated more than 500 taels to help rebuild the monastery in its original location and [saw to it that] it continued to be known by its previous name.[38]

This long inscription shows that Shu was a pragmatic prefect, asking the natives about the local situation and showing an understanding of the importance of the integrity of the dike. As the inscription notes, during the reconstruction, Shu even lived on the worksite. This kind of dedication made him famous for his achievements in water control.[39] If such a pragmatic person believed in the mighty power of the deity and wanted to place the dike under divine protection, it is easy to imagine that the Jianghan people were perhaps even more convinced of the power of the deity. Evidence of this is the temples the people built on dikes—in the 1870s, for example, thirty-four temples stood on or along the Wancheng Dike alone.[40]

The Jianghan people believed not only in water deities but also in the power of bodhisattvas (particularly Guanyin, the goddess of mercy) to control floods and protect dikes. In Qianjiang, for instance, the Xinzhou Si (New Islet Monastery) was built by a county magistrate under Emperor Yingzong (r. 1457–1464) of the Ming dynasty. The magistrate decided to build the monastery—dedicated to preventing future floods—on an islet that emerged

38 *Jingmen zhili zhou zhi* 1868, vol. 6: 13b–14b.

39 Will 1985: 315n43.

40 Ni Wenwei 2002 [1885]: 90–91.

as floodwaters receded.[41] Another temple, a Guanyin nunnery, was also built during the late Ming dynasty, but by a retired county magistrate who once saw that the Han River dike in his hometown was about to rupture and hurriedly recruited and supervised laborers to reinforce it. That night, he claimed to have seen Guanyin and soon the waters stopped rising, and in the end the dike held. He thus donated funds to build a nunnery by the side of the dike, which has remained safe ever since.[42] From their names, however, it is impossible to discern the relationship between these two temples and water control or dike safety. In any event, there must have been more of such kinds of temples that were not recorded in local gazetteers.

Some of these water deity temples might have been intended to do more than ensure the integrity of dikes. In Jingmen, for example, in the spring of 1867, rebels associated with the anti-Qing Nian movement chased with government troops on the bank of the Han River in the Shayang area. The troops had great difficulty stopping the rebels from crossing the river due to its length. A commander of the government troops had heard about the effectiveness of a local temple (i.e., the Zhenjiang wang miao) and brought his soldiers inside to worship. At that moment, the Han River water rose four chi within two days. The Nian rebels were unable to cross the river and had to withdraw. The water also soon receded. Military officials believed that the government forces were under the protection of a mighty deity.[43] This, however, might have been a coincidence—the Han River water often rises rapidly during the high-water season when floodwaters flow down from the upstream mountains. On the twenty-eighth day of the eighth lunar month in 1742, for example, at Shayang the Han River suddenly rose eight or more chi; on the fourteenth day of the ninth lunar month in 1753, at Qianjiang it even rose 15 chi overnight.[44]

Nevertheless, the local people believed in water deities. In another example, Wang Baixin, a local elite who was famous for his views on water control, once wrote an essay about the worship of water spirits:

> In the summer of 1859, the river dike in Luoshan [in Jianli County] burst. By the ninth lunar month, the river water rose again, and two islets outside the dike suddenly split in the middle; the fissure was 1 li or more long and several zhang deep. There was a spirit hidden inside and it wreaked

41 *Qianjiang xian zhi* 1694, vol. 7: 6ab.

42 *Qianjiang xian zhi* 1879, vol. 7: 5ab.

43 *Hubei tongzhi* 1921: 980.

44 Shuili dianlibu 1991: 251, 327.

> havoc by riding the waves. ... I wrote an essay about worshiping, and the spirit [then] furtively disappeared.[45]

Clearly, Wang also believed in the mighty power of the spirits. The splitting of the two islets, however, may actually have been the collapse of the islet into the river, a common phenomenon in the middle Yangzi River valley. Wang was not a government official, but he was influential nonetheless.[46] In his essay, he went on to explain that the local people had already suffered greatly, and thus he, along with them, would henceforth more devotedly worship water deities in the hope there would be no more disasters and that the people would farm diligently to pay the taxes. This is the standard official tone on such matters: carry out water control to protect the people and guarantee the payment of taxes. It is no wonder that the local gazetteer reproduced his entire essay, which would mean that the local authorities wholeheartedly agreed with him.

The Revered Mr. Xiao (Dan), the Revered Mr. Yan (Shuzi), the Revered Mr. Ni (Fuke), and the Revered Mr. Xie (Lin)—all mentioned in Table 5.1—were worshiped by the local people because of their divine power to control floods or their contribution to the construction of dikes.[47] The same is true of He Liaogong of Jianli and Ding Agong of Jiangling (neither of whom is listed in Table 5.1).[48]

Although it seems that in the Qing the inhabitants of the Jianghan area believed in the power of water deities when it came to high-water control and dike protection, the beliefs of the emperors are not as well known. The following section explores what we do know about the thinking of the emperors.

2 Iron Oxen, Imperial Edicts, and Water Deity Worship

As past scholarship has shown, the Qing emperors, whether they functioned as managers or engineers, had to take the management of rivers—particularly the Yellow River, the Grand Canal, and the Yongding River—seriously. Some went so far as to express their concern over river management by frequently visiting water-control sites in person.[49] Most river dikes in the Jianghan area

45 *Jingzhou fu zhi* 1880, vol. 27: 28a.

46 Wang had a close relationship with some important Xiang Army generals such as Hu Linyi (who once served as the governor of Hubei), and thereafter was very influential (*Jianli xian zhi* 1994: 833).

47 Yang Guo and Chen Xi 2008: 335–45.

48 *Jianli xian zhi* 1872, vol. 2: 8b; Fang Ming and Dong Yumei 2003: 60.

49 Leonard 1996; Dodgen 2001; Lillian Li 2007; Michael Chang 2003.

were not directly managed by the central government, but that does not mean the emperors completely ignored them. On the contrary, they often issued edicts directing officials how to manage Jianghan's dikes.

From the perspective of ideological or spiritual control, the Qing dynasty emperors, who were Manchus and not Han Chinese, favored the construction of temples and the worship of various (traditional Han Chinese) deities to win the support of their Han Chinese subjects and to show their legitimacy as the new rulers of China. That was also true in the Jianghan area and was particularly manifested in the placement of more iron oxen on dikes.

The Chinese had a long tradition of putting statues of animals along riversides to ward off water or water demons. One of the most common statues was the iron ox.[50] The Jianghan people began placing iron oxen on dikes at least as early as the Song dynasty. In Jianli, for example, a Song dynasty monastery was named Tieniu si (Iron Ox Monastery) because an iron ox was placed in front of the monastery to protect the dike.[51] In the same county in the Qing dynasty, at Tieniu zhou (Iron Ox Islet) the locals cast a 1,000 jin iron ox and put it on the islet to ward off "water spirits" in the river.[52]

In the record-breaking flood on the Jianghan Plain in 1788, the Wancheng Dike broke in twenty-two places, and the city of Jingzhou was inundated. The central government allocated two million taels of silver to rebuild the dike and the city. After the reinforcement of the Wancheng Dike, the Hunan and Hubei governor-general Bi Yuan, who oversaw the rebuilding project, was ordered by the emperor to cast nine iron oxen and put them on the most dangerous segments of the Wancheng Dike. In the eleventh lunar month of the same year, the Qianlong emperor issued an edict:

> I thought it had always been that iron oxen were cast and put on the dangerous segments to contain water. That is because the flood dragon fears iron, while [the related element (of the Five Elements) of] the ox is earth, and earth can hold back water, and hence [people] cast iron into the shape of oxen to control [flooding]. This time Jingzhou suffered grievously from a flood, which was thought to have been caused by the flood dragon. Now that a new dike project has been built in this place, [I] wrote an edict ordering Bi Yuan to follow up and cast and erect iron oxen on the

50 Huang Zhigang 1970 [1934]: 25–27; Wang Peijun 2012.

51 *Jingzhou fu zhi* 1880, vol. 28: 39a.

52 *Jianli difang zhi* 1991: 167.

> dangerous and important segments of the Wancheng Dike in Jingzhou and Shashi to protect the dike. This is also a way to prevent inundation.[53]

In 1859 the prefect of Jingzhou cast another iron ox and had it put on the segment of the Wancheng Dike near the town of Haoxue, bringing the total of iron oxen on this dike to ten. This "iron ox," as the local people called it, was a statue of the *jiaoduan* (a mythical auspicious animal in ancient Chinese legends).[54]

The Qianlong emperor apparently believed that casting iron oxen and putting them on the dike brought protection against inundation. After the 1788 flood, however, he accused the officials in charge of dike maintenance of mismanagement. About twenty of those officials—from the governor-general to the deputy county magistrate—were fined, dismissed from office, and jailed, but none of them was blamed for being insufficiently devout in their worship of the water deities.[55] In other words, here the emperor seems to have valued regular dike management over water deity worship. No one actually ever attributed that year's huge flood to the failure of the responsible officials to earnestly worship the water deities. After the repair of the ruptured Wancheng Dike, however, the emperor reiterated his belief in the role of divine protection and ordered officials to place iron oxen on the dike.

The Jiaqing emperor (r. 1796–1820), the son of Qianlong, also seems to have believed deities could control floods. In the eighth lunar month of 1796 he issued an edict reflecting this belief:

> There was another memorial that talked about how [the Wancheng Dike] remained intact after the autumn high water and the solidity of river dike projects. This was all due to the protection of the deities. [Thus I] specially sent out twenty big sticks of Tibetan incense, [to be] handed over to the governor of Hubei and used in worshiping in the River God Temple in thanks for the god's effectiveness.[56]

Here he clearly attributed the safety of the Wancheng Dike to divine protection. During the reign of his son, the Daoguang emperor, the Jianghan Plain frequently suffered from water calamities—the Wancheng Dike alone sustained one to four ruptures in each of the seventeen years between 1822 and

53 *Jingzhou fu zhi* 1880, vol. 17: 6ab.
54 Ni Wenwei 2002 [1885]: 90; *Jiangling xian zhi* 1990: 656.
55 Ni Wenwei 2002 [1885]: 210–14.
56 Ni Wenwei 2002 [1885]: 31.

1850.[57] Daoguang issued eighteen edicts on dike management in Jingzhou, but none of them mentioned divine protection; instead, when the dikes were breached, he tended to blame government officials and clerks for not doing their job.[58] In other words, he realized the importance of practical (and also effective) dike management in the protection of the integrity of dikes. His successor, the Xianfeng emperor (r. 1851–1861), faced a similar if not a worse situation (as the Taiping rebels frequented the Jianghan area) and did not mention divine protection of dikes either.

The next emperor, Tongzhi (r. 1862–1874), however, once again issued edicts that acknowledged the power of the water deities to protect dikes. In the tenth lunar month of 1870, he wrote: "This year the summer high water has risen and is close to overflowing the [Wancheng] dike; the situation is very dangerous. The circuit intendant, the prefect, et al. visited temples and worshiped piously, and thus relied on the hidden protection of the deities. [The locals] turned danger into safety." He then sent two *bian'e* (horizontal inscribed boards) to the governor-general of Hunan and Hubei, and commanded him to hang them in the River God Temple and the Zuzhang God Shrine to pay homage to the divinities.[59]

This practice was continued by the next ruler, the Guangxu emperor (r. 1875–1908), who in an 1884 edict declared that the deities of "the River God Temple and the Zuzhang God Shrine are usually efficacious." He then also sent out two bian'e and directed the local officials to hang them in temples to thank the deities for their protection of the dike, and did almost the same thing again in 1887.[60]

Unlike the Qianlong emperor, who responded to floods by not only ordering that iron oxen be placed on dikes but also punishing the responsible officials for poor dike management, both the Tongzhi and Guangxu emperors praised officials who, in the face of potentially disastrous floods, worshiped in temples instead of supervising high-water control. Does this mean that these emperors really believed that divine protection was more important? To answer this question, it must be noted that both Tongzhi and Guangxu were children when they were enthroned. The Tongzhi emperor died at the age of nineteen; the Guangxu emperor was only three years old when he ascended to the throne in 1875. Thus, it is essential to uncover what officials in the central government thought about this issue.

57 *Jingjiang dadi zhi* 1989: 71–73.
58 Ni Wenwei 2002 [1885]: 31–38.
59 Ni Wenwei 2002 [1885]: 39.
60 Shu Hui 2002 [1896]: 379.

In the Qing, there were regulations pertaining to dike management in Jingzhou: the responsible official would be promoted for keeping the dike safe or dismissed or imprisoned if it broke.[61] In 1883, however, the Ministry of the Personnel stipulated that the maintenance of dike safety was mandatory, and the responsible officials therefore did not deserve to be rewarded for simply doing their job. In 1886, some provincial officials insisted that the Wancheng Dike was so important that the incumbent prefect of Jingzhou, who ensured its integrity, deserved promotion, but the Ministry of the Personnel flatly refused. The emperor agreed.[62]

Thus, from 1788 to 1887, most Qing emperors issued edicts in recognition of the role water deities supposedly played in maintaining dike safety in the Jianghan area. If inundations or dike ruptures occurred, however, the responsible officials would be punished for mismanagement, not because their worship of the water deities was insufficient pious. This implies that the emperors and/or central governmental officials might not really have believed that dikes held because of divine protection; most likely they turned to religion as a form of psychological consolation. Or they may have believed in both divine protection and regular maintenance, but different people had different ways of ranking their relative importance. Nevertheless, water control and particularly dike management were pragmatic matters.

Qing society seems to have echoed this approach. In an 1888 memorial, someone reported that Lin Zuo, who had been the magistrate of Jiangling for two years, was impeached for never having visited a dike project site and for being overly fond of the opera. According to the person (an official) who investigated these charges:

> Magistrate Lin has served for more than a year. He managed dike affairs particularly well; it was a bad year last year and [the government] did not tax much, but he did what he could to divert funds for must-do projects. … The high water was unusually high last year, but all the dikes are [still] solid. … As for going to the opera and [patronizing] actresses, according to my investigation, after ensuing the integrity of dikes, it had always been the custom to stage opera performances to repay the deities, not toy with actresses.[63]

61 Zhang Youyu and Gao Chao 1991: 850–52.

62 Shu Hui 2002 [1896]: 470–72.

63 Shu Hui 2002 [1896]: 461–62.

Whether the charge raised in the report was accurate or not, the investigator clearly believed that it was absolutely correct to stage operas—as part of a celebration that could last for three months in the Yellow River valley[64]—to repay the deities for maintaining the safety of dikes. In other words, the commoners also acknowledged the existence of water deities and their efficacy, a tradition that continued into the Republican period.

3 The Case of Xu Guorui

Early in the twentieth century, with the fall of the Qing dynasty (as well as the end of the imperial system) and the establishment of the Republic of China in 1912, and thereafter the anti-tradition and anti-imperialist New Culture Movement of the 1910s–1920s, many traditions were denounced or abandoned. In the 1920s–1930s, as a part of a nation-building movement, the Republican government shifted its water-control practices from old customs to new technology, with the formation of some modern hydraulic government departments, and efforts to unite administration of water conservancy under the National Economic Commission in 1934.[65] In Hubei many dike bureaus were created, such as the Bureau of Dike Work on the Jing River and the Bureau of Engineering Work on the Yangzi and the Han Rivers. After a flood in 1931, about thirty governmental officials (include twenty-two county heads) of Hubei were punished (some were, for example, suspended from office, downgraded in rank, or hit with a salary cut) by the Nationalist central government.[66] All of them were punished for their mismanagement of the dikes, not because they failed to worship water deities piously, nor, on the other hand, because they devoted too much time to worshiping the deities.

At the same time, many traditions became targets of the Nationalists' anti-superstition campaign.[67] But it is not easy for people—including those who were in charge of dike management—to turn their back on tradition. Xu Guorui is a good example. Like many youths at that time, Xu attended a military academy and later participated in the Wuchang Uprising, which eventually overthrew the rule of the Manchus and brought the new Republic to power in 1912. From 1924 to 1937 he headed the newly established Bureau of Dike Work on the Jing River. This was a technical department, but he was not trained as

64 Fang Mu 1995.

65 Pietz 2002; LS1-5-606.

66 LS1-4-452.

67 Nedostup 2009.

a hydraulic engineer. Before 1924, Xu had served in a local police department and dealt with disputes involving boats that shipped stone for dike projects,[68] which provided him with some expertise in dike affairs and presumably qualified him to serve in this bureau. After leading the bureau for ten more years, it seems he had come to understand dike affairs quite well—at least in the eyes of Zhang Qun, the governor of Hubei at the time.[69]

According to some sources, Xu, appreciating the importance of the Wancheng Dike, and investigated the entire dike to examine dangerous sections. He reformed some dike regulations to fight corruption; he often personally supervised the laborers who reinforced the dike; and several times he lived on dikes for weeks and worked closely with the laborers to control floods. It was said that he was always the first person to appear on the scene during an emergency. He advocated the dredging of the riverbed and opposed the blockage of flood-diversion outlets. With the support of the population of ten counties, he killed a proposition to block the four southward outlets on the Jing River into Lake Dongting. He also edited *Jingjiang di zhi* (Gazetteer of the Jing River Dike). Xu was also a devout Buddhist. After the Great Jingjiang Dike ruptured in 1935, he prayed to the deities for help, "providing sacrifices, setting up an incense alter, worshiping the river gods on the dike, and forbidding butchering within the city for three days." After the flood, he was punished for being lax in leading the rescue work.[70]

Much of the above information about Xu is favorable. Other sources, however, describe Xu as a very unprofessional manager—he was condemned as greedy and corrupt, and the 1935 flood was blamed on his incompetence. According to these sources, the Bureau of Dike Work on the Jing River had ample funds, but Xu spent only a fraction on dike repair and high-water control; the dike was in poor condition due to corruption and jerry-building; and high-water control was also performed in a perfunctory manner.[71] Although Xu was punished for negligence after the 1935 flood, these completely negative comments made by PRC informants seem one-sided, as is typical of the official Communist attitude toward Nationalist bureaucrats.

According to these sources, Xu was not only a devout Buddhist, but also a social climber. The sources claim he peculated dike fees to fund his cultivation of social contacts—including the elites in the Jingzhou-Shashi area as well as in Wuhan (the capital of Hubei province), and various Buddhist monks and

68 LS1-5-852: 4.

69 LS1-5-851.

70 *Hubei shuili zhi* 2000: 1512–13; *Jingjiang difang zhi* 2012: 784.

71 Qin Zhongxiang 1998; *Jiangling difang zhi* 1984: 21.

nuns. Xu was the chief patron of some important local Buddhist temples and attended all the local temple fairs. When there was a serious flood, Xu would ask Buddhist monks and nuns for help—hundreds of them would then chant Buddhist scriptures along the Yangzi River. After the waters receded, Xu would hire them to chant for a few more days to thank Buddha for his protection.[72] These activities were expensive.

Combined, these sources, positive or negative, reveal that Xu apparently had an extensive knowledge of dike affairs and also believed in divine protection. From the Communist perspective, if Xu really worked diligently on dike affairs, he should not have asked Buddha for help at the last moment—one Communist critic in fact argued that if Xu had not done what he did, the Great Jingjiang Dike in 1935 would have been held.[73]

Why was Xu, who was the head of the Bureau of Dike Work on the Jing River, so devoted to Buddhism? It is said that when, at the age of forty, he fathered a son, he thought it was "a heavenly bestowal." He chanted Buddhist scriptures during the high-water season to ensure the protection of heaven. During the 1931 flood, he received desperately needed high-water control materials just in time (see Chapter 2), which he attributed to "heavenly help" and thereafter he became even more devoutly Buddhist.[74]

According to revolutionary ideology (of both the Communists and the Nationalists), worshiping water deities was superstitious. In the Republic, however, some government officials such as Xu may have just followed tradition without necessarily being convinced of its efficacy. In any case, respecting tradition might also have offered some psychological comfort. Thus we see that even the pragmatic prefect of Jingmen in the Qing also rebuilt a monastery to thank deities for protecting the dike. In another example, in 1933 in the battle against the Communist forces in their Honghu Base Area, the Nationalist Hubei government allotted special funds to build and repair some dikes in Honghu—this was viewed by the compiler of a gazetteer in the 1990s as a Nationalist attempt to "psychologically conquer" the people of the Communist base area.[75] Thus for the populace who had faith in tradition, officials' worship of water deities proved that they took high-water control seriously. At the same time, however, that did not mean that they would not take some practical steps to control floods. Officials may well have taken some measures eventually, but, according to their critics, these were too little and too late. It is no wonder that

72 Qin Zhongxiang 1998.
73 Qin Zhongxiang 1998.
74 *Shashi shuili difang zhi* 1994: 179–80.
75 *Honghu xian zhi* 1992: 13.

dikes broke. Although these officials lived in the early twentieth century, their education was traditional and they thought their measures were effective, or at least might be emotionally or psychologically beneficial.

According to an on-the-spot investigation, the real reason for the rupture of the Great Jingjiang Dike in 1935 was mismanagement. At noon on July 5, the river overflowed one segment of the dike and soon a small fissure burst open. The local inhabitants rushed to the site and tried to repair the break. But Wu Jingtang, the director of dike management of the Bureau of Dike Work on the Jing River, who was responsible for this section, fled, claiming that he was going to pray to Guan Yu. The local people who came to help had to leave because there were no materials and no one to lead their work. The small fissure then widened into a breach more than 165 meters long and 3 meters wide. The water surged through the break and flooded more than ten counties on the Jianghan Plain, causing millions of people to become refugees.[76]

After the disaster, the Bureau of Dike Work on the Jing River argued that this was an "inundation," a nature disaster beyond human control. The bureau also claimed that the Jiangling County government had failed to provide enough laborers to deal with the break and was therefore partly to blame. The county government, however, insisted that the dike had burst and that the personnel of the bureau had been negligent and had to take full responsibility.[77] Both the county administration and the technical bureau wanted to blame each other, but neither mentioned that the opposite side had failed to appease the water deities.

Lei Xiaoqin, head of Jiangling County at the time, in fact had a very different attitude toward the worship of water deities. Early in the spring of 1935, Jingzhou was struck by a severe drought. The residents from the outskirts of Jingzhou went to the King Yu Temple to pray for rain. According to legend, a relic of Yu the Great in this temple, *xirang* (growing earth), was buried inside an earthen well, and if anything inside the well was damaged, a huge flood would follow. Lei dismissed this legend and hoed away one layer of topsoil over the well in May. In July, after the dike broke, many people surrounded the county government office and said that Lei had caused the disaster by desecrating the xirang.[78]

Before the flood, Lei was extremely dissatisfied with what Xu Guorui had done and threatened to kill Xu if the dike broke. Sources claim that Xu was so afraid of a huge flood that his hair turned white within a few days. After the

76 *Jiangling difang zhi* 1984: 16; Li Zinan 2001.

77 Qin Zhongxiang 1998; Li Zinan 2001.

78 Qin Zhongxiang 1998.

flood, however, Xu was supported by the garrison troops and was only mildly punished, while Lei was forced to leave his position and flee in fear.[79] As the county head, Lei also should have taken responsibility for the rupture. It is said that when he first heard the news of rupture, he thought that the dike was in another county.[80] Needless to say, the conflict between local county governments and the dike bureaus in general is one of the main reasons for the ineffectiveness of high-water control in the Republic.[81]

Since he was a devout Buddhist, the worship of Buddha had become an inseparable part of Xu's daily life. No one perhaps would object to honoring Buddha, but it was equally important to send laborers to watch the dike. Sacrificing animals is not the first thing a pragmatic dike manager should do in the middle of a flood. At the same time, it is unfair to claim that Xu did not care about the integrity of the dike or that he alone was responsible for the rupture of the Great Jingjiang Dike.

Some scholars have argued that traditional states used religious practices—particularly rainmaking—as a means of controlling rural society. On the Jianghan Plain, however, this seems at most to have been only a minor objective of the local government. Rather, the example of Xu Guorui shows that local officials were just doing what was expected of them. According to recent research, when drought hit a region, the local officials were expected to practice rainmaking rites, which also provided a rare chance for these officials to publicly display that they were benevolent and model officials.[82] Local officials' worship of water deities in the Jianghan Plain should be viewed in this light. That is to say, they did what they were supposed to do, and they also likely wanted to take advantage of the opportunity to showcase that they were diligent, responsible, and competent.

4 "Human Beings Can Overcome Nature"

On the one hand, worshiping water deities, putting iron oxen on dikes, and paying respect to Buddha all show the limited capacity of humans to withstand natural calamities. On the other hand, people to this day still believe they can use their own resources to control floods.

---

79 Qin Zhongxiang 1998.
80 *Jiangling difang zhi* 1984: 16.
81 Jiayan Zhang 2014: 45–74.
82 Snyder-Reinke 2009: 19, 173.

This tradition is also deeply rooted in Chinese culture, as seen in the legendary Yu the Great's taming of the rivers and in Xunzi (ca. 313–238 BCE), one of the most important Confucian scholars, who "firmly believed in man's ability to control nature."[83] The gazetteers of Jiangling—a county where the all-important Wancheng Dike was located—repeatedly pointed out that dike management was essential to the people's safety and security and that human beings were capable of acting to ensure their safety. What happened in Baiju polder late in the eighteenth century is a good example.

> [In Jiangling] each polder has always had a head and laborers. But since there is a difference in residents' power and the distances from their homes to dike, there have been many faulty practices. ... Recently [the residents of] Baiju polder and Hulu polder have made vigorous efforts to end these malpractices. In years when there was a bumper harvest, [the residents of] the whole polder have cooperated in building the polder dike, no matter if one was strong or weak, or if one lived far [away from] or close [to the dike]; and there was no specially appointed polder head. The dike is divided into sections and the laborers are assigned according to the amount of land each one has; the rich contribute money, and the poor contribute manpower. [The residents of] each polder recommend an experienced person to supervise the work. The residents watch each other, and if a weak place is found, all the residents come together to punish [the responsible person] and order [him] to rebuild it quickly. ... In 1788, [the area suffered from] a huge flood, the Yangzi River and the Han River flowed together, and farmland and crops in low-lying areas were inundated. Only the dikes of Baiju polder and Hulu polder remained secure. ... Human beings can overcome nature (*ren ding sheng tian*), is that not so?[84]

A main point of this paragraph is that if a dike was well maintained, it would not break even during a huge flood. It is from this point of view that the compilers concluded that "ren ding sheng tian"—which I translate as "human beings can overcome nature." There is nothing in the paragraph about water deity worship or divine intervention.

Many later gazetteers, including the gazetteer of the Wancheng Dike, reproduced this paragraph. One can assume that the compilers of all these gazetteers agreed with its main idea, particularly "human beings can overcome

83 Lillian Li 2007: 2.

84 *Jiangling xian zhi* 1794, vol. 8: 10b–11b.

nature." Li Hongzhang, one of the most important Qing statesmen in the late nineteenth century, had the same idea when he wrote the preface to the gazetteer of the Wancheng Dike in 1876: "The dike should be solidly built, the funds should be raised from various sources, the government's responsibilities should be checked, and high-water control should be carried out diligently. If all of this is done, then [the damage caused by] huge floods can be reduced and minor floods can be eliminated. This is how human beings can overcome nature."[85] Li thus emphasized that floods can be controlled, at least to a point. He reiterated the official attitude toward natural disasters: people are not powerless. Otherwise, what was the point of encouraging people to build and repair dikes? Again, Li did not mention the worship of water deities.

The idea of "ren ding sheng tian" has been carried forward in the PRC—but has been more commonly interpreted as "man can conquer nature," or even "man must conquer nature."[86] Like the imperial government, the new PRC government spent a great deal of time and money on water control. In the Jianghan area that included the reinforcement of the Great Jingjiang Dike, the building of the Jingjiang Floodwater Diversion Zone, and most importantly, emphasizing annual high-water control. In the high-water season of 1952, there were hundreds of emergencies on the major dikes of Hubei, but all of them (except one) were eventually overcome because people believed that "man can conquer nature."[87] In that year, to mobilize as many people as necessary for high-water control, the Hanyang County government launched a political campaign to publicize among the peasantry the idea of "man can conquer nature" and "to build dikes to protect life, and to carry out high-water control to protect households."[88] The county government wanted the peasants to understand the relationship between ensuring the integrity of dikes and their personal survival. Here the idea of "man can conquer nature" is more a platitude than a policy. If the county government told the people that man cannot conquer nature, there would have been no way to mobilize them to repair and maintain the dikes.[89]

In the late 1950s, the CCP promoted the notion of "ren ding sheng tian" to increase production.[90] It was also a guideline for socialist construction in the

85 Ni Wenwei 2002 [1885]: Li's preface.

86 Luo Jianjin 2001; Shapiro 2001: 66.

87 SZ34-2-414.

88 SZ113-2-35A.

89 After the 1931 flood, a reporter argued that people should have "the confidence of ren ding sheng tian" in facing calamities, but it was found that most refugees from the flood lacked that confidence (Xie Qianmao 1931: 194).

90 Murphey 1967.

following two decades. That, according to Judith Shapiro, amounted to "Mao's war against nature," which caused many environmental disasters, particularly during the GLF and the Cultural Revolution.[91] But this idea has a very long history in China and was not original to the CCP and Mao.

As a native of Hunan (a province adjacent to Hubei), Mao must have known that the Jianghan Plain was subject to frequent flooding. At a social gathering in Wuhan in 1959, Mao asked a performer from Mianyang whether he had heard the saying "In Shahu and Mianyang, nine years out of ten are lean years" (*Shahu Mianyang zhou, shinian jiu bushou*) (because of frequent floods on both the Yangzi and the Han Rivers). Although Mao might have just been making small talk, the performer interpreted the comment to mean that Mao had a deep understanding of the suffering of the people of Mianyang.[92]

In another case, one day early in the PRC Mao reportedly burst into tears while reading a document. Surprised, his bodyguard asked what was wrong. Mao answered that he had just read about how the people in the middle Yangzi River valley suffered from a flood. He even said that he wanted to resign from his position as chairman of the Communist Party to assist hydraulic experts who were trying to contain the Yangzi River. He declared to many people: "You can despise anybody and anything, but you must under no circumstances despise the Yellow River and the Yangzi River."[93]

In these two examples, it seems that Mao in fact rather respected the power of nature and recognized the difficulty in conquering it. He particularly mentioned that he wanted to "assist hydraulic experts," implying, in other words, that he recognized the importance of science and technology in the control of floods—or more broadly, the power of science and technology to help human beings conquer nature. Indeed, almost all newly compiled PRC gazetteers emphasize how up-to-date science and technology and management (particularly the leadership of the CCP) have contributed to dike safety in the PRC.

This is not an apology for environmental disasters caused by the propaganda that "man can conquer nature." Rather, as the above discussion has shown, Mao's promotion of this idea was rooted in a historical tradition—his understanding of hydrology and his intention to harness rivers for the benefit of the people represent a continuation of the tradition of the imperial rulers. But the phrase "ren ding sheng tian" has been interpreted literally to mean that human beings are so powerful that they really can "conquer" nature, which is not at all what the compilers of traditional gazetteers meant.

---

91 Shapiro 2001.

92 Yang Tong 1977: 4.

93 Zou Dongting and Zheng Zehua 1998: 149.

5 Water Watching, the Dragon Dance, and New-Style Tourism

After the founding of the PRC, the government vigorously promoted the idea that "man can conquer nature," advocated atheism, and openly criticized all "superstitious" practices, including the worship of water deities. Water deity worship was even blamed for contributing to the frequent floods in the pre-1949 era. The author of a commemorative article asked rhetorically why floods were so frequent before 1949, and then supplied the answer himself: because the rulers in the past placed their hope for high-water control in the deities. Taking the Haoxue iron ox as an example, the author continued: even if it had been a living ox, it could not have guaranteed its own safety under the old social system—the iron ox was unearthed from the rive sludge early in the PRC.[94]

Unlike the local officials in the Qing and the Republic who visited temples to pray for divine protection of the dikes, officials of the PRC government at different levels who are responsible for high-water control most certainly do not worship water deities. Instead, they have been required to go to dikes in person, especially when there is an emergency, as many imperial and Republican officials also did. In the early years of the PRC, because everyone was subject to the authority of the Communist Party, conflicts over dike affairs between administrative organs and technical bureaus as in the Republic became virtually impossible. Moreover, responsible officials could not run away and avoid punishment. After floods were contained, the government would have a celebratory meeting, which is functionally similar to what the Qing did by staging operas.

In the PRC era, most inhabitants of the Jianghan Plain no longer worship water deities either. Instead, they enjoy "water watching" in Shashi. The segment of the Great Jingjiang Dike in Shashi is one of several dangerous sections, particularly during the high-water season when the water of the Yangzi River almost reaches the top of the dike. At this point, the river water extends so far that the river looks like a sea. In the 1980s, during the high-water season the water level of the Yangzi River was ten to fourteen meters higher than the ground on the northern side of the dike along Shashi.[95]

Under these circumstances, any tiny breach of the Great Jingjiang Dike would quickly lead to a disaster, with floodwater immediately submerging the whole city of Shashi and most of the Jianghan Plain. The citizens of Shashi, however, now know that the government will do its best to maintain the

94 Cheng Qijing and Xu Yiping 1987: 14.

95 *Jingjiang dadi zhi* 1989: 14.

dike—the Great Jingjiang Dike has not ruptured since 1954.[96] They are so confident in fact that they consider the top of the dike an ideal place to cool down after a hot summer day. Thus, every night in the high-water season, thousands of people go to the dike to look at the water and enjoy the cool air.[97] This would have been unthinkable in the Qing and the Republic.

In the past three decades, with economic development, social stability, and the improvement of people's standard of living, water conservancy projects have also undergone changes. Their major functions still include high-water control, irrigation, and electricity generation. Since the 1990s, however, with the increase of personal income, tourism has become an important part of the life of the people of China, particularly in urban areas. Water conservancy projects, which usually are in remote mountains or in the countryside, allow visiting city dwellers to enjoy fresh air and a picturesque landscape. Thus many water conservancy projects have now become tourist destinations.

In traditional China, each county or prefecture had its so-called "eight scenes" or "ten scenes," which usually included famous local mountains, lakes, and buildings. In contemporary China, the number of touted tourist destinations has been greatly expanded. Many of the locales depicted in the *One Hundred Scenes of Jingzhou*, published in 2002, are water conservancy or related projects, such as the Great Jingjiang Dike (see Table 5.2). Nationwide, the most famous is the Three Gorges Dam, which attracts tourists from all over the world. A major purpose of this dam, which is not far from the Jianghan Plain, is to protect the plain from severe floods.

In the Qing dynasty, "to commemorate the hydraulic achievements of the emperors, numerous temples and shrines were erected along the major rivers."[98] In the PRC, however, after a successful high-water control project, the government built not temples or shrines but monuments. In the Jianghan area, a representative example is the Monument to the Jingjiang Floodwater Diversion Project. The Hubei provincial government built the monument to celebrate the completion of this huge project in 1952 and to showcase how the new government was superior to the previous governments.[99] It is now one of the new scenes of Jingzhou (see Table 5.2).

---

96 Another good example would be the successful control of the 1963 Tianjin flood due to "campaigns involving mass mobilization, commanding flows of resources and quick decision-making" (Paltemaa 2011: 25).

97 *Shashi shuili difang zhi* 1994: 110.

98 Lillian Li 2007: 71.

99 *Jingjiang fenhong gongcheng zhi* 2000: 310–11.

TABLE 5.2 Water conservancy projects in the *One Hundred Scenes of Jingzhou*

Order in the source	Name of the scene	Note
20	The Iron Ox Crag	Iron Ox Park (in Haoxue) on the Great Jingjiang Dike
21	Guanyin Rock	
45	Honghu (Lake Hong) Scenic Zone	In Honghu
46	Tian-e-zhou David's Deer Reserve	In Shishou
47	Weishui (the Wei River) Scenic Zone	In Songzi
48	Scenic Zone of the Jing River Flood-Diversion Zone	In Gong'an
49	Great Jingjiang Dike Scenic Zone	
50	Jingsha River Scenic Zone	
59	Pagoda of Longevity	On the Great Jingjiang Dike
65	Monument to the Jing River Flood Diversion Project	On the Great Jingjiang Dike
80	The old course of the Jing River	

SOURCE: JIANG YONGWU AND ZHANG JIAWEI 2002

Recently, the local county/township government has built a public park on the site of the iron ox on the Haoxue segment of the Great Jingjiang Dike. I have visited this park several times in past years. The iron ox (jiaoduan) is still there, facing the Yangzi River, but the local people no longer rely on it to control floods. Instead they come here for exercise, relaxation, and sightseeing. Another iron ox, on the Lijiabu segment of the Great Jingjiang Dike, which was destroyed during the Cultural Revolution, was re-created in 1982.[100] In Wuhan, on the site of the Dragon King Temple—the temple itself is long gone—a public recreation and sightseeing area was built in 2001.[101] Similar developments have been undertaken in many other parts of China as well.[102]

At the same time, the PRC government no longer insists that religion is the opium of the masses, and many traditions that had been criticized or abandoned have been revived since the beginning of the economic reform in 1979. For example, the state now sponsors an annual ceremony to worship

100 *Jiangling xian zhi* 1990: 656.

101 Tu Deshen and Yang Zhichao 2002.

102 Wang Peijun 2012.

the Yellow Emperor in his mausoleum (in Huangling, Shaanxi, which was destroyed during the Cultural Revolution but restored recently) in the spring and one to honor Confucius (in Qufu, Shandong) in the autumn. Many such state-sponsored restorations of old religious sites or traditions are actually aimed at developing the local economy via tourism.[103] In the Jianghan area, religious believers are free to visit temples to pay their respects, many temples that were destroyed during the Cultural Revolution have been rebuilt, and ancestor worship is practiced all year. The tradition of the dragon dance has returned, but the people perform it not to beseech the Dragon King to control the water but simply for recreation.[104]

The tradition of water deity worship in the Jianghan Plain, in short, seems to have died out. After the founding of the PRC, particularly in the era of collectivization, the state deeply penetrated and controlled rural society (down to the village or even household level), but since 1979, the state has gradually withdrawn from the village level. Nonetheless, it still has the power to mobilize people for high-water control. But just as local government officials responsible for high-water control will not visit temples to pray for the safety of dikes, and no longer mention water deity worship in their management of dikes, the Jianghan people rarely mention it in their daily lives either. At the end of the nineteenth century, thirty-four temples stood on or along the Wancheng Dike. Like many other temples elsewhere, these were destroyed or abandoned in the twentieth century. One of these temples, Jiuhua Temple, has recently been rebuilt by the local people. When I visited the temple on May 17, 2013, an informant (the temple's public liaison person) told me that it is a normal Buddhist temple, not a water deity temple. He also informed me that in fact three other temples—Xuxian, Jinguo, and Yangsi—were also rebuilt along the Great Jingjiang Dike recently by the local residents, but none of them is a water deity temple. I visited Xuxian Temple and agreed with him. It seems the achievements of water control by the PRC government have convinced the local people that they no longer need to pray to the deities to ensure the integrity of dikes or pray to the Dragon King for rain, even though they are free to do so in the post-Mao era.

Recently, the local government has rebuilt the city wall and some other buildings in Jingzhou and even sent officials abroad to promote Jingzhou as a tourist destination. In the summer of 2006, a delegation from Jingzhou visited Atlanta. The chief present they brought with them was a small portrait of Guan Yu—a representative of the culture of the Three Kingdoms and the first mayor

103 Oakes and Sutton 2010.

104 *Jiangdi xiang zhi* 2006: 233.

of Jingzhou, according to them. In their presentation, however, I heard nothing about Guan Yu's divine power to control floods.[105] Official reports indicate that the worship of Guan Yu now is a major part of Jingzhou's tourism and "tourism souvenirs of Jingzhou mainly centre on the Guan Gong culture,"[106] but there is no mention of his mighty power to subdue the waters. On May 17, 2013, when I visited the recently rebuilt Guan Yu Temple in Jingzhou City, I found that in this temple Guan Yu actually was worshiped as the God of Wealth; the whole temple looks more like a place of money-making than a place of religion.

This had happened before. The Xuzhou area, in northern Jiangsu, in the middle of the Qing, when the Yellow River was still merged with the Huai River, suffered frequent water calamities, and thus the worship of water deities was important to both government officials and the populace. After the Yellow River again changed its course in 1855, it flowed further north, through Shandong, not Jiangsu. The frequency of water calamities in Jiangsu declined, and the local residents ceased to worship water deities. Many of the temples dedicated to them fell into disrepair and disappeared.[107] In Shandong, however, the frequency of floods increased after 1855, and many temples were built along the Yellow River (which now runs through the province) for people to worship river deities for the protection of dikes.[108]

6 Conclusion

In response to the frequent flooding of the Jianghan Plain over the past three centuries, the local inhabitants built temples to worship water deities and put iron oxen on dikes in the hope of containing the water. This is a natural reflection of the environmental situation in worship culture. In the Qing and the Republic, governmental officials stressed the role of water deity worship in dike management—at least its psychological function.[109] At the same time,

105 What Yan Gao (2012: 92) observed was different. She claimed that water deity worship had been revived in Jianli-Honghu border region in the 1980s, and she visited a rebuilt temple that enshrined Guan Yu, and that "the major expectation for Guan Di is to prevent [*sic*] the communities from floods."

106 "Jingzhou Folk Custom: Temple Fair at Guan Yu Temple" 2011.

107 Hu Qiwei 2008.

108 Dong Longkai 2001.

109 A similar episode occurred in Georgia in 2007. To overcome a record-setting drought, the governor of Georgia held a ceremony in front of the state capitol to openly pray for rain. It brought no rain. But his action at least demonstrated to the public that the governor was doing his job—even if he was providing no more than psychological comfort. From

they also had to fulfill their duty to maintain the safety of the dikes, particularly river dikes. If the dikes proved to be safe, the water deities were rewarded for their protection; if the dikes ruptured, the responsible officials were punished, whether they had worshipped water deities or not.

After the establishment of the PRC, worship culture was criticized as a feudal superstition and many practices and beliefs were abandoned and replaced with revolutionary ideology. The state particularly revived the idea that "human beings can overcome nature," which emphasized the human ability to control floods. This idea, however, long predates the PRC, though it has been interpreted in different ways. The PRC government has invested great time and energy in harnessing rivers to reduce the frequency of dike ruptures and inundations, and in response the populace no longer feels the need to rely on water deities for protection.

Since the economic reform launched in 1979, the people's revolutionary ideology has faded away. The development of the economy has become the first priority of local cadres, who mostly have ignored the requirement of environmental protection. As a result, the damage to the environment caused by economic construction is startling. It has argued that this environmental degradation originated in the Communist Party's environmental policy during the Mao era and in the idea that "man must conquer nature."[110] It is undeniable that in the past half century China has developed its economy at the expense of the environment. In particular, the nationwide pollution caused by rapid rural industrialization and urbanization has exacted a terrible toll on the environment. To struggle with nature for survival, however, is nothing new, whether in the Jianghan area, in other parts of China, or elsewhere in the world—the landscape of the Netherlands, for example, is almost all man-made, as expressed in the famous saying, "God made the world, but the Dutch made Holland." People in water-rich areas—be in Holland or the Jianghan Plain—had to build dikes to protect the land from dangerous floods.

Many temples that had been neglected or destroyed in the first thirty years of the PRC, particularly during the Cultural Revolution, were rebuilt after 1979. Some of their sites have even been turned into public parks or tourist destinations. While some old traditions such as the dragon dance have been revived, they now have nothing to do with the control of water and everything to do with entertainment. At the same time, most dikes are now managed by the governments at different levels. During the high-water season, it is hard to

---

the reports in the media it was clear that the governor was more mocked than thanked (Emerson 2007).

110 Shapiro 2001.

imagine that any responsible government official would visit a temple to pray to water deities for the safety of dikes. It seems this tradition is gone for good.[111]

In sum, before 1949 water deity worship was important in the Jianghan Plain in high-water control due to frequent water calamities that plagued the area for centuries, but after the CCP took power, the frequency of floods and dike ruptures was dramatically reduced. This plus the Communists' campaign during the Mao era against "superstition" and religion, contributed to the tradition of water deity worship gradually dying out. Even when religious activities were revived in the post-Mao era, there was no return of the worship of water deities. Instead, some temples became money-making places and water conservancy projects became tourist destinations.

---

111 In Taiwan, however, officials still perform some religious rituals such as rainmaking (praying for rain) (Snyder-Reinke 2009: 260-61n1).

CHAPTER 6

Polders and Lakes

Polders or *yuan*, also called *yuanzi* or *wei* in the local terminology of the Jianghan Plain, are usually surrounded by dikes, with sluice gates on the dikes and canals for irrigation and drainage inside the polder. In some places, polder dikes also function as river dikes, thus *di* (dikes) and *yuan* (polders) are frequently referred to by a single term, *diyuan*. Depending on the location from which a polder was reclaimed, it could also be referred to as a beach polder (*tanyuan*), an islet polder (*zhouyuan*), or a lake polder (*huyuan*). According to its size, location, and the time of enclosure, a polder could also be referred to as an upper polder, lower polder, large polder, small polder, old polder, or inner polder. Finally, in the Qing and the Republic, polders could generally be categorized according to their legal status: official polders, which were built with government funds and officially registered; people's polders, funded by the residents but recognized by the state; and private polders or illegal polders, which were not supposed to exist. In this chapter our discussion focuses on polders in general; in the next chapter we will take a deeper look at private polders in particular.

1 Reclamation and Enclosure of Polders

In a region vulnerable to water calamities, the reclamation of wasteland and the enclosure of polders had a close relationship with local geography and human activities to adapt to such an environment. A late Qing gazetteer of Jingzhou noted that "even though [the area] suffered water calamities and drought, no place had not been reclaimed."[1] That is to say, the local people did everything they could to reclaim land no matter where it was located. This they did to survive, but also because they were responding to the age-old craving of Chinese peasants for land and their tradition of industriousness.

Three major factors contributed to wasteland reclamation and polder enclosure on the Jianghan Plain over the past several centuries. First, geography. The Jianghan Plain is a fluvial plain formed by silt deposited by river waters and lakes. These silts were gradually deposited and formed strips of land along

1 *Jingzhou fu zhi* 1880, vol. 5: 2a.

 | DOI:10.1163/9789004739482_008

rivers at a relatively high elevation. The elevation became even higher after the construction of river dikes, particularly along the Yangzi and the Han Rivers. For this reason, the Jianghan people built polder dikes to protect their land from both river water and lake water. Thus, polders on the Jianghan Plain were initially enclosed from relatively high places along rivers, then gradually moved to relatively low places adjacent to lakes. Second, state encouragement. The Ming state not only encouraged reclamation by taxing newly opened land very lightly, but it also ordered local officials to engage in water conservancy (particularly diking) to boost the local economy. To restore the economy torn by war at the end of the Ming, the early Qing state also encouraged people to reclaim lakeside wasteland. Later, the Qing state particularly required local officials to pay close attention to diking and followed this with a series of regulations on annual repairs and rewards and punishments. Hence in the middle and late Qing, many lakes were reclaimed. Third, population pressure. To some extent, the spread of polders was an inevitable outcome of land reclamation pushed by a growing population. One of the most effective means of feeding ever more people was bringing virgin land under cultivation. The largest category of such land available in the Jianghan Plain was lakeside wasteland, which, in the Ming-Qing era, attracted millions of immigrants. Their arrival further increased pressure on land, and in response more wasteland was reclaimed. In this spiraling process, more and more land was opened, and more and more polders were enclosed.[2] In the Republic, though the political situation was unstable, the Jianghan people continued to enclose polders by reclaiming land from lakeshores, river beaches, and islets. In the 1950s–1970s, even though it was discouraged by the PRC state, the Jianghan people still reclaimed a great amount of land from lakesides, river beaches, and islets, due mainly to the pressure to feed an exploding population.

Depending on the type of land that had been reclaimed and enclosed, there were two general patterns of polder enclosure: the blockage of river courses and the enclosure of lakeside wasteland. The former can be divided into two sub-patterns: the reclamation of river courses that were abandoned because their outlets were intentionally blocked, and the reclamation of river harbors that had been gradually silted up. The polders that were enclosed from the land between dikes and rivers also fall in this pattern.[3] The enclosure of lakeshore also can be divided into two sub-patterns: the enclosure of preexisting lakeside farmland by building polder dikes to protect the land from floodwater, and the enclosure of a part of a lake and draining it to turn it into

2 Zhang Jiayan 2001; Jiayan Zhang 2014: 33–40.

3 Mei Li, Zhang Guoxiong, and Yan Changgui 1995: 98–100.

farmland. Many polders enclosed over the past three centuries belong to the second sub-pattern, but it took some time—or in some cases, a long time—for these enclosures to become regular polders, due to the slow process of land improvement. Most private polders enclosed in the Qing also belonged to this sub-pattern. "First [the locals] block water in many places along the banks and in the middle of the lake to form silt, then use taxation as an excuse to build dikes all around to form polders."[4] The pattern of enclosure varied from location to location, and the benefits derived from different patterns also varied: "high land [is enclosed] as polders, low land remains a marsh; [the locals can] harvest grain annually from high land, [and they] can get water chestnuts, wild ducks, and fish and shrimp from low land."[5]

At first glance it may seem that these enclosures, or polders, are the same as the enclosures call *wei* in the Yangzi delta. According to historian Peng Yuxin, the dikes and polders (diyuan) in Hunan and Hubei were "transplanted" wei from the Yangzi delta.[6] If we look closer, however, we will find that although they shared some similarities, they also differed in many ways.

Similar land in the Yangzi delta was called *wei* or *yu*. The two terms were usually used interchangeably, though the meaning was not the same: wei land consisted of "enclosed fields" "reclaimed [from] lake- or riverfront land (polders)," while yu land comprised "embanked fields," that is, "fields surrounded by built-up embankments on low-lying land."[7] In other words, wei consisted of land that was first enclosed from shallow swamps, river beaches, and lakeshores, and then was reclaimed and farmed. Yu, on the other hand, was land that was reclaimed and farmed first, and only later was surrounded by dikes to protect it. Such land was found on large plains. Wei were smaller and had a relatively simple water conservancy system; yu, on the other hand, had a relatively sophisticated water conservancy system. People of the Yangzi delta started to reclaim wei land as early as the late Zhou dynasty, while the yu land system did not appear until the early Tang dynasty and was not fully developed until the Song dynasty. After another century or so the terms of *yu* and *wei* were used interchangeably.[8]

According to the renowned anthropologist and sociologist Fei Hsiao-tung, in Kaixiangong (Kaihsienkung) Village in the Lake Tai region, the core area of the Yangzi delta, "*Yu* is the local term for the unit of land surrounded by water.

4 Shuili dianlibu 1991: 298.
5 *Hanchuan xian zhi* 1873, vol. 9: 6a.
6 Peng Yuxin 1990: 171.
7 Philip C.C. Huang 1990: 30.
8 Miao Qiyu 1982: 20; Yao Hanyuan 1986; Liang Jiamian 1989: 329–33.

Each *yu* has its own name. Its size is determined by the distribution of streams and thus varies." There were eleven yu in Kaixiangong, varying from 8.55 mu to 986.4 mu.[9] Based on Fei's description, it appears that there were at least four major differences between yu or wei in the Yangzi delta and yuan (polders) on the Jianghan Plain. First, the yu in Kaixiangong were surrounded by water, while on the Jianghan Plain, some polders were adjacent to lakes, although most were not adjacent to rivers. Only polders built on islets in the middle of a river course were surrounded by water. Second, in terms of water conservancy, the priority in Kaixiangong was irrigation and drainage, hence the importance of canals and channels or ditches. On the Jianghan Plain, on the other hand, the priority is high-water control, and thus dikes are of paramount importance. Third, there is only arable land and a water conservancy system inside a yu or wei, while inside most Jianghan polders, there is arable land, wasteland, a water conservancy system, and villages. Fourth, a polder is far larger than a yu. The largest yu in Kaixiangong Village was only 986.4 mu, but a polder of the same size on the Jianghan Plain would be considered small (see Table 6.2).

Enclosing a polder on the Jianghan Plain, however, was not easy, particularly if it was done by a clan. For example, it took eight generations of the Jiang clan, from the late Ming to the mid-Qing, to reclaim a sandbar in Tianmen and turn it into a polder.[10] In another case, when the Qu clan of Honghu in the Republic needed more land to feed its growing number of members, one plan was to reopen a polder that had collapsed to a lake in 1832. But it took decades for the organizer to get the agreement of every household. Some people worried about poor fengshui, and others were concerned about infringement on their property rights. The death of the organizer's wife from heatstroke while she was irrigating the fields in 1939 reinforced his determination to enclose a new polder, and agreement was reached the next year and work was started in 1941. The residents enjoyed good harvests every year after the completion of the enclosure of this polder in 1943. For this reason, it was named Minsheng (people's livelihood) polder.[11]

2 Growth and Decline of Polders

The number of polders continued to grow from the Qing to the Republic, but in many places in the twentieth century, people also began to merge their

---

9 Fei Hsiao-tung 1962 [1939]: 17–18.

10 *Tianmen shuili zhi* 1999: 116.

11 *Qujiawan zhi* 1986: 30, 181–83.

polders. A drastic decline in the numbers of polders, however, only occurred in the PRC, when state-sponsored water conservancy projects eventually rendered most polder dikes superfluous. At that point they were transformed into farmland.

Although this was the general trend, there were variations from county to county. In some places people wanted to merge their polders because of the blockage of river branches. In Mianyang, for example, many branches of the Han River were silted up or blocked in the early nineteenth century, reducing or eliminating the threat from these branch rivers. Residents of some polders merged their polders to reinforce their dikes in order to collectively protect them from the Han River floods and the backflow of the Yangzi River floods. In the process some small polders disappeared.[12] The situation was slightly different in Tianmen where even though some polders were merged in the late Qing and the early Republic, more polders were reclaimed at the same time, thus the number of polders in Tianmen increased from 76 in 1840 to 257 in 1917.[13] In Songzi, before the formation of the Songzi River in the 1870s, there were only 34 polders in the county's lake area. But after that, silt carried by river water was deposited, and many new polders were enclosed. The increase in the number of polders, however, not only created more conflicts between the residents of different polders, but also increased the task of annual repairs and high-water control. Thus, people started to merge their polders, reducing the number from about a hundred to thirty-three by 1936.[14] In 1946, residents of tiny Sanxing polder, consisting of only 200 mu of land but 7 li of dikes, wanted to merge their polder into Tongxin polder.[15] In Gong'an, there were more than a hundred polders at the end of the Qing, but they were gradually consolidated into 51 in the Republic.[16]

It was not always easy to merge neighboring polders. In a case in Hanyang in 1947, a township head planned to merge around ten neighboring polders into one, to reduce residents' heavy task of high-water control. Surprisingly, some residents of three small polders firmly rejected his proposal, and even accused him of violating "democracy" in decision-making,[17] a quite new idea. In 1938, residents of Daxing polder in Hanyang also refused to merge with a neighboring polder.[18] In Songzi, in a move initiated by the county government,

12 *Mianyang xian zhi* 1989: 131.
13 *Tianmen xian zhi* 1989: 141.
14 *Songzi xian zhi* 1986: 87–88, 351.
15 100-8-38: 47.
16 *Gong'an xian zhi* 1990: 75.
17 LS31-4-1404.
18 LS31-4-1403.

four polders were merged in 1942, but the new polder was too large and too complicated to operate and so it was broken up into four polders again two years later, a step that left the residents "very satisfied."[19]

Although it was challenging to merge polders in the Republic, the task became far easier in the PRC, due mainly to the state's high level of despotic power and the abolishment of private landownership. To better manage rivers and dikes, the new state issued orders in 1950 to nationalize lake lands that hindered water conservancy and to relocate the residents, to nationalize all old polders that had been turned into lakes, to nationalize all designed flood diversion zones, and, finally, to nationalize all land within 100 meters of a dike foot on the river side and within 50 meters on the inner side.[20] Even so, the wishes of the locals still counted. In Hanyang and Hanchuan in the early 1950s, the county government announced that three basic principles should be followed when merging polders: the merger must be voluntary, the benefits must be greater than the harm, and the original water system must be maintained. Furthermore, the consolidation of large polders required the approval of the provincial government.[21] People benefited tremendously from these consolidations. In Hanchuan, for example, before three small polders were merged into one large polder in 1957, the locals frequently fled because they could not cope with the floodwater. Although the peasants received government relief after 1954, they still suffered: "70 percent of the peasants tended to relocate, … five couples are filing for divorce, and four families called off engagements because of life's difficulties." But after consolidation, they enjoyed a stable income, and many marriages were saved.[22] In short, the number of polders was drastically reduced in the PRC. In Qianjiang, the number of polders did not change much throughout the Qing dynasty, but most of them disappeared in the 1960s.[23]

For all these reasons, it is difficult to be certain about the number of polders in any particular county. In the Qing and the Republic, neither the size of polders nor their total number was stable. Some polders or a part of a polder could be destroyed by floodwater while new polders might be reclaimed, and some old polders might be merged into one at the same time.

This has been true in the PRC as well. After 1949 some new polders were enclosed, while more disappeared. The 257 polders in Tianmen in the Republic

19 100-6-1060: 52–53; 100-1-99: 7–8.

20 *Jingjiang dadi zhi* 1989: 258–59.

21 30-1-4b; 40-1-55a.

22 50-1-50b.

23 *Qianjiang shuili zhi* 1997: 130–39.

were consolidated into 72 in 1957 (11 of which had been created after 1949), and further consolidated into 24 in 1985.[24] Although it was not easy to merge them, the benefits were clear: a lightening of the burden of high-water control, an increase in farmland, a growth of agricultural output, a reduction in the spread of snail fever, and so on.[25] In fact, it is widely believed that the consolidation of dikes and polders in lake areas was the only effective means to overcome water calamities.[26] At the same time, due to the large scale of water conservancy projects beginning in the early PRC, particularly the building of many new river dikes and the reinforcement of the existing river dikes, many polder dikes were razed and turned into farmland. Table 6.1 offers a rough picture of the changing numbers of polders in some counties on the Jianghan Plain from the early Qing to the PRC.

The most dramatic changes occurred in Mianyang. One source mentions that in the 1930s Mianyang had around 500 large polders, but if one counted the small polders located inside these large polders, the total number would reach an astonishingly 3,900.[27] That would have been the largest number of all time in any county on the Jianghan Plain.[28] The situation began to change in 1951 when a part of Mianyang was separated and turned into a new county, Honghu, instantly causing Mianyang to "lose" a great many polders. Another source indicates that after the separation, Mianyang still had 93 polders, but most of them disappeared in the following decades due to the constant construction of water conservancy projects. In a township in Mianyang, some large polders were formed by the consolidation of 10 or 15, or even 83 small polders after 1949. By the early twentieth-first century, only six individual polders (plus some small polders) still existed.[29] The great divergence among the sources on the numbers of polders in Mianyang also reminds us of the difficulty of

24 *Tianmen xian zhi* 1989: 141–42.

25 50-1-42.

26 50-1-50b.

27 *Jianghe xiufang jiyao* 1971 [1936]: 258.

28 A memorial from 1745 stated "there are no less than several thousand polders in Mianyang" (Shuili dianlibu 1991: 267). This number was not based on an actual count, but it indirectly substantiates the point that the number of polders in Mianyang in the Republic was not an exaggeration. According to a new gazetteer, under the private ownership system people built small dikes everywhere before 1949 to protect their land; one village had no fewer than three lake polders (*Mianyang Chenchang qu xiangtu zhi* 1987: 69). This account also indirectly backs up the assertion that polders proliferated in Mianyang before 1949.

29 *Huangjing qu zhi* 1986: 50; *Xiantao shuili zhi* 2008: 203–12. According to one source, there were three small polders in the Xiantao portion of the Dujiatai flood diversion zone (*Xiantao shuili zhi* 2008: 212); but according to another source, that number was five (*Xiantao shi zhi* 2011: 254).

TABLE 6.1 Number of polders in selected counties on the Jianghan Plain at different times

County	Early Qing	Mid-Qing	Late Qing	Republic	PRC
Xiaogan	56			201 (in 1934), 222 (in 1935)	46 (in 1956)
Yingcheng				22 (in 1947)	8 (in 1960)
Hanchuan	8		265	c.280	
Tianmen	72		76	257 (in 1917)	72 (in 1957), 24 (in 1985)
Qianjiang			154	173 or 117 (in 1946)	210 (in 1955)
Mianyang		1,393	1,813	c.3,900 (in 1937)	6
Jiangling			179		
Jianli			491 (in 1859), 498 (in 1873)	520 (in 1935)	256 (in 1959), 210 (in 1962)
Gong'an			c.100	c.51 or 78	
Shishou			75	116	29 (in 1985)
Songzi	c.100		38	33 (in 1936)	22 (in 1985)
Dangyang				c.20	17
Zhijiang		11	17	43	28 (in 1985)

Note: The numbers in this table roughly reflect changes in the number of polders, but do not reflect changes in administrative jurisdictions over time.

SOURCES: XIAOGAN (*XIAOGAN XIAN JIANZHI* 1959: 41–42; *HUBEI XIANZHENG GAIKUANG* 1934: 643); *YINGCHENG XIAN ZHI* 1992: 211; HANCHUAN (*HANCHUAN TUJI ZHENGSHI* 1895, VOL. 4: 4B; *JIANGHE XIUFANG JIYAO* 1971 [1936]: 267); *TIANMEN XIAN ZHI* 1989: 141–42; QIANJIANG (*QIANJIANG XIAN ZHI* 1990: 287; *QIANJIANG SHUILI ZHI* 1997: 131); MIANYANG (ZHANG JIANMIN 1999: 311T4–14; *MIANYANG XIAN ZHI* 1989: 131; *JIANGHE XIUFANG JIYAO* 1971 [1936]: 258; *XIANTAO SHUILI ZHI* 2008: 205); JIANGLING (*JINGZHOU FU ZHI* 1880, VOL. 20: 3A–5A); JIANLI (*JIANLI XIAN ZHI* 1959: 16; *JIANLI DIFANG ZHI* 1991: 71, 80); GONG'AN (*GONG'AN XIAN ZHI* 1990: 75, *HUBEI XIANZHENG GAIKUANG* 1934: 1001); SHISHOU (*JINGZHOU FU ZHI* 1880, VOL. 20: 7B–12A; *SHISHOU XIAN ZHI* 1990: 129); SONGZI (*JINGZHOU FU ZHI* 1880, VOL. 20: 17B–18B; *SONGZI XIAN ZHI* 1986: 351–52); *DANGYANG XIAN ZHI* 1992: 242; ZHIJIANG (*JINGZHOU FU ZHI* 1880, VOL. 20: 19A–20A; *ZHIJIANG XIAN ZHI* 1990: 165)

ascertaining an accurate count. The consolidation of small polders, the renaming of old polders, and the fact that the same polder could be known by different names all contributed to the uncertainty surrounding their numbers.

We also must keep in mind that all these numbers are those recorded in the sources. But there is no way of knowing how many polders may not have been recorded. A Qing gazetteer reported that Jiangling had "forty-eight large polders and a hundred or so small polders."[30] These numbers are unrealistically low, as memorials reported that 203 polders had been flooded in Jiangling in 1781, and 236 in 1840.[31] A 1934 survey of Hanchuan shows it had about 200 polders, but a report mentioned that it had more than 280 in 1935.[32] It is true that one can find the names of hundreds of polders in the documents—mostly in gazetteers, which are also the main sources of Table 6.1. But it is also true that many polders had never been recorded in gazetteers. In Tianmen, for example, many polders appeared or recorded on land deeds of the Qing were not mentioned in its gazetteers.[33] Another common category, private polders, was usually not recorded in gazetteers either. Thus, it is fair to say that in the Qing and Republic there were more polders on the Jianghan Plain than what can be found in the records, but it may be impossible to know exactly how many there were.

The size of polders, as mentioned earlier, also varied. Some were tiny, while others were enormous. According to a 1959 survey in Jianli, Hegang polder had only 7.48 mu of land (4.33 mu of paddies and 3.15 mu of dry land), but Da polder had 129,471.69 mu of land (83,156.29 mu of paddies and 46,315.4 mu of dry land).[34] The most numerous groups of polders in Qianjiang (in the early Qing) and Hanchuan (in the late Qing) ranged from 1,000 to 10,000 mu; but in Songzi (in the 1930s) and Jianli (in the early PRC), the number of polders with an acreage of 1,000 to 10,000 mu almost equaled those with either less than 1,000 mu or more than 10,000 mu (see Table 6.2).

Some large polders, as mentioned earlier, may have contained many small polders, and some also contained one or more lakes.[35] Most land inside polders consisted of rice paddies, with dry farmland second in acreage.[36] In some lake polders, however, the major part (before 1949) was covered by water (i.e.,

30 *Jiangling xian zhi* 1876: vol. 8: 2a.

31 Shuili dianlibu 1991: 468, 836.

32 *Hubei xianzheng gaikuang* 1934: 782; *Jianghe xiufang jiyao* 1971 [1936]: 267.

33 Zhang Jianmin and Tang Gangmao 2014.

34 *Jianli xian zhi* 1959: 116, 118.

35 Jiayan Zhang 2014: 32.

36 Jiayan Zhang 2001.

TABLE 6.2 Number of polders of different sizes in Hanchuan, Qianjiang, Jianli, and Songzi

County	Number of polders	Size of polders (mu)				Dates
		Less than 1,000	1,000–5,000	5,000–10,000	More than 10,000	
Qianjiang	161	25	59	36	41	Early Qing
Hanchuan	34	3	20	7	4	Late Qing
Songzi	31	8	10	5	8	The 1930s
Jianli	256	69	92	41	54	Early PRC

SOURCES: *QIANJIANG XIAN ZHI* 1694, VOL. 3: 11A–26A; *HANCHUAN XIAN ZHI* 1873, VOL. 9: 7A–33B; *SONGZI XIAN ZHI* 1982 [1937]: 118–20 (NOTE: THE ORIGINAL NUMBER OF POLDERS WAS 33, BUT THERE IS NO DATA ON THE SIZE OF TWO POLDERS); *JIANLI XIAN ZHI* 1959: 109–19

lakes), and the residents mostly made their living by fishing, only farming a bit of high land adjacent to their homes.[37]

The length of polder dikes also varied greatly: some were as short as a couple of li while others stretched for hundreds of li. The circumference of Tongcheng polder in Mianyang in the Qing was a remarkable 200-plus li.[38] On average, each of the thirty-five official polders in Hanchuan in the late Qing had 2,501 zhang (nearly 8 kilometers) of dikes.[39] In Zhijiang, the dike of Zhangjia quanzi polder was only about a kilometer long, while the dike of Lower Bailizhou polder spanned nearly eighty kilometers in 2002.[40]

As explained earlier, with the building of ever more river dikes and the reinforcement of existing river dikes in the early PRC, many polder dikes were rendered useless and were razed and then farmed. Most polders that still exist today are located outside river dikes, and some of them are more stable than others. When Yonghe polder in Shishou, for example, was first reclaimed from the Yangzi River beach in 1830, it had fewer than a hundred households; but by 1986, it had 3,451 households and 18,034 people, a sign of its stability. River beach polders that were still standing along river courses in the high-water season, and had been reclaimed long in the past, were essentially no different

37 *Mianyang Chenchang qu xiangtu zhi* 1987: 34.
38 Shuili dianlibu 1991: 460.
39 *Hanchuan xian zhi* 1873, vol. 9: 7a–33b.
40 *Zhijiang shi zhi* 2012: 178.

from regular old polders. Although the local people did their best to protect their dikes from being breached, where relatively unstable river beach polders existed, the government might order that their dikes be broken open to discharge high-water every year. People who lived in these polders found a way of coping: if the high water was not too high, they would grow crops; otherwise, they would make a living by fishing. Every household, it was said, had a boat.[41]

One source indicates that the fourteen people's polders along the Great Jingjiang Dike were illegal even in the late Qing and the Republic. In 1951, the PRC government turned these polders into a high-water-discharge-and-farming area. That is, if the high-water was low, the land would be farmed, but if there was a great volume of high-water, the land would be used to discharge high-water. In the 1980s, Jingzhou Prefecture still had fifty-one such polders.[42]

In Hubei in 1998, there were 204 beach polders located on the outside of the Yangzi River dikes, 101 of which had been reclaimed after 1954. These polders caused many problems. They blocked waterways in the high-water season and threatened the integrity of the Yangzi River dikes; their upkeep consumed a great deal of manpower in the high-water season; and when they were breached, the costs of public relief were burdensome. Thus, after the 1998 flood, the central government decided to raze them and relocate their residents to inside the Yangzi River dike. In all, 50,000 households and 217,700 people from 133 beach polders were relocated.[43]

As the number of polders shrank, the overall length of polder dikes contracted, which greatly reduced the burden on the local people of high-water control and annual repairs.

3 The Vicissitudes of Lakes

The names of many polders in the Jianghan Plain include the character for *hu* (lake), suggesting they had been reclaimed from lakes. During the Ming-Qing period, lakes in the Jianghan area covered about 26,000 square kilometers. This decreased to around 10,000 square kilometers by the end of the Qing and 8,330 square kilometers by the end of the Republic.[44] More dramatic changes

---

41 Wang Minfu 1989: 5, 80.

42 *Jingjiang dadi zhi* 1989: 84, 426.

43 Hubei sheng shuiliting and Hubei sheng fangxun kanghan zhihuibu bangongshi 2000: 122, 125.

44 Guan Qingtao 1983.

continued in the PRC, including a reduction in the number and surface area of lakes, and a change in their functions.

First, as for the reduction of the numbers of Jianghan lakes, in Qianjiang, sixty-one lakes disappeared since the early Qing. Seven of these lakes with a surface area of a thousand or more mu disappeared from 1936 to 1956, and thirty-four from 1956 to 1986. Many smaller lakes also vanished at the same time. This loss was entirely due to reclamation. Thirty-one lakes were reclaimed in the years 1957 to 1960 alone, and many were reclaimed by state-owned farms.[45] For example, in 1975 Wuchang carried out eleven key lake-enclosure projects. To complete this work required a great deal of machinery and materials: 38 transformers, 42 motor generators, 42 pumps, 737 poles, 788 meters of steel pipe, 1,969 tons of cement, and others.[46] These numbers, from one perspective, demonstrate the huge investment in lake enclosure in the PRC, which was almost impossible in earlier eras.

In the Jianghan area from the early 1950s to the 1980s, 150 lakes with a surface area of 5,000 or more mu disappeared (Table 6.3).

Numerous small lakes also disappeared. Records show that Hubei had 1,332 lakes with a surface larger than 100 mu in the early 1950s, but the number was reduced to 844 in the 1980s—that is, 488 lakes with a surface of 100 or more mu had vanished.[47] It is unclear how many of them were on the Jianghan Plain, but since that is where most of Hubei's lakes were found, it seems safe to assume that most of these onetime lakes were located there as well.

Second, over time the surface area of Jianghan's lakes shrank. The disappearance of 150 lakes with a surface area of 5,000 or more mu amounted to a loss of 1,923.4 square kilometers of lake-surface in total (see Table 6.3). Among Jianghan's lakes, Lake Hong (*Honghu*) is the largest lake in Hubei today and is shared by two counties (Jianli and Honghu), but is mostly located in the county of Honghu, giving that county the largest surface area of lakes in Hubei. In the early 1950s, Honghu County had three large lakes (Lake Hong, Lake Datong, and Lake Dasha) and forty-four middle-size and small lakes. Lake Hong had a surface area of 736.56 square kilometers in the early 1950s, but this had been reduced to 653 square kilometers by 1959 and 348.36 square kilometers by around 1986. At the same time, most of Lake Datong and Lake Dasha were

---

45 *Qianjiang shuili zhi* 1997: 65–66.

46 Wuchang 49-1-61; Wuchang 49-1-65.

47 *Hubei shuili zhi* 2000: 136.

TABLE 6.3 Number and area of reclaimed lakes of 5,000-plus mu on the Jianghan Plain, 1949–1980s

County	Number of disappeared lakes	Acreage of disappeared lakes (in the early 1950s)(km^2)	Reclaimed acreage (in the 1980s) (1,000 mu)
Jianli	24	231.5	347.7
Jiangling[a]	23	374.2	556.9
Honghu	19	372.0	491.5
Qianjiang	17	158.3	236.6
Xiantao	17	146.3	219.7
Hanchuan	11	162.4	258.3
Yingcheng	10	73.5	114.4
Yunmeng	5	40.6	58.0
Zhongxiang[b]	4	46.4	82.5
Shishou	3	15.8	31.9
Zhijiang	3	23.9	79.3
Wuhan	2	38.7	58.1
Caidian	2	16.6	21.2
Jingmen	2	44.7	67.0
Shashi	2	31.5	46.3
Songzi	2	13.1	19.7
Tianmen[c]	2	93.4	174.3
Dangyang	1	22.1	45.6
Xiaogan	1	18.4	27.6
Total	150	1,923.4	2,936.6

a In Jiangling, one lake (Lake Cao) was turned into a reservoir, and thus is not included in the calculation.

b The reclaimed acreage of Lake Guanzhuang includes land reclaimed from Lake Linjia, which still exists.

c Lake Chen was in Tianmen and Hanchuan—the source provided data on the lake as a whole. For the sake of calculation, I have assigned the entirety of this lake to Tianmen.

SOURCE: *HUBEI SHUILI ZHI* 2000: 137–48, TABLE 1-4-2. COUNTED AND CALCULATED BY THE AUTHOR

reclaimed. In total, two-thirds of the lakes in Honghu County were reclaimed in the early PRC.[48]

Outside the county of Honghu, Lake Chen was the largest of Hubei's disappeared lakes. Lake Chen was shared by Tianmen and Hanchuan, and still had 180,000 mu of water surface in 1958, but by 1970 it had been completely reclaimed and ceased to exist.[49] At the beginning of the PRC, Songzi still had fifty or so lakes with a surface larger than 100 mu, but many of them were soon reclaimed, and by the 1970s, 95,000 mu of lake had been reclaimed. The acreage of lakes in Songzi shrank by more than 80 percent.[50]

The most dramatic changes occurred in Hanchuan, a county famous for its history of water calamities and grain shortages. After 1949, to reduce water calamities, polders in Hanchuan were merged, rivers were harnessed, and lakes were separated from rivers. At the same time many lakes were reclaimed. By 1979, Hanchuan had become self-sufficient in grain, a historic achievement. This, however, was accomplished at the expense of Hanchuan's lakes: the lake surface area shrank from 1,120,771 mu in 1957 to only 111,843 mu in the early 1980s, a reduction of about 91 percent in less than three decades.[51] Lake Diaocha, which is situated mostly in Hanchuan, shrank from about 1,000 square kilometers before 1949 to only 38.6 square kilometers in 2005.[52] In addition to Hanchuan, many other counties in Xiaogan Prefecture witnessed the same outcome: from the 1950s to 1996, Xiaogan lost 83.75 percent of its lake surface due to reclamation.[53]

The so-called "Four-Lake Area," in the core of the Jianghan Plain, contained Lake Hong, Lake Chang, Lake San, and Lake Bailu. By the end of the twentieth century, both Lake San and Lake Bailu had been reclaimed out of existence,[54] leaving just two lakes (Table 6.4).

As Table 6.4 shows, the total surface area of selected lakes on the Jianghan Plain in the late 1970s was only about a third of that in the early PRC. This sort of shrinkage is true of all the Jianghan area: in total, the surface area of Jianghan's lakes shrank by two-thirds between 1949 and the end of the twentieth century. In fact, so many lakes were reclaimed within so short a time in

48 *Honghu xian shuichan zhi* 1986: 29–30. A source states that Lake Datong was completely reclaimed by the 1980s (*Hubei shuili zhi* 2000: 137).

49 *Tianmen xian zhi* 1989: 112.

50 *Songzi shuili zhi* 2008: 3, 23.

51 *Hanchuan xian zhi* 1992: 2, 81.

52 *Hanchuan shi zhi* 2010: 94.

53 *Xiaogan diqu shuili zhi* 1996: 260.

54 *Hubei shuili zhi* 2000: 128.

TABLE 6.4 Changes in the surface area of selected major lakes on the Jianghan Plain, 1950s–1970s

Lake	Location (county or city)	Lake-surface (1,000 mu)		Percentage remaining
		In the early PRC	In the late 1970s	
Lake Hong	Honghu, Jianli	990	603	61
Lake Diaocha	Hanchuan	480	100	21
East-West Lake	Wuhan	398	13	3
Lake Chen	Hanchuan, Tianmen	255		
Lake Lu	Wuchang	113	68	60
Lake Chang	Jingmen, Jiangling, Qianjiang	225	184	82
Lake Datong	Honghu	166	13	8
Lake Pai	Mianyang	195	49	25
Lake Houguan	Hanyang	148	98	66
Lake Dasha	Honghu	144	50	35
Lake San	Jiangling	141	6	4
Lake Bailu	Jianli	110	18	16
Lake Tangsun	Wuchang	87	55	63
Great Wangjia Lake	Songzi	89	12	13
Total		3,541	1,269	36

Note: The original table also includes several lakes in counties outside the Jianghan Plain as defined in this research. They are excluded from the present table; also excluded are lakes shared by one Jianghan county and one or more counties outside Jianghan. Thus, the total percentage is slightly different from the percentage in the source.

SOURCE: *HUBEI NONGYE DILI* 1980: 88, TABLE 3–6

the early PRC that there was high pressure on laborers to farm and drain at the same time.[55]

The shrinkage of lakes in Hubei in general, however, was not as dramatic as on the Jianghan Plain. From the early 1950s to the 1980s, Hubei lost 488 lakes with a surface larger than 100 mu, or 65 percent of its lake acreage. At the same time, about 8.7 million mu of lakes were reclaimed, which increased

55 *Hubei sheng zhi dili 1* 1997: 1154–55.

farmland by about 5.2 million mu and contributed to a reduction of population pressure.[56] Since 1978, with the development of the non-farming economy and the growing emphasis on environmental protection, both the number and acreage of lakes on the Jianghan Plain are on the rise (see Chapter 7).

Third, the functions of lakes changed over time. Before 1949, many of Jianghan's lakes were linked with the Yangzi and the Han Rivers, and thus the surface area of any lake could have varied greatly. For example, Lake Chen in Tianmen encompassed an area of only 17 square kilometers in 1925, but in 1948 it spread over 246 square kilometers.[57] With the separation of lakes and rivers and the continued reclamation of lakes in the early PRC, the surface of Jianghan's lakes shrank, and since they no longer took in or stored floodwater from rivers, many of them became enclosed lakes. Up until the 1980s, many of Jianghan's lakes also gradually grew shallow or became marsh wetlands.[58] Since 1979, as local governments have gradually shifted their priority from agriculture to the non-farming economy and GDP growth, the drive to reclaim land has declined. With the growth in people's income, travel and recreation have become increasingly important in their daily lives, which has led to new functions connected with water conservancy projects and lakes, particularly recreation and tourism.

The changes in Lake Diaocha are a prime example. It is unclear when Diaocha polder was formed, although it was probably before the Ming dynasty.[59] In 1767 the polder flooded and reverted to a lake. After that, it underwent dramatic changes since the late Qing and particularly since 1949 when the new state invested great efforts in taming rivers and lakes. Historically speaking, Lake Diaocha was an open lake (that is, one linked with a river), and its water level and size were closely related to the amount of floodwater of the Han River. Its surface could grow to about a thousand square kilometers in years with floods, but then shrunk after floodwater retreated.[60] During the high tide of the GLF in the late 1950s, the Hanchuan County government believed it could reclaim around four hundred thousand mu of land from Lake Diaocha, which would not only meet an age-old demand of the local people but also transform Hanchuan from a county with a shortage of grain to one with a surplus. It also felt that such a reclamation would display the superiority of socialism over private ownership before 1949, since reclamation of Lake Diaocha was impossible

56 *Hubei shuili zhi* 2000: 132.
57 Zhongguo kexueyuan et al. 1957: 29.
58 *Hubei nongye dili* 1980: 80.
59 *Xiaogan diqu shuili zhi* 1996: 42.
60 Zhongguo kexueyuan et al. 1957: 29–30; Wang Qinzeng 1982: 70.

before the PRC. In a word, this was a project with both economic and political significance.[61] The same was true of the building of the Honghu Gedi—a dike to separate Lake Hong—in 1955. Such a dike was something the locals had dreamed of for two centuries, but it was claimed only the CCP managed to accomplish the task.[62] The same claim was made regarding the reclamation in Dongxihu in 1957, which the government asserted could only have been done under socialism. In any case, all attempts in earlier eras had failed.[63]

From the early 1950s to the 1980s, Lake Diaocha was broken up into many smaller lakes, and many parts of it were reclaimed. By the 1980s, 252,000 mu of land had been reclaimed.[64] As a result, Lake Diaocha changed from a gigantic lake with several million mu of surface area into a controlled lake of less than a hundred thousand mu. At the same time, the ownership of the water surface changed from private to public. The PRC state first enclosed the lake to reclaim some land, and then, since the 1980s, it has returned some land to the lake mainly because farming proved ineffective. The number of full-time fisherfolk has also drastically declined, from 158,330 in 1948 to 1,416 by 2005.[65] These changes unavoidably affected the socioeconomic functions of Lake Diaocha, gradually transforming it from an open lake that accepted outside waters to an area of fishing and farming. The function of its aquatic products also changed, shifting from hunger relief to healthy food. As people's incomes increased after the 1979 reform, Lake Diaocha has also become a recreational and tourist destination.[66]

4 Land Reclamation and State Farms

Much of the land reclaimed from lakes in Hubei in the PRC era was turned into state farms. On the Jianghan Plain, many of these farms were established on lakeside wasteland that in the past had been ravaged by floods or abandoned due to schistosomiasis.

Let us look at a few cases in point. Jingzhou Prefecture's first state farm was launched in 1953. By 1993 the prefecture had eighteen state farms (and 101 branch farms), with 970,000 mu of farmland.[67] These farms, most of which had

61 50-1-71.
62 D35-1-175.
63 50-1-67.
64 *Hubei shuili zhi* 2000: 137.
65 *Hanchuan xian zhi* 1992: 176; *Hanchuan shi shuichan zhi* 2010: 101.
66 Wu Shisi 2001; *Hanchuan nianjian* 2008: 305.
67 *Jingzhou diqu zhi* 1996: 197.

been reclaimed by 1963, ranged from 15.6 square kilometers (Zhouji Farm) to 366.67 square kilometers (Wusan Farm). Furthermore, most of the farms also built dikes to protect the land.[68] In Qianjiang, from 1950 to 1985, twenty-three state farms were established, accounting for 24.18 percent of the county's acreage.[69] The reclamation of all these farms was organized by governments at the county level or above, but many small reclamation projects were also organized by production brigades or even production teams. In one instance, a brigade reclaimed only 21 mu from a lake.[70]

Some state farms were reclaimed not from lakes but from wasteland. Jianghu Farm, for example, was reclaimed from wasteland formed after a rupture of the Han River dike in 1930.[71] Wusan Farm was also reclaimed from a huge stretch of wasteland created by a rupture of the Han River dike in 1935.[72] People's Dayuan Farm, reclaimed from a long expanse of beachland on the Jing River, was slightly different because it was established on what had been sparsely populated wasteland. Within a circumference of about thirty miles in 1956, there were only 423 households, most of which were dirt-poor woodcutters. During the high-water season, the area became a veritable sea, forcing the locals to rely on small boats to eke out a living.[73] Other state farms were established on land that had been abandoned because of bandits and diseases, such as Guanzhuanghu Farm, reclaimed from land abandoned due to the 1935 flood, the Japanese invasion, and rampant banditry; Lingjiaohu Farm, reclaimed from wasteland plagued by banditry and schistosomiasis; and Taihugang Farm, reclaimed from land abandoned because of floods, pestilence, and bandits.[74]

The early PRC state exempted virgin land brought under cultivation from agricultural tax for five years, and land reclaimed from abandoned farmland for three years.[75] The goal was to encourage land reclamation, which is precisely what the imperial dynasties sought in their early stages.

According to Qing memorials, the total acreage in some counties was far larger than the registered acreage. A 1761 memorial stated that the total acreage in Mianyang exceeded the registered acreage by 2,212,631 mu; in Hanchuan, by

68 *Jingzhou shi nongken zhi* 2010: 2–6, 27, 35.

69 *Qianjiang xian zhi* 1990: 247.

70 *Longdian qu zhi* 1989: 45–46.

71 *Jianghu nongchang zhi* 1988: 41.

72 *Wusan nongchang zhi* 1987: 35.

73 *Renmin dayuan nongchang zhi* 1988: 15–16.

74 *Guanzhuanghu nongchang zhi* 2000: 1–2; *Lingjiaohu nongchang zhi* 1991: 37; *Taihugang nongchang zhi* 1988: 1–2.

75 *Qianjiang xian zhi* 1990: 247.

277,327 mu; and in Jiangling, by 283,737 mu.[76] An aerial survey conducted during the Republic also demonstrated that the actual acreage in many of Jianghan's counties far exceeded the acreage reported by the local government.[77] This also partly explains why the PRC state was able to reclaim so much wasteland and establish so many state farms on the Jianghan Plain. It was widely believed that all reclaimable land in inland China had been brought under the plow by the end of the eighteenth century, and thus many Chinese in search of land migrated to the Northeast in the nineteenth and early twentieth centuries. However, the Jianghan area still had a huge amount of reclaimable wasteland. In the early PRC Hubei ranked third in the nation in reclaimable land, next only to the frontier regions of Heilongjiang and Xinjiang.[78]

Creating new land is arduous, whether it is wrestled from wasteland or from lakes. On the Jianghan Plain, most wasteland was densely covered by reeds, with deep and twisted roots. Reclamation in the early PRC therefore was mostly undertaken with tractors. Another problem was that the water was infested with *Oncomelania* snails, exposing workers to infection. Many of the farms were reclaimed from wasteland during the GLF, when workers' rations were reduced from no limit to one jin (half a kilogram) per day, then to half a jin per day. Workers survived on vegetables, wild lotus roots, and reed roots; no meat was available. They even drank lake water. The result was malnutrition so serious that many suffered from dropsy and hepatitis, and some even lost their lives.[79]

The reclamation of lakes was equally tough. According to a memoir, when the protagonist participated in the building of a dike in Dangyang in 1963, he walked to the worksite carrying with him everything he needed for survival. He stayed in a peasant's home, ate meat only once every two weeks and continued to work even when it snowed. He toiled under these kinds of conditions for several months, "turning yellow and skinny."[80] Why would people push themselves—or be pushed—to such extremes? A Jianli County gazetteer

76 According to this memorial, the total acreage in these counties were: Hanchuan, 886,258 mu; Mianyang, 4,065,373 mu; and Jiangling, 2,592,876 mu (Shuili dianlibu 1991: 362). The registered acreage was 608,931 mu in Hanchuan (*Hanyang fu zhi* 1747, vol. 24: 1a–2a), 1,852,742 mu in Mianyang (*Mianyang zhou zhi* 1894, vol. 4: 5ab), and 2,309,139 mu in Jiangling (*Jingzhou fu zhi* 1880, vol. 14: 8b–9a).

77 He Bingdi 1988: 111.

78 Chen Guoxiao 2000.

79 *Zongkou nongchang zhi* 1986: 102–03; *Zhouji nongchang zhi* 1991: 29; *Huanghu nongchang zhi* 1988: 33, 38.

80 Wan Zhaonong 1991.

provides at least part of the answer: lakes were enclosed to "develop food grain production" after the "three years of natural disasters" (i.e., the GLF).[81]

Unlike in the Qing and the Republic, when almost all reclamation and enclosure was done by individual households or clans, in the PRC such projects were directly or indirectly organized by the government at different levels, or at least organized by production teams. The enclosure of Dongxihu region in 1957, for instance, was even a national project, bringing together a hundred thousand laborers, of whom seventy-one thousand came from the province of Henan. A great number of Wuhan residents also participated as "volunteers," and seventy thousand laborers worked on the reclamation site even during the Spring Festival. Thus, the enclosure was completed within five months.[82]

Another impetus behind the PRC's drive to reclaim wasteland was that some of this land had a special connection with the Communist revolution, stretching from the 1920s to 1949. The land in question was once controlled by the Communists, or at least the local people supported them. After the Communist Party came to power in 1949, the new government felt it had a moral obligation to help lift the locals from poverty. For example, the PRC was deeply concerned about poverty in the Lake Zhongzhou area in Hanchuan, which was the center of a Communist guerrilla area during the Anti-Japanese War. In fact, the PRC government considered poverty alleviation in this area "an obligatory political task." One way to address this concern was to reclaim land from lakes.[83] A similar calculus was involved in the reclamation of the floodplain on the lower reaches of the Dongjing River valley in the Hannan region, south of Wuhan.[84]

Competition among neighboring administrative units also stimulated villagers to reclaim lakes and enclose polders in the PRC. In Jingmen, most new people's polders were enclosed during the 1960s–1970s, and one of its gazetteers claimed that virtually everyone was competing to enclose lakes. Many of these new polders thus were very small—eighty of them had only 15 to 95 mu of land.[85] Eighty-nine polders were enclosed from Lake Chang (at the junction of Jingmen, Jingzhou, and Qianjiang) due mainly to population pressure—they were reclaimed during the collective era when agriculture was valued and population growth was at a peak, resulting in pressure to reclaim land. The last reclamation in Jingmen was in 1984.[86]

---

81 *Jianli xian nongye zhi* 2011: 149.

82 *Wuhan shi zhi: nongye zhi* 1989: 521; Wang Pingsheng 2004: 59–62.

83 *Xiaogan diqu shuili zhi* 1996: 261; *Zhongzhouyuan nongchang zhi* 1989: 234.

84 XX000084-WS03-0137-0004.

85 *Jingmeng shi shuili zhi* 1989: 202.

86 *Jingmen shi shuili zhi* 2012: 76–80.

In Jingzhou, most state farms were reclaimed from shallow swamps, waste lakes, waste beaches, and waste hills,[87] all of which were infertile and not attractive prospects for reclamation before 1949. Some wasteland, however, was potentially very productive. In People's Dayuan Farm, the land, reclaimed from river beaches, was so fertile that in the early years after reclamation, crops flourished even without the application of fertilizer. But after decades of cultivation, the land lost its fertility and there was an increasing resort to chemical fertilizer. At the same time, many wild animals also disappeared.[88]

The reclamation of wasteland in the Jianghan area not only helped support a growing number of local people, but also reduced population pressure outside the region. At the end of 1957, the central government deployed fifty-seven thousand people from hunger-stricken Henan to Jingzhou to participate in land reclamation and water conservancy construction, and, according to a Jingzhou gazetteer, another thirteen thousand came with them voluntarily. Later all of them were settled in Jingzhou.[89]

Most of these lake reclamation and polder enclosure projects were clearly encouraged by the state, either to expand farmland or to eradicate *Oncomelania* snails. In many cases, however, the locals were also eager to reclaim lakes and enclose polders since this would improve their lives in countless ways. In Xiaogan in 1956, for example, it was reported that the local people were "extremely excited" when they heard the news that a dike was to be built to enclose a lake, as this was their "age-old wish." The GMD, the report continued, had once tried to build a dike to enclose the lake, but failed.[90] In Yingcheng, the reclamation of Lake Xiaojiafan in 1957 not only opened 500 mu of new land but also freed 26,000 mu of existing land from water calamities, a result, it was said, that local people had dreamed of for centuries. The sources claim that they were so enthusiastic that it only took forty days to finish the project, thirty days ahead of schedule. After reclamation, according to the sources, the locals escaped a life of poverty. They had been so poor that men had trouble finding a wife, but after the reclamation this problem evaporated.[91] In Yingcheng County, the reclamation of Swan Lake (Tian'edang) in 1957–1959 would add 100,000 mu of land and a potential harvest of 70 million jin of grain, enough to support the county's entire population for half a year. The land in this region was originally fertile but turned to wasteland after the 1839 flood. After that,

---

87 *Jingzhou diqu 40 nian* 1989: 112.

88 *Renmin dayuan nongchang zhi* 1988: 80.

89 *Jingzhou diqu zhi* 1996: 25.

90 81-1-31.

91 Qi Lafa 2009; Yingcheng 49-1-61.

the local people tried five times to enclose the lake, but always in vain. Thus, reclamation of Swan Lake was, according to the sources, supported by both the state and the people.[92] The same was true in Songzi, where in the 1940s the GMD troops stationed in the area and the local military forces were ordered to reclaim the Great Wangjia Lake to drive out bandits who haunted the area, but had no success due to various reasons, including the Japanese invasion. To eliminate snail fever and to promote high-water control (by freeing about 100,000 mu of farmland from water calamities and reducing the pressure of high-water on 60 kilometers of dikes) and agricultural production, from the 1950s to the 1970s about 47,000 mu of the lake was reclaimed by Songzi and Gong'an, the two counties that shared the lake. By 1985, there were still twenty-two polders in Songzi, fifteen of which were enclosed in the 1960s–1970s. And several of these polders had been reclaimed from the Great Wangjia Lake.[93] To mobilize people to enclose a large lake area on the outskirts of Wuhan in 1957, a Communist leader told representatives of the laborers that after the completion of the enclosure, the crops and livestock raised on it could support 45,000 to 60,000 people.[94]

Thus, in the early PRC, in many cases the local people strongly favored reclaiming lakes. It was not, as some critics want us to believe, that a despotic Communist state forced the local people to reclaim lakes. Quite the opposite—in many places the locals wanted to reclaim lakes even before 1949 but were unable to. In the early PRC a top priority was the development of the economy, and there was little concern with protecting the environment. Nowadays critics charge that lakes were over-reclaimed. This criticism, though valid, is also ahistorical.

5 The Battle against Snail Fever

Although from the Qing all the way to the PRC the main purpose of the Jianghan people in reclaiming wasteland and lakes was to expand farmland to feed a growing population, there was, however, another crucial factor in the early PRC that motivated reclamation on a large scale: to wipe out snail fever.[95]

---

92 Yingcheng 49-1-90ab.

93 100-8-32: 83; *Songzi xian zhi* 1986: 367; *Songzi shuili zhi* 2008: 26–27; *Weishui liuyu zhi* 2010: 187–88.

94 30-1-16.

95 *Hubei sheng zhi nongye 2* 1999: 407.

In fact, eliminating this scourge became a national public health campaign directly pushed by Mao and the central government.[96]

The Jianghan people had suffered from snail fever for more than two millennia. The earliest known victim was found in a tomb dating back to 167 BCE in Jiangling.[97] Some have argued that the troops of the state of Wei were defeated in the famous Battle of Red Cliffs in the year 208 not because they were outwitted by the resourceful Zhuge Liang or were defeated by the alliance of the Shu-Wu states (the other two kingdoms of the Three kingdoms era), but because of a "southern disease"—snail fever—to which the soldiers of the state of Wei had no resistance.[98] Almost two thousand years later, when PLA troops were preparing to cross the Yangzi River in 1948, many northern soldiers got sick after they were infected by snail fever.[99] The former is a new interpretation deduced from historical data, the latter is a fact; both cases reveal how devastating this disease can be. For example, in Zongkou Farm, founded in 1955 in Qianjiang, 90 percent of the residents were suffering from snail fever when the farm was established, and many adults were disabled.[100] According to a 1956 survey, 763,400 people on the Jianghan Plain were infected with snail fever, and 4.1 million mu of land were infested with snails.[101]

There are countless heartbreaking stories related to snail fever in this region before 1949. In Mianyang, for example, ravaged by schistosomiasis, a polder with more than twenty villages had only around twenty people in eight households in four villages who were still alive right before 1949.[102] In Jiangling, from 1920 to 1949, snail fever caused 31,800 or more deaths. As a result, 3,260 households and 735 villages were wiped out.[103] In Qianjiang, from 1938 to 1948 alone 520 villages were eliminated by snail fever.[104] In Zhongxiang in the first half of the twentieth century, snail fever also wiped-out hundreds of villages and killed thousands of people.[105] Tragedies continued after 1949. In 1957, twenty-five

96 Gross 2016; Xun Zhou 2020.

97 *Jiangling xian zhi* 1990: 619.

98 Li Yousong 2007.

99 Tang Zheng 2007. Many PLA soldiers were also infected with snail fever in the Shanghai area in 1950 when they were preparing to take Taiwan. Later it was found that many new enlistees from the countryside had snail fever and were too sick to drill. Both were factors that contributed to the new government launching the campaign to stamp out snail fever (Gross 2016: 15–40).

100 *Zongkou nongchang zhi* 1986: 103.

101 *Hubei sheng zhi shuili* 1995: 342.

102 *Mianyang xian zhi* 1989: 580.

103 *Jiangling xian zhi* 1990: 619.

104 *Qianjiang shuili zhi* 1997: 232.

105 *Zhongxiang xian zhi* 1990: 804.

people in Tianmen were infected with acute schistosomiasis when they were gathering waterweeds to be used as green manure; eighteen of them died.[106] And in 1965 on a water-conservancy project worksite, even though they were well-protected, 1,881 laborers were infected with snail fever, and some died.[107]

A local gazetteer summarized several factors that made this disease endemic in the Jianghan region before 1949. First, social factors. There was no specific organization responsible for preventing and treating snail fever. In addition, the disrepair of water conservancy projects spread *Oncomelania*, the intermediate host of the blood fluke that causes schistosomiasis, everywhere. Second, natural factors. The environment of the land between rivers and dikes is conducive to the growth and spread of *Oncomelania*. Third, the nature of local people's work. People spent a long time in infected water fishing, harvesting lake weeds, swimming, and so on; they also had to rely on lake water for daily use (drinking, washing, etc.). And fourth, untreated human and animal waste polluted waters and spread the disease.[108]

One of the most effective ways to prevent the spread of snail fever is to reclaim riverside and lakeside wasteland and thus reduce the breeding ground for *Oncomelania*.[109] As early as 1957, the central government ordered regions plagued by *Oncomelania* to carry out water conservancy projects.[110] Governments on the Jianghan Plain, one of the nation's major areas affected by *Oncomelania*, actively responded. Since residents who lived along Lake Tongjia (a lake shared by Xiaogan and another county) had long been subjected to both frequent inundation and snail fever, after many investigations and extensive dialogue with the local peasants, in 1956 the government decided to enclose the lake.[111] In another case, after the PRC unsuccessfully attempted to eradicate snail fever in the area around Lake Guanlian, a small lake (with a surface area of 4 square kilometers in 1960) on the lower reaches of the Han River in Hanyang, the government of Wuhan launched multiple studies and concluded in 1960 that the only way to stamp out the disease was to reclaim the lake. This would also protect 4,950 mu of land from frequent inundation and reclaim 6,000 or so mu of new land. The project was reportedly welcomed by the local residents.[112] In still another case, a lake in Hongxigang polder in Zhijiang was

106 *Jingzhou weisheng zhi* 1990: 11.

107 *Mianyang xian zhi* 1989: 134.

108 *Songzi xian weisheng zhi* 1986: 124.

109 *Jingzhou weisheng zhi* 1990: 226–27.

110 *Xiantao shuili zhi* 2008: 373.

111 81-1-31.

112 30-1-24c.

reclaimed in 1970 to wipe out the host of snail fever.[113] And, in a final example, a brigade in Shishou was created by reclaiming the Yangzi River beach in the early 1970s specifically to eradicate snails.[114] In the local dialect, this was called *mieluo yuan* (lit., an eradicating-snail polder).

Other than wasteland reclamation, methods to eliminate snails include building irrigation channels and leveling the land. In 1971, Zhongzhouyuan State Farm became the first snail-free farm in Hubei, and by 1985, no live snails were found in eleven state farms.[115] By the end of the twentieth century, all across Hubei, large water conservancy projects also gradually eliminated snails in many counties and elsewhere greatly reduced the area in which they were endemic.[116]

In short, after more than three decades of hard work, particularly the reclamation of wasteland and the construction of numerous water conservancy projects, snail fever was eliminated by the 1980s in Tianmen, Jingmen, and Dangyang, as well as many other places.[117] Thus, snail fever in the Jianghan area has largely been controlled, though the disease has not yet been completely eradicated.

The water conservancy projects launched during the Mao era have been criticized as a "war against the nature,"[118] but this is definitely not the case within the Jianghan Plain. If it was indeed a war, it was mostly a war against hunger and an unending struggle against a "little worm"—the blood fluke, the parasite that causes snail fever.

113 *Zhijiang xian zhi* 1990: 169.
114 *Hubei sheng Shishouxian diming zhi* 1983: 155.
115 *Hubei sheng zhi nongye 2* 1999: 408–09.
116 *Hubei shuili zhi* 2000: 25.
117 *Tianmen xian weisheng zhi* 1984: 75; *Jingmen weisheng zhi* 1990: 107; *Dangyang shi zhi* 2008: 179.
118 Shapiro 2001.

CHAPTER 7

Studies on the Reclamation of Private Polders

Although almost all studies on the agrarian history of Hunan and Hubei—the location of most polders—in late imperial times mention or focus on polders, few have explored private polders (*si-yuan*) in detail.[1] Furthermore, most of these studies treat the Lake Dongting Plain in Hunan and the Jianghan Plain in Hubei as a single unit, though there are many differences between them.[2] Since the present book focuses on the Jianghan Plain, it does not cover the Lake Dongting Plain except for occasional comparative purposes.

The earliest reclamation of polders on the Jianghan Plain probably occurred in the Song dynasty, while the enclosure of polders became common and reached its first peak in the Ming dynasty, and then peaked again in the mid-Qing dynasty. The local people continued to enclose polders throughout the Republic. In the first three decades of the PRC, a great number of new polders were enclosed from lakes, while villages or production teams along the Yangzi, Han, and Dongjing Rivers also reclaimed many new polders from river beaches or islets.

Past scholarship has tended to argue that China's government during the Qing and the Republic was incompetent, decrees were not followed, corruption among officials and underlings was rife, and officials and their staff colluded with the local powerholders. These were the major reasons the enclosure of private polders never ended despite the frequent decrees banning the practice. The powerful new government of the PRC, according to this narrative, was able to ensure the strict enforcement of its orders. There is no question that the PRC in many cases vigorously and speedily implemented policies whose results were obvious, such as Land Reform and the reform of social customs—for example, the closing of nearly all brothels, gambling houses, and opium dens almost overnight. People thus had good reason to believe that the illegal reclamation of river beaches and lakeshore would also end immediately after the new state banned it. But, as we have seen, there was in fact an upsurge of reclamation of river beaches and lakes in the early PRC. Thus, it should be

1 The term *si-yuan* is ambiguous. It can mean "private polder," or "illegal polder," or even "illicit polder." Saying that "illegal polders were banned" does not make much sense—if something is "illegal," then of course it is banned. In this sense, "private polder" is better, though it does not convey the ambiguity in the term *si-yuan*, particularly its inherent notion of illegitimacy.

2 Jiayan Zhang 2014: 12.

 | DOI:10.1163/9789004739482_009

apparent that past scholarship has oversimplified the issue. It is unfair to say that the state before 1949 did not or was not willing to take any action. Most dikes, however, were locally managed. Furthermore, Chinese peasants' hunger for farmland also contributed to the continual reclamation of private polders despite government bans, even under the strict rule of the PRC.

Research on this topic has focused mostly on the Qing, rarely on the Republic, and almost never on the longue durée spanning the Qing, the Republic, and the PRC. The Qing dynasty has been gone for more than a century, and the GMD rule of mainland China ended more than seventy years ago, but the competition for land between humans and water continues. If it is true that the Qing ban on the reclamation of private polders failed because the so-called feudal government represented the landlord class and its bureaucrats colluded with local powerholders, both the GMD government in the Republic and the CCP government in the early PRC boasted that they were the people's government. Thus, it may have been reasonable to believe that enclosing private polders would be easily controlled, at least in theory. In reality, however, the enclosure of private polders continued in the Republic and in the early PRC as well. Although the term *si-yuan* was not used in the early PRC, the fact was that these polders were reclaimed illegally, just as in the Qing and the Republic. In other words, the reclamation of private polders apparently had no connection to changes in China's political system.

All of this should prompt a rethinking of the nature of "Old China," particularly the supposedly hostile relationship between the people and the bureaucrats, as well as the personal integrity and motives of those bureaucrats. In fact, the bureaucrats of "Old China" and "New China" shared much in common—in both eras one of their primary responsibilities was to see that people under their jurisdiction were fed. If they could accomplish this, it would be considered a major political achievement. Land reclamation clearly was one of the most important ways to foster food security, particularly when the level of agricultural technology was relatively low. Both Lillian M. Li and William Rowe, among others, have mentioned the responsibility and pressure on the Qing state to feed the population;[3] one of the ways it sought to achieve this was to bring more land under the plow.

Population pressure was undoubtedly a major force driving people to enclose private polders, which has also been regarded as a main reason the Qing's ban on them failed.[4] In the Ming-Qing era, the Jianghan Plain was an important immigrant destination. The population of the PRC, however, has

3 Lillian Li 2007; Rowe 2001.

4 Peng Yuxin 1990: 189.

nearly tripled in the past seven decades—does that mean that people today must open far more land (and private polders) than before to survive? Again, this is not the case. The reclamation of river beaches and lakeshore reached its peak in the 1960s–1970s. Since 1979, though population pressure still exists, the steady growth of the non-farming sector of China's economy has provided many opportunities for young villagers who no longer wish to stay in the countryside. Land reclamation has no attraction for such people, and thus the impetus behind the enclosure of private polders has evaporated.

Therefore, if we place the phenomenon of the enclosure of private polders in the context of a longer period, we will move beyond such clichés as governmental incompetence and official corruption, and begin to see aspects of China's past that have been overlooked by previous scholarship. This chapter explores the phenomenon of private polder enclosure over the span of three centuries and probes its complicated implications.

1 What Are Private Polders?

According to Ni Wenwei, "The polders reported to the Board [of Works] with names are official polders, [but] … those built illegally by the people are private polders. [Some people] want to hide the truth about *si* [private or illegal], and thus claimed [their polders] are people's polders."[5] Here he mentioned that some polders were called "people's polders" but in fact were private polders. That is, all polders enclosed by the people themselves without official permission and without being officially registered were private polders.

However, a late Qing gazetteer mentions that the designation depending on whether a polder was an obstacle to discharging floodwater: "all polders that block river streams and are a hindrance to the major dike are private polders."[6] Here "the major dike" refers to the Wancheng Dike, highlighting that these private polders were blocking waterways, but the gazetteer does not make it clear whether the private polders were enclosed legally or illegally. Since the state usually would not allow the enclosure of these polders, they were most likely enclosed illegally. Moreover, they were enclosed from beaches on the water side of river dikes. According to the same gazetteer, there were fifty private polders in total in Jingzhou, all of which were enclosed from river beaches. Many of them had been surveyed by the state and were recorded, and some

5 Ni Wenwei 2002 [1885]: 322.
6 *Jingzhou fu zhi* 1880, vol. 20: 4b.

were even on the records of the central government.[7] These fifty private polders would have included only those on the record; the number of unrecorded private polders would have been far larger.

Private polders were not necessarily located on the water side of river dikes; they also could be on the land side. As a late Qing official pointed out, "private polders located outside [river] dikes are admittedly worrisome; even those inside [river] dikes and along river branches and lakes are also harmful, since if one foot of land is occupied, one foot of stream will be hindered."[8] To him, private polders referred to all polders enclosed without permission and could be situated on either the water side or the land side of river dikes. Some private polders were even built inside existing polders.[9] From this perspective, it did not matter whether they were built legally or illegally; all that mattered was whether they blocked waterways. There was a reason for this position. One of the main causes of the 1788 flood was that waterways were blocked by a sand islet in the middle of the Yangzi River course that had been reclaimed to grow reeds.[10] Hence, after the flood, the state ordered those polders to be destroyed.

As we have seen, these definitions are those of Qing bureaucrats. But to this day scholars' understanding of private polders is mostly based on these definitions and are thus very much the same. According to Peng Yuxin, "polders built with state funds are official polders, those built by the people without state funds but recognized by the government are people's polders, and those built by the people but not recognized by the government are private polders."[11] Zhang Jianmin further elaborated that "the *si* in *si-yuan* refers to illegality," because these polders blocked waterways and hindered the discharge of floodwater and "thus the state regarded them as illegal." It is certain that no state funds were used in high-water control and annual repairs of these polders—all the costs were borne by the locals.[12] Morita Akira and Pierre-Etienne Will also mentioned that private polders were illegal and blocked waterways.[13]

Mei Li and her colleagues concur that *si* in the term *si-yuan* meant "illegal," since, unlike official polders, *si-yuan* were not registered and reported to the Board of Works, the central government agency in the imperial era responsible for state-sponsored building projects. Whether any particular polder was legal or not depended on whether it was a hindrance to discharge and drainage, and

7 *Jingzhou fu zhi* 1880, vol. 20: 4b-20a.
8 Ni Wenwei 2002 [1885]: 242.
9 Shuili dianlibu 1991: 461.
10 Ni Wenwei 2002 [1885]: 25.
11 Peng Yuxin 1990: 173.
12 Peng Yuxin and Zhang Jianmin 1993: 210.
13 Morita Akira 1974: 35–42; Will 1985.

whether "it had not yet been transformed into taxpaying land."[14] This interpretation is supported by a PRC gazetteer: "private polders are those built by the people but not officially registered, and they are 'illegal.'"[15]

In sum, we can conclude that the two main features of private polders were that they were illegitimate (i.e., they were enclosed without official permission) and they blocked the discharge of floodwater. They could be reclaimed from river beaches or islets, or enclosed from lakes, and could be located on either side of river dikes. It is possible that some of them were actually people's polders. Most private polders were illegal, but only in theory, since the state tacitly consented to the reclamation of many of them, because they could be taxed. These taxes, however, were not paid to the central government and were not recorded in tax registers. It is for this reason that we should also consider those polders enclosed by the local people without permission in the early PRC as private polders.

2 The Enclosure of Private Polders in the Qing and the Republic

In China's long history, the transition from one dynasty to another was usually violent and chaotic. The governments of the new dynasties understood that restoring agriculture was absolutely crucial to stability and legitimacy. Thus, they encouraged migration and incentivized the reopening of farmland and the opening of new land by offering tax reductions or exemptions. The most extensive wasteland available for reclamation on the Jianghan Plain was lakeside wasteland, which people enclosed as polders. Therefore, the enclosure of such polders on the Jianghan Plain was encouraged and supported by the state at the beginning of both the Ming and the Qing dynasties, and the Jianghan area hence became a famous immigrant destination.[16] With the constant growth of the population, however, more and more wastelands were reclaimed, and less and less land was left to receive discharged floodwater. By reclaiming polders without permission and not reporting them to the Board of Works, people could evade the obligation to pay tax to the central government. These were private polders.

14 Mei Li, Zhang Guoxiong, and Yan Changgui 1995: 115–16.

15 *Honghu xian zhi* 1992: 98. The situation in Songzi at the end of the nineteenth century, however, was different. There the locals built small polders with their own money; these polders were called *si-yuan* (*Songzi xian zhi* 1982 [1937]: 80). In this case, *si* has nothing to do with legal or illegal. It is not even clear whether these polders were subject to the land taxes. The only thing that is clear is that they were built by local people.

16 Jiayan Zhang 2014: 33–38.

That the main force driving the enclosure of private polders was population pressure was frequently mentioned by Qing officials in their memorials, as Chang Jun (governor of Hunan) did in 1766.[17] Later, Ni Wenwei explicitly explained the relationship between population pressure and diking:

> Jingzhou is located in the upper reaches of the Yangzi River. ... Since medieval times, the situation of the Yangzi River has changed frequently, the population has grown, and the productivity of land has almost been totally exhausted. Areas that formerly stored floodwater have become land, graveyards, towns, and cities. [But the locals] do not want to leave their hometowns where they have long lived, [and thus] they have had to think about how to defend themselves. To deal with the clash between land and water, [they] turned to the job of high-water control and annual repairs. Nowadays thousands of miles of surging water are constrained within dikes. [If] one foot of a dike is not secure, one thousand miles of land will be inundated. [We] safeguard the dikes and worry about ruptures, so how can [we] have the heart to say we will give them up?[18]

As a pragmatic official, Ni criticized the idea that dikes should be abandoned as utterly absurd. As far as he was concerned, there was no other choice but to protect the people and the land with dikes. At the same time, he also pointed out that the growing population increased the pressure on the land, and that in turn drove people to reclaim still more land. He did not blame the locals for doing what was necessary to feed a growing population: that is, reclaiming land. As a prefect, his greatest concern was the integrity of river dikes (more precisely, the Wancheng Dike), but for the average people, the logic of reclaiming private polders was similar: more people need more farmland, and reclaiming river beaches and lakeside wasteland was essential for survival.

In a 1748 memorial aimed at banning private polders, Peng Shukui, the governor of Hubei, briefly explained how the locals reclaimed lakes step by step. They first blocked water along the bank of a lake or in the middle of the lake to form silt, then built dikes to enclose the deposited silt.[19] After most available lakeshores were reclaimed, the locals turned to reclaiming river beaches, a phenomenon that probably emerged at the end of the Ming dynasty but became very common in the late eighteenth and the early nineteenth

17 Shuili dianlibu 1991: 399.

18 Ni Wenwei 2002 [1885]: 6.

19 Shuili dianlibu 1991: 298.

centuries.[20] This was also a peak time of population growth. The pressure to feed ever more mouths forced the locals to reclaim ever more river beaches. A batch of recently discovered land deeds from the Qing (mostly in the reign of Daoguang) confirms that a Xiong clan in Tianmen controlled extensive beachland along the Han River.[21] Although these beaches were not supposed to have been reclaimed, not only were they reclaimed but they were recorded in land contracts. The deed holders, in other words, owned this land and their ownership was officially recognized. In a similar situation, in Ouchi in the 1940s a lawsuit was filed in a fight over the ownership of some legally purchased silted beaches.[22] The Ouchi River was formed only in the 1850s, and hence these silted beaches only came to exist about a century before the lawsuit. And, as in other cases, beachland was not supposed to be reclaimed in the first place. Practice, as we can see, was often divorced from theory.

These river beaches were reclaimed also because their soil was fertile. The governor of Hubei once complained in 1839 that whenever there was new silt, some "ignorant" people would reclaim the deposits as private polders and then farm them.[23] Regulations on the treatment of silted beaches changed in the reign of Qianlong. In 1748, if a household reported that it lost some farmland because it had collapsed into a river, the household was allowed to take an equivalent amount of nearby river beaches. If all the households with land that had collapsed into rivers were compensated in this way, then the rest of the silted beaches would be opened for landless people to reclaim.[24] This was still a continuation of the early Qing policy, under which people were encouraged to reclaim wasteland. Fifteen years later, however, a memorial of the governor of Hubei forbade the reclamation of newly silted river beaches.[25]

Because of its strategically important position, Jingzhou had been a hotly contested battleground since ancient times. Therefore, the Qing state stationed a considerable number of elite Eight Banners troops there. Part of the pasturelands for its cavalry were located on the river beaches.[26] In the middle of the nineteenth century, the local government recruited people to reclaim about twenty thousand mu of these pasturelands in Shishou, Jianli, Qianjiang, and Zhijang, and collected taxes on the reclaimed land. Twenty thousand mu, roughly the area of a relatively large polder in the Jianghan Plain, is hardly a

---

20 Mei Li, Zhang Guoxiong, and Yan Changgui 1995: 120.

21 Zhang Jianmin and Tang Gangmao 2014.

22 LS31-4-63.

23 Shuili dianlibu 1991: 823.

24 Chen Zhenhan et al. 1989, vol. 1: 290.

25 Chen Zhenhan et al. 1989, vol. 2: 406–07.

26 See Yan Gao (2022) for more details about these troops and their pasturelands in Jingzhou.

vast swath of land. The important thing is that these beaches should not have been reclaimed according to the previous regulations, but now the government officially recruited people to reclaim them. This land was thus subject to taxation, the main lure that attracted the local authorities. In 1861, the local government stipulated that the newly silted beaches on the southern bank of the Yangzi River would be opened for reclamation, and would be rented out, with the revenue to be used for local education and other expenses. Furthermore, it noted that there was "no need to report to the Ministry of Revenue."[27] Although this reclamation was not technically "illegal," it is obvious that it violated the previous regulations. The other side of the coin is that the Taiping Rebellion (1850–1864) was raging and threatening the very existence of the imperial state. In that context, it is understandable that the government would seek to exploit every possible source of revenue.

Despite the ban, the reclamation of private polders continued and in some instances was done quite openly. For instance, in 1859, a gentryman in Zhongxiang asked for permission to build a dike and form a private polder.[28] Though his request was rejected, it is likely that some private polders were openly reclaimed. The private polder in this case, however, might refer to a people's polder. Otherwise, how dare a gentryman flagrantly ask the state to approve of something illegal? Some people who openly built dikes to enclose private polders claimed that they were building a road. And if no official discovered the work, they would extend the road year after year and eventually turn it into a dike.[29] Sometimes even a little floodwater would lead people to enclose private polders. In Songzi, for example, from 1900 to 1902 the Yangzi River had less water than usual, and the local powerful households took the chance and enclosed many private polders.[30] Those who built private polders in these circumstances might come from the same community or the same clan. Jiangzheng polder in Zhijiang, for instance, was originally a private polder enclosed from a lake by the Cai and the Zhang clans.[31]

Because private polders were not considered legitimate, statistics are hard to come by. According to an 1880 gazetteer, there were 179 regular polders and 9 private polders in Jiangling; these numbers were 47 and 1 in Gong'an, 41 and 34 in Shishou, 498 and 4 in Jianli, and 36 and 2 in Songzi, respectively. But the number of official and private polders in Songzi is uncertain, as the gazetteer

27 Xi Yuan et al. 1990 [1879], vol. 6: 38a–39a.

28 *Xiangdi cheng'an* 1969, vol. 1: 148a–149a.

29 Ni Wenwei 2002 [1885]: 237.

30 *Songzi xian zhi* 1986: 351.

31 *Zhijiang xian zhi* 1990: 167.

mentions that Songzi had 36 official polders, 2 private polders; or 8 official polders, and "all the rest are private polders."[32] The gazetteer also mentions that only Shishou had a large number of private polders, with the number in other counties too small to be counted. This hardly seems credible. If there were a great number of regular polders but only a few private polders, there would have been no need for the state to repeatedly issue orders to ban them. In any case, the gazetteer was an official history, while private polders were not reported to the state and were thus not officially recorded. Their number was most certainly far larger than what was recorded. One source implied that some private polders were recorded because they had been surveyed and should have been razed but still existed.[33]

All these sources recorded only the number of private polders, not their size. Despite these polders being illegal, they were not necessarily small. In Jianli, the dike of one private polder was "about ten or so li long." If the private polders around Xiaozhiyuan in Shishou were linked together they would have been "about 60 li wide …, and about 80 li long"[34]—a very large swath of land. According to a 1915 statistic, there were about one million mu of private polders along the Yangzi River in Jingzhou alone.[35] If one were to include private polders in the Han River valley, the total area of private polders on the Jianghan Plain would be even more impressive.

3 Banning Private Polders: Bureaucrats of the Qing and the Republic

According to Ni Wenwei, private polders blocked waterways, caused the rise of riverbeds, and created huge waves that scoured river dikes. The damage private polders caused was, in Ni's words, "the greatest."[36] The existence of private polders, in short, seriously threatened the safety of river dikes. Official policy, as we have noted, was to ban them, but the state constantly compromised and yielded to the local people. Furthermore, those who stuck to the ban were usually high-ranking bureaucrats at the provincial level and above; prefects and magistrates did not always hold identical views on this issue.

32 *Jingzhou fu zhi* 1880, vol. 20: 3a–18b.

33 Ni Wenwei 2002 [1885]: 218–20.

34 Ni Wenwei 2002 [1885]: 219.

35 *Jingjiang di zhi* 1937, vol. 4: 10. In 1915, the new government wanted to legalize private polders along the Yangzi River, encompassing more than one million mu, in order to tax them, but the dike administration was strongly opposed, noting that this step would negatively affect more than twenty counties (*Jingjiang di zhi* 1937, vol. 4: 10–13).

36 Ni Wenwei 2002 [1885]: 217.

As early as 1734, a Hubei provincial official had advocated "prohibiting the illegal building of polder dikes," which has been regarded as the first use of the term *si-yuan*.[37] This implies that the Qing state had already tended to no longer tolerate over-reclamation. One of the most influential memorials on this subject, however, was the one submitted to Emperor Qianlong in 1748 by Peng Shukui:

> Small polders enclosed from unoccupied lakeshore and riverside lands are always strictly prohibited. … Unfortunately, the floodwater of the Han River carries a high content of sand and very easily becomes silted, and powerful people thus are quickly drawn to take advantage of this. They first block water in many places along the banks and in the middle of the lake to form silt, and then use taxation as an excuse to build dikes all around to form polders. The villagers are concerned about survival; thus, the more the land the better, but they forget that if humans seek benefit by contended for space with water, water certainly will contend with humans, which brings disaster … We propose that where there are no polders, illegally building polders must never be allowed, and the rebuilding of broken polders [dikes] must not be allowed.[38]

This memorial explicitly sets out the causes, process, and current situation of the formation of private polders and proposes a strategy to deal with the issue. Peng's strategy, however, only makes sense on paper. Faced with a growing population, how were people to survive without continuing to open more land? Because of Peng's unrealistic strategy, his proposal to ban private polders was doomed to fail. Later, when discussing the ban on private polders, Qing bureaucrats usually started with the words "since the memorial of the revered Mr. Peng, … private polders have always been banned." This cliché alone is evidence of the failure of Peng's proposal and the continuing enclosure of private polders in the Qing. For more than a century and a half, from 1748 (the year Peng submitted his memorial) to 1911 (the end of the Qing dynasty), the reclamation of private polders continued unabated, despite being banned. In other words, on this issue, the local people were always playing games with the Qing bureaucrats.

The bureaucrats' view on private polders is crystal clear: all private polders had to be destroyed since they blocked waterways. This was a natural conclusion after taking everything into consideration. In 1763, when Fude, the

37 Mei Li, Zhang Guoxiong, and Yan Changgui 1995: 116.
38 Shuili dianlibu 1991: 298.

governor of Hubei, reported on how to deal with dike affairs he listed six "ought to" do things, one of which was banning private polders: "no matter if the polder is new or old or if its soil is fertile or infertile, as long as it blocks waterways, it should be razed immediately. … If newly silted beaches reappear, reclaiming them [as private polders] must not be allowed."[39] The private polders Fude referred to even included some old polders, apparently legally enclosed. They too should have been razed because they blocked waterways. Later, in the reign of Daoguang, in discussing the water-control situation in Hunan and Hubei, Wei Yuan, an important scholar, went further to argue that even if official polders blocked waterways, their dikes should not be rebuilt after a rupture, but private polders that did not block waterways and were flooded should not be razed.[40]

Both Peng and Fude were high-ranking provincial-level bureaucrats. When Wei Yuan wrote his essay, he was not a top official, but served in the office of the governor of Jiangsu, also at the provincial level. All three argued that all polders, as long as they blocked waterways, should be razed. The thinking of officials at the prefectural and county levels, however, was divided. In 1796, Gugeng polder, located on the river side of the Wancheng Dike, came to the attention of the government of Jingzhou. The chief of the marine battalion in Jingzhou argued that "although [this small polder] has been submerged under a flood, its old dike still exists." If it were to be rebuilt, he argued, it would protect the Wancheng Dike from being flooded. The official who was sent to survey the situation reported back that since there was no polder there before, this polder was not an official polder, but an illegal one, which blocked river streams. He asked that the polder be left alone and "let it be destroyed on its own," after which it would never be rebuilt. Ni Wenwei disagreed because the foot of the Wancheng Dike would be slowly submerged every year in the high-water season, which helped to disclose any potential problems to be solved in annual repairs. If there was a polder dike on the river side, once it broke, the floodwater would then rush to the Wancheng Dike, endangering its integrity. Therefore, he mocked the military chief as having good intentions but not really understanding the situation.[41]

However, the local people certainly would not "let it be destroyed on its own." On the contrary, they would continue to rebuild it. Later prefects thus had to repeatedly reiterate the ban. They even blamed lower-level governments for instigating people to enclose private polders, thereby "sympathizing

39 Chen Zhenhan et al. 1989, vol. 2: 406–07.

40 *Zaixu xingshui jinjian* 2004: 968.

41 Ni Wenwei 2002 [1885]: 223–25.

by granting small favors [but] forgetting great harm."[42] But since so many river beaches and/or islets were enclosed as private polders, even the emperor had to tacitly consent to the status quo. In an edict in 1833, Daoguang declared,

> Show leniency for those who occupied and reclaimed [river beaches/sandbars] in the past, but strictly prohibit occupying and reclaiming in the future. ... From now on, if there are newly silted beaches, all of them should be taken as government wasteland and their cultivation must never be allowed. ... If anyone dares to engage in illegal reclamation and grows reeds on newly silted beaches and islets, they must be harshly punished.[43]

Daoguang continued the policy on newly silted river beaches initiated by Emperor Qianlong: that is, they should not be reclaimed. However, this ban was also ineffective.

After suppressing the Taiping Rebellion, many generals of the famous locally organized Xiang Army of Hunan, which played a leading role in the defeat of the Taipings, became high-ranking provincial officials. They were tough, pragmatic, and resolute. When dealing with the ban on private polders, however, it seems they were not tough enough, or simply could not be tough. Zeng Guoquan (a younger brother of Zeng Guofan, the leading organizer of the Xiang Army), when serving as the governor of Hubei, once got very angry about the enclosure of small private polders, and harshly ordered local officials to cooperate with village leaders to investigate the regions along the Yangzi and the Han Rivers, and "if [they] find illegally built [polder] dikes, immediately destroy all of them." Severe as this may sound, it is more like an official gesture than a real order; Zeng probably understood that no official or village leader would be so naïve as to follow his order to destroy all private polders. Thus, in the following sentence, he spoke of "leniency for the past [illegal reclamation]" and stated that reclaiming private polders would not be allowed in the future.[44] "Leniency toward [construction] in the past, but a strict ban on [construction in] the future" became a common practice among provincial and prefectural officials for dealing with the issue of private polders.

County magistrates, however, had different views. Some supported the construction of polders on the water side of river dikes by arguing that they could protect the dikes. In 1839, both the governor of Hubei and the prefect of

42 Ni Wenwei 2002 [1885]: 217.

43 *Zaixu xingshui jinjian* 2004: 135.

44 Ni Wenwei 2002 [1885]: 238–39.

Jingzhou ordered the destruction of all private polders on the river side of the Wancheng Dike.[45] But the magistrates of Dangyang and Jiangling, who surveyed the area in 1842, disagreed. They tactfully argued,

> After a careful survey, [we have found that] since Baozhang polder served as outside protection for the Wancheng Dike, it does not block waterways, and its existence seems beneficial. Moreover, in the past it was named "Baozhang" [safeguard], which apparently refers to safeguarding the Wancheng Dike. There is a great deal of military land and farmland crisscrossing this polder. Since there is military land there, it [the polder] would have been there for a long time, which is different from private polders built on river beaches that block river streams. Furthermore, people's taxes and garrison troops' grain are involved. Abandoning [Baozhang polder] will both reduce national tax revenue and imperil the people's livelihood.[46]

In all of Qing officialdom, the county magistrates knew best the situation in their jurisdictions. They had to take local people's opinions, reactions, and interests into account, and therefore could not obey their higher-ups to the letter, even though that improved their chances for promotion. Here the two magistrates particularly mentioned that there was military land in this polder, which meant that the reclamation must have been approved by the state and hence it could not be a private polder. They also warned that the destruction of the polder would bring harm by "reducing national tax revenue and imperiling the people's livelihood," which instantly made it a national moral issue. Since no provincial or prefectural bureaucrat could object to this, these county magistrates achieved their goal without openly offending their superiors.

A magistrate of Shishou even directly challenged his superior by openly favoring the enclosure of private polders. He not only supported the construction of small polders within river dikes by arguing that these polders would protect the river dikes, but also recorded them in the new county gazetteer to legitimize their status. Furthermore, he even collected fees to build private polders.[47] In other words, he turned a completely blind eye to the ban on the enclosure of private polders. In the above discussion, it was mentioned that most private polders were in Jiangling and, particularly, Shishou. Although

45 Ni Wenwei 2002 [1885]: 226–27.
46 Ni Wenwei 2002 [1885]: 228.
47 Ni Wenwei 2002 [1885]: 239–40.

that was related to their specific geographical features (i.e., they had many river beaches), the support of their magistrates was also an important reason.

Thus, the reclamation of private polders never stopped. In the late Qing, Ni Wenwei hit the mark with a single comment: "County bureaucrats benefit from tax collection, yamen runners and clerks benefit from surcharges, dike officials benefit from gifts, … therefore the more bans on private polders, the more private polders there will be."[48]

The enclosure of private polders continued to be a vexing issue in the Republic. The General Bureau of the Wancheng Dike was established in 1912 and soon sent staff to investigate the private polder situation, asking the government to take action to destroy old private polders and ban the enclosure of new ones.[49] That the new state started to deal with private polders immediately after it was established demonstrates the seriousness of the issue. The strategy it applied, however, was the same as that of the Qing state, and the results were the same as well. The locals simply would not obey this ban either. Thus, in 1942, when the Republican government issued its Law on Water Conservancy, it stipulated that "Waterways, sandbars, and river beaches cannot be enclosed and reclaimed."[50] If this had to be included in the legal code, it must have been a very common practice.

4 Banning Private Polders in the Qing and the Republic: Gentry and Villagers

Needless to say, unlike most government personnel, who supported the ban on private polders, most local gentry and villagers favored the enclosure of private polders. For example, in 1867 in Haoxue, a prefect in Jingzhou witnessed many people building polder dikes on the river beaches on the other side of the Yangzi River. Later it was found that some gentry were leading these actions. The prefect particularly emphasized that he had witnessed this in person and the locals could not deny it. The dike was destroyed.[51]

This prefect was tough, but other prefects might have had a soft heart and could not bear to see the well-being of the locals threatened by their private polders being immediately destroyed. They tried to find a middle ground that

48 Ni Wenwei 2002 [1885]: 323.

49 *Jingjiang dadi zhi* 1989: 351.

50 100-2-52: 102.

51 Ni Wenwei 2002 [1885]: 237–38.

could be accepted by both sides. In 1809, a circuit intendant (*daotai*) issued an order to destroy private polders in Shishou:

> Since you people have already spent money on the building [of a polder dike], if this dike were razed, [I am] afraid the land will be abandoned, which would be a waste of labor. … Now, at my discretion, I plan to let you open one or two or three places on the restored broken polder [dike] facing the river and following the stream, with each opening being two or three hundred zhang wide. Thus, after the river water rises, it can flow freely without being blocked, and the unbroken dikes will gradually fall into ruin.[52]

He went on to tell the people of Shishou that they should let these polder dikes fall into ruin on their own and should not surreptitiously rebuild them. If they were caught doing so, they would be harshly punished.[53] This circuit intendant may have sounded tough, but he actually was sympathetic. As an official, he had to make a gesture in his order and demonstrate to his higher-ups that he had done what he was supposed to do. As a circuit intendant, on the other hand, he also had to be mindful that the locals' labor, time, and money should not go to waste. Thus, he only asked them to open the dike at several places. If there was not excessive floodwater, the locals could still have a harvest, and then they could let the dike simply waste away. The locals, however, had a very different agenda in mind. First, they definitely were not so obedient as to actually open the dike in several places to let floodwater flow freely and eventually destroy the entire dike. Second, even if they were forced to open the dike when government personnel were present, it would be easy for them to fill up those openings after the government personnel had left.

In another case, a prefect of Jingzhou in the late Qing heard that some people were constructing dikes on river beaches on the pretext of building a dividing road between two plots. He sent representatives to investigate in the spring, but all the local leaders avoided meeting with them. He then ordered the magistrates of Jiangling and Shishou to attend to the problem in person. Surprisingly, both magistrates warned the prefect to be cautious.[54] Here the local leaders turned a deaf ear even to the call of the prefect's representatives, which demonstrated that the prefect's control over local society was very weak. The local leaders clearly knew the prefect's call was not a good sign, but since

52 Ni Wenwei 2002 [1885]: 225.
53 Ni Wenwei 2002 [1885]: 226.
54 Ni Wenwei 2002 [1885]: 342.

they did not want to jeopardize their own interests, they simply refused to meet his representatives. The prefect thus turned to the magistrates, but they too did not want to offend the locals. They quibbled that there were too many polders, and that they could not be destroyed by force without sparking an incident. This may or may not have been true, but nonetheless it was a possibility everyone had to take seriously. Besides, the county magistrates would not benefit in any way from this ban. In fact, they would lose tax revenue, offend the locals, and provoke resistance. It is no wonder they remained passive.

This certainly encouraged the local people to persevere in their illegal diking. The policy of "leniency for the past, a strict ban in the future" was tantamount to encouraging people to enclose private polders—the state might order a private polder be destroyed, but rarely was one actually demolished. Thus, it is readily understandable that the ban on private polders remained a mere formality. This was hardly because government personnel were greedy and corrupt; rather, it is because they took the local people's livelihood into consideration. After the 1935 flood, for example, the state decided to destroy private polders around Lake Dongting, but it was not easy. According to the 1947 report of an investigation mission,

> Since the building of one dike easily took tens of thousands of laborers, the project to destroy it also needed the same number of laborers. Because the government had no budget for such an undertaking, in the past while carrying out the task of destruction, only one opening was dug. But after the people who carried this out left, the locals instantly filled up the opening. Moreover, there are a great number of houses located inside the polder, and the crops there grow luxuriantly. Breaking down the dike and letting the water flow in will cost the people their means of livelihood. Since this is a matter of life and death, [the people thus will] fight to the death to protect the dike. Those sent to knock down dikes also cannot bear [the thought] of completely destroying them. Thus, they usually deceive their superiors. Sometimes the residents of a polder have even opened the dikes themselves. They took pictures and handed them to those who were sent to break down the dikes so that they could report to the government that they had carried out their duty. Therefore, in reality, no dike had ever been destroyed by obeying [government] orders.[55]

55 *Hunan sheng binhu zhoutu shichatuan baogaoshu* 1947: 111.

This investigation mission found fifty-five private polders ranging in size from 500 to 30,000 mu in seven counties, all of which were built between 1939 and 1944.[56] The explanation this investigation mission offered is convincing: that's the way of the real world. In other words, the fact that many private polders were not destroyed during the Republic was not always because corrupt bureaucrats colluded with local powerholders, as past scholarship has led us to believe.

Even the law in the Republic itself had loopholes that could be exploited. According to Article 65 on Water Conservancy, "Waterways, sandbars, and river beaches cannot be enclosed and reclaimed. But if the responsible agency has reported to the upper-level responsible agency that [such reclamation] will not block waters and is not a hindrance to storing floodwater, and if the upper-level agency concurs, then [this stricture] shall not apply."[57] The first part of the article repeats the rule in the Qing, but the Qing state was very strict about the ban and countenanced no exception, at least on paper. The second part of the article, however, clearly left room for people to manipulate the rule, and in fact they would try to find any possible excuse to justify enclosure.

Although, unlike on the Lake Dongting Plain during the Republic, there was no large-scale reclamation on the Jianghan Plain, the enclosure of private polders was still common. It was found that the locals usually reclaimed newly silted land but did not report it to the government. As a result, a great deal of land or an entire polder escaped taxation.[58] Thus, an expert in water control urged that all private polders along the Yangzi and the Han Rivers be banned since they blocked waterways and caused frequent water calamities, particularly in the counties of Gong'an, Shishou, Jiangling, and Jianli. According to this expert, the Republican government did not stop the local people from reclaiming river beaches and islets, because it wanted to increase its tax revenue.[59]

This is not a groundless statement. In Shishou in 1946, residents of Yonghe polder, which was reclaimed from newly silted river beaches, asked upper-level authorities to upgrade the status of their polder dike to a major dike to ensure its safety, as upstream rapids constantly scoured the dike base and caused the dike to collapse.[60] First, this polder was basically an illegal, since the law prohibited newly silted river beaches from being reclaimed. Second, the fact that the rapids washed out the dike base indicates that this dike blocked waterways

56 *Hunan sheng binhu zhoutu shichatuan baogaoshu* 1947: 111–18.

57 100-2-52: 102.

58 Xiong Daorui n.d.: 60.

59 *Jingjiang difang zhi* 2012: 711.

60 LS36-2-18.

and thus should be abandoned. Not only did the locals not abandon it, but they went so far as to ask for more investment from the upper-level government, because that would have a bearing on "national taxation." The same situation occurred in Zhijiang, where the locals implored the county government to list the dike of their private polder as a major dike.[61] Many river beaches of the Dongjing River were also gradually reclaimed in the Republic.[62]

5 Reclamation of River Beaches, Islets, and Lakeshores in the Era of Collectivization

The enclosure of polders took several forms in the early PRC. In the 1950s, private landownership was abolished, ownership was transferred to the state, and the right to manage land was handed to collectives or villages. Individuals thus lost their motivation for reclamation. But population pressure still pushed the collectives, or even production teams, to continue to reclaim land. On the one hand, the PRC state launched many large-scale water conservancy projects to reduce the frequency of water calamities, to reclaim lakeside and riverside wastelands, and to eliminate the breeding ground of *Oncomelania*. On the other hand, the reclamation of more and more lakes and river beaches also increased the pressure on high-water control.

Right after the establishment of the PRC, the new state nationalized all rivers, lakes, and river beaches and islets. On February 4, 1950, the government of the Center-South Region issued its first regulation on water conservancy.

> Everywhere [people] must not contend for land with water and must not enclose [land]. All lakes and water systems along the rivers must not be changed or infringed on or shrunk, no matter their size. Any alteration must be approved. It is absolutely prohibited to enclose newly silted sandbars in the middle of the Yangzi River. Old silted lands as well may not be enclosed.[63]

In the same year, "to eliminate water calamities forever," the new state stipulated that:

61 LS1-5-855B.

62 *Yanglinwei qu zhi* 1989: 53–56.

63 *Jingjiang dadi zhi* 1989: 259.

> Sandbars or islets in rivers and lakes must be entirely preserved, even though sandbars give rise to morbid waters. They also block floods but at the same time hinder shipping. In the old society only the interests of some were considered. Capitalists and evil landlords wantonly engaged in enclosing and reclaiming. Thus, sandbars are growing larger and larger, and the morbid state of rivers is getting worse and worse. If this is not corrected in time, it will eventually lead to serious consequences. The state must preserve these lands, draw up a comprehensive plan of water control, and restrict or undo reclamation.[64]

This regulation bears a very clear ideological imprint, attributing as it does the enclosure of private polders in the Republic entirely to the encroachment of landlords and capitalists. Past scholarship has taken basically the same position. This subjective judgment implies that enclosure would not be tolerated in the new society. Although the state realized sandbars are problematic, it still regulated to preserve them as part of "a comprehensive plan of water control." The nationalization of river beaches was indeed a revolutionary step. But aside from its ideologically loaded language, this regulation is not greatly different from those of the Qing and the Republic. More importantly, at the very beginning of the PRC, the new state issued regulations on water control, particularly banning private polders, just as the imperial state had done.

The tone of this regulation conveyed determination, but it was immediately challenged by reality. In Gong'an in 1951, the local residents immediately restored the dike protecting Nanwuzhou, an islet with about twenty thousand mu of arable land, which had collapsed. According to the new regulations, the islet should have been abandoned, and the government of Hubei repeatedly ordered the local county government to raze it. But the fertile land of this islet supported seven thousand people. In other words, before abandoning this islet, the upper-level government had to first find a place to relocate these people or find another means of supporting them.[65] As a result, not only was the islet preserved, but its dikes were repeatedly reinforced.[66]

Baili Islet is another example. In 1951, after receiving an order from the Hubei provincial government to turn the polder on Baili (a 228 square kilometer island in the middle of the Yangzi in Zhijiang) and six other polders into a flood diversion zone, the leaders of the Zhijiang County government reported

64 SZ37-1-9.

65 SZ34-2-223.

66 Nanwu Islet originally belonged to Jiangling but was separated from it and transferred to Gong'an in 1950 (*Gong'an xian zhi* 1990: 59–60).

that they certainly would obey the order, but that the locals had different ideas: the islet could not restore much floodwater, and so would not be very useful in safeguarding the Great Jingjiang Dike. Furthermore, if the residents were to be relocated, then the labor they had invested in the dikes and farmland would be wasted. Finally, where could one relocate 119,000 people, and how could one arrange for them—particularly craftmakers and merchants—to continue their work? Therefore, the residents asked the upper-level government: Is there a way to avoid turning these polders into a flood diversion zone? The response was to postpone the plan for a flood diversion zone.[67] In this case, the upper-level government clearly acted with undue haste.

To a certain extent, "leniency for the past, a strict ban in the future" continued in the PRC. In 1953, the state emphasized that "building new dikes along the banks of all rivers and lakes in Hubei is prohibited. If dikes illegally built after 1938 are dilapidated, they may not be restored without the permission of the provincial government; if they are not dilapidated, they may not be reinforced." Building houses or growing crops on dikes was also prohibited, but it seems that these proscriptions were not well enforced.[68] During the GLF, the central government encouraged peasants to "stick in a pin wherever there is room" or "make use of every bit of space" (i.e., grow crops on any available land) to cope with China's grain shortage. Since river beaches and islets were readily available, they became ideal targets of reclamation, and the governments of counties along the Yangzi River in Jingzhou asked their higher-ups to allow them to enclose beaches and islets "in order to bring the potential of this land into full play, and increase industrial and agricultural production." This was a timely appeal, one the Jingzhou prefectural government had to come to terms with. In fact, it gave in.[69]

In the Qing, as we have seen, if a polder—no matter whether it was old or new, or whether its land was fertile or infertile—blocked waterways, it was supposed to have been immediately destroyed. But, to repeat, this ban was impractical and was thus later amended to prohibit the reclamation of newly silted river beaches only. The PRC state, though powerful, was also pragmatic. It not only differentiated between uncultivated and cultivated land, and between old and new polders, but it also adjusted its policy according to changing circumstances. Since the PRC government billed itself as a people's government, it had to take the livelihood of the peasants into consideration. Thus, as we have seen, it allowed the reclamation of river beaches. These cases

67 SZ34-2-225.

68 SZ34-2-404.

69 39-1-36-1.

may seem to have involved merely local issues, but in fact they were a matter of national importance as well, since instability in one region could threaten the stability of the whole country.

The blockage of Xiaozhiyuan in Shishou is another good example. As mentioned earlier, there were a large number of private polders in Shishou, many of which were located along the northern bank of the Yangzi River. Shishou and Jiangling had battled for about two centuries over the opening/closing of Xiaozhiyuan, a rivulet running across the region. It was blocked in 1939 due to war. In 1950, in response to an order from the Hubei government, the Shishou County government reopened it, but just a year later it was reclosed. Sources usually provide rich information and discussion about historical conflicts, and how the new state skillfully solved such problems, but they only mentioned the blockage in 1951 in passing. What is unspoken is that the state had little alternative but to block the rivulet. The reason is simple: hundreds of thousands of mu of fertile land in this area were too valuable to be ignored. In 1951 and 1952, a state farm was established here, which turned 140,000 mu of river beaches into cotton fields. And in 1952 the state started to build the Jingjiang flood diversion project in Gong'an, making the area a prime place to resettle relocatees.[70] Simply put, population pressure forced the PRC state to be realistic.

In theory, since there was to be no more privately owned land in the PRC, there would be no more private polders. But it has not been unusual for the locals to enclose polders without permission. Both provincial and prefectural governments have opposed the enclosure of beach polders, but the position of county governments has varied: some have supported enclosure, some have opposed it, and some have taken no consistent position. Although in Jiangling, for instance, enclosing river beaches (particularly along the Great Jingjiang Dike) was formally prohibited, it still repeatedly occurred. From 1960 to 1974, fourteen small polders were enclosed from river beaches in Jiangling alone. The county government ordered that they be destroyed, but some in fact went untouched. One such polder was enclosed in 1960, and it was not until 1974, after twenty-seven people died when the dike ruptured, that it was abandoned.[71] In the 1980s, there were still fourteen polders outside the Great Jingjiang Dike; many of them had been repeatedly destroyed and rebuilt since the Qing era.[72]

The situation was similar along the Han River and the Dongjing River. In 1957 in Zhongxiang, the Bureau of Water Conservancy asked the county government

70 *Shishou xian zhi* 1990: 159; *Jingjiang fenhong gongcheng zhi* 2000: 230–31, 237; SZ34-2-117.

71 *Jiangling difang zhi* 1984: 93–94.

72 *Jingjiang dadi zhi* 1989: 84–91.

to enclose nine polders from the Han River beach, as the county had about 330,000 mu of farmland along the Han River but only could get one harvest a year, in the summer. The enclosure of those polders would guarantee one more harvest in the fall and substantially improve the local people's standard of living. It is unclear whether the county government approved the request or not, but in any case, these polders likely were enclosed anyway.[73] Tianxinzhou polder, which was enclosed from the Dongjing River beach in Mianyang, collapsed in the 1931 flood but likely was not rebuilt due to a shortage of funds.[74] Knowing the Mianyang County government planned to reenclose this polder, the Hubei government sent an express telegram on February 2, 1962, to the government of Jingzhou Prefecture (which governed Mianyang at that time) and copied it to the government of Mianyang and ordered Mianyang to stop the enclosure.[75] The government of Mianyang nonetheless carried out the enclosure in 1963.[76] In other words, the county government simply ignored the order from the provincial government.

A 1964 survey turned up 133 large beach polders in Hubei, of which 70 were along the Yangzi River and 25 along the Han River. More than half had been enclosed after 1949, particularly after 1954, and most of them were enclosed without permission. Although the provincial government frequently issued orders to destroy them, most of them remained unaffected.[77] In Jingzhou Prefecture, one hundred polders were enclosed from Yangzi River beaches before 1975. In that year the Jingzhou government ordered that these polders be destroyed, but three years later, only three of them were demolished. At the same time, eighteen new polders had been enclosed. Also in Jingzhou, eleven beach polders on the Dongjing River were ordered to be destroyed in 1975, but by 1978, no action had been taken.[78] According to the standards of the Qing and the Republic, these polders were private polders and should have been destroyed. They still existed in the Mao era because most local governments were reluctant to take action.

Mainly because of the gradual silting-up of the Yangzi River beaches in Shishou, the local people planted cotton on these beaches and in normal years got a harvest. After several years they decided to enclose the beaches as polders, and by 1964 had built eleven of them. These polders, however, blocked

73 54-1-35b.

74 LS1-3-0236-005.

75 SZ34-5-210.

76 *Honghu xian zhi* 1992: 121.

77 SZ113-3-285-006.

78 00046-001-0144.

waterways and were ordered to be destroyed.[79] Some of them were in fact demolished, but more survived. By the 1980s there were still nine small polders on the water side of the river dikes, all of which had been reclaimed from river beaches from 1952 to 1974.[80] In other words, they were like the typical private polders in the Qing and the Republic, though some of them were reclaimed to eliminate *Oncomelania*.

Both the Jingzhou prefectural government and the local county governments, however, opposed the reclamation of these polders. In other words, they were reclaimed by the sub-county (township or people's commune) government, or even by brigades or production teams. This was precisely the case in Jiangling: all its fourteen beach polders, ranging from 80 mu to 1,400 mu, were enclosed by brigades or production teams, although both the Jiangling County government and the government of Jingzhou Prefecture had repeatedly issued orders to raze them. Some were in fact demolished, while some others were razed but then restored.[81] Several cases reveal that governments at various levels seemed to be unable to stop the building of polders. In Shishou, Fuxingzhou polder was reclaimed from an islet in the middle of the Yangzi River by a commune in 1975 without permission; the Jingzhou prefectural government ordered it to be destroyed, but it still existed until 1998, when it was flooded once again and the state ordered that it be abandoned and its residents relocated into the town of Tiaoguan.[82] Fanxing polder, in the same county, had been reclaimed, destroyed, and restored many times from 1924 to 1975. Each time the polder was destroyed, the locals simply rebuilt it.[83] In Songzi, though the governments at the county, prefectural, and provincial level all banned the enclosure of river beaches and islets, many new polders were nonetheless enclosed from river beaches/islets in the 1950s. In 1963, the Songzi County government sent a note to its sub-county governments to demolish polders enclosed from river beaches before early May, that is, before the coming of the summer high water. To help cadres persuade the villagers, the note suggested "leniency for the past," but for those who continued to enclose river beaches without permission, it called for "severe punishment."[84] It is clear from the note that the enclosure of all these beach polders was organized by townships

79 D35-1-412a.
80 *Shishou xian zhi* 1990: 133.
81 *Jiangling difang zhi* 1984: 93–94.
82 39-1-122-1; *Tiaoguan zhen zhi* 2001: 79, 175.
83 Wang Minfu 1989: 80.
84 33-1-80.

or even villages, but the tone of the note echoes documents from the Qing. The locals certainly would not readily obey.

The same situation occurred along the Han River. In Hanchuan in 1957, residents of a township who lived in poverty mainly because of a land shortage, repeatedly asked the county government to enclose a lake, but all their requests were rejected with the excuse that an enclosure would affect state policy and the people living on the opposite bank of the river. One resident was so upset that he went to the county government compound to submit a petition to the authorities. Likely because of public pressure the county government finally approved the enclosure of the lake. Also in Hanchuan in 1957, residents of another township petitioned to reclaim a polder that had been destroyed by the 1931 flood. Their petition was denied, but they reclaimed the polder anyway. After reaping a good harvest, they wanted to reclaim more, but the Bureau of Water Conservancy stopped them. Three hundred or more people from 114 households went to the county government twice to demonstrate. The county officials eventually gave in, and the land was enclosed.[85] In 1964 in Hanchuan, after several brigades enclosed beaches along the Han River, they were ordered to entirely raze the dikes they had built. It is unclear how the locals responded, but in Qianjiang in the same year, residents were loath to demolish nine polders enclosed from river beaches.[86] The upper-level government repeatedly ordered that all these polders be destroyed, but in vain. On the one hand, we see that the locals were assiduous about reclaiming river beaches, but on the other hand, we should also recognize that the upper-level government, intentionally or not, left room for doubt about its determination. For example, in 1975, the Hubei government issued an order that polders enclosed without authorization be "resolutely" destroyed. At the same time, it declared that "reclamation without enclosure" was permissible.[87] The tone of this order was harsh, but the principle of "reclamation without enclosure" was in fact a compromise and offered an opportunity for local governments to manipulate the order, just as in the Qing.

Not only river beaches, but also lakeshores and entire lakes were reclaimed. In 1964, the Fourth Production Team of Tanglin Village in Jingmen reclaimed 40 mu from a lake.[88] In Yingcheng, from 1959 to 1975, many lake beaches were reclaimed by communes or brigades along the lakes, which increased their

85 50-1-52ab.

86 D35-1-412bc.

87 42-1-169.

88 *Tanglin cun zhi* 2008: 76, n1.

farmland by 45,784 mu.[89] In Qianjiang, from the 1950s to the 1960s about sixty lakes disappeared due to reclamation, adding 380,000 mu of farmland. Reportedly, this contributed greatly to feeding and clothing a growing population.[90] In Honghu, from the end of the 1950s to the end of the 1970s, 262,500 mu of land were reclaimed from Lake Hong by production teams along the lake.[91] Most of the 177 polders in Jingmen in 1985 had been newly reclaimed after 1949, particularly in the 1960s–1970s. Eighty of these polders were less than 100 mu, with the smallest only 15 mu, a clear indication that they were enclosed by villages or production teams.[92] As late as 2006, in the lake region of Shayang, there were still eighty-nine polders larger than 100 mu, most of which had been reclaimed from lakes in the 1960s–1970s by production teams.[93] On the Jianghan Plain, 150 lakes with a surface area of 5,000 or more mu disappeared between the 1950s to the 1980s.[94] Some people in Jiangling even reclaimed reservoirs, despite it being illegal, dangerous, and absolutely prohibited.[95]

In short, although the early PRC state tightly controlled rural society, some peasants still turned a deaf ear to the government's orders and continued to reclaim river beaches without permission. It would be unfair, therefore, to blame the local government alone for the over-reclamation of river beaches in the early PRC. In fact, most governments at the county level and above were opposed to this kind of reclamation, which was for the most part organized by the communes, brigades, and even production teams. Yet, since China's population almost doubled from 1949 to 1979, and since agriculture was the backbone of the economy at the time, more farmland meant more people could be supported. Facing mounting population pressure, the PRC government at all levels had little choice but to give in.

6 Changes in the Reform Era

Up to the early 1980s, the over-reclamation of river beaches was still a problem. In 1981 it was found that thirty-one polders with an acreage of 73,930 mu

89 *Yingcheng xian zhi* 1992: 230.
90 *Qianjiang shuili zhi* 1997: 64–66.
91 *Honghu xian zhi* 1992: 98.
92 *Jingmeng shi shuili zhi* 1989: 202.
93 *Shayang xian zhi* 2013: 324–27.
94 *Hubei shuili zhi* 2000: 137–48.
95 *Jiangling xian shuili zhi* 1984: 103.

were enclosed from beaches in the Jingzhou section of the Han River.[96] In 1983, there were 132 such polders in Jingzhou alone. Although the provincial government ordered that 79 of them be completely demolished, and the remaining 53 to be opened for the discharge of floodwater, the reality was that only eight had been completely demolished by May, with all the rest either untouched or opened but very soon reblocked. The 35 beach polders of the Han River dikes occupied almost half of the river's beaches and blocked waterways, which seriously threatened the integrity of the dikes.[97]

It was only after the 1979 rural reform that the enclosure of private polders started to slow down and eventually came to a stop. A key component of the reform involved the central government calling on local governments to adjust their agricultural structure and give individual households the right to grow whatever they wanted, and in fact encourage peasants to develop the rural economy according to local conditions.[98] As a result, more and more lowlands were returned to lakes, fishponds, and lotus, and a growing number of villagers turned from farming to aquaculture. An example is Wagouzi Village, adjacent to Lake Hong, where there were only about 200 mu of farmland in 1949, but that increased to 1,510 mu by 1970. After 1978, however, villagers gradually returned the land to the water—there was almost no farmland in the village by 1993, and most villagers had become aquaculturists. The result was that they enjoyed a skyrocketing income since in recent decades the commercial value of aquatic products has been generally far greater than that of crops such as rice.[99] In the county of Honghu, 120,000 mu of farmland were returned to lakes from 1979 to 1992.[100] Returning land to lakes occurred almost everywhere across the region: Gong'an returned 30,000-plus mu of land to lakes in the 1980s, Jianli returned 63,399 mu from 1982 to 1985, Wuchang returned 66,899 mu from 1979 to 1985, and Hanchuan returned 152,646 mu to lakes or fishponds from 1983 to 2005.[101] From the 1980s to 2000, in all of Hubei about 1.8 million mu of land were returned to lakes.[102] After the 1998 flood, between the years 1999 and 2012, Hubei abandoned 320 polders, which provided 1,379 square kilometers (or 2,068,500 mu) of land to discharge floodwater.[103]

96 *Hanjiang difang zhi* 2015: 186.

97 40-1-532ab.

98 *Wuchang xian zhi* 1989: 289.

99 *Wagouzi zhi* 2013: 24, 93–94.

100 *Honghu xian zhi* 1992: 98.

101 *Gong'an xian zhi* 1990: 172; *Jianlin xian zhi* 2013: 153; *Wuchang xian zhi* 1989: 276; *Hanchuan shi shuichan zhi* 2010: 175.

102 *Hubei shuili zhi* 2000: 734.

103 *Hubei sheng hupo zhi 1* 2014: 19.

This is understandable from a historical perspective. Based on the level of agricultural productivity in the early PRC, the reclamation of wasteland was one of the most obvious ways to increase agricultural output, particularly in the 1960s–1970s, when the population surged. The desire of peasants in a pre-industrial society for land was also a powerful force driving them to reclaim riverside and lakeshore wastelands, which had continued for centuries on the Jianghan Plain and was not something that occurred only in the collective era. The difference is that in the early PRC, the nationalization of rivers and lakes, the completion of many water conservancy projects, and the power of the local administrative units all made the reclamation of river beaches and lakeshore easier than before, and thus far more river beaches and lakes were reclaimed.

It is against this macro background that an increasing amount of low-lying land on the Jianghan Plain has been returned to lakes or turned into fishponds. Although beach polders reduced population pressure and contributed to the local economy, they blocked waterways and raised the water level of the Yangzi River, which not only threatened the integrity of nearby river dikes but also consumed a huge amount of manpower and materials for high-water control. Thus, after the 1998 flood, the state ordered that many beach polders be either razed or abandoned. In Jingzhou, from 1999 to 2005, 138 beach polders were removed or abandoned, and more than two hundred thousand people were relocated to the land side of river dikes.[104] At the same time, a great deal of land was returned to water. In Jingzhou alone, 10,000 hectares (or 150,000 mu) were returned to rivers.[105] Recent research has found that both the number of lakes on the Jianghan Plain and their surface area are on the rise: slowly from 1978 to 1998, and drastically from 1995 to 2010.[106]

7 Conclusion

Corruption among officials, which past scholarship has often blamed for the failure of the ban on private polders in the Qing and the Republic, was largely restricted in the early PRC. Nonetheless, the enclosure of private polders continued. This was because China still faced the age-old problem of how to feed a growing population. In a pre-industrial society, the best way may have been to reclaim land. But the PRC state in its early years officially banned many polder enclosures. Nonetheless, the prohibition ran up against the unremitting need to

104 *Jingzhou shi zhi* 2015: 286–87.

105 *Jingjiang difang zhi* 2012: 163.

106 Fang, Jingyun, Sheng Rao, and Shuqing Zhao 2005; Wang Wenli et al. 2012.

feed a burgeoning population. The clash between the dictates of the state and the demand for survival shaped the actions not only of the peasants but also of government officials. Objectively speaking, reclamation in many instances was reasonable and therefore was supported, overtly or covertly, by lower-level government and tolerated by high-level government. Thus, the enclosure of private polders continued from the early Qing all the way to the end of the twentieth century. This had nothing to do with changes in the China's political system, and everything to do with the unchanged demand for food. But the ways the people enclosed private polders and how the state handled the issue varied from era to era.

As discussed earlier, although in the Qing and the Republic the enclosure of private polders was a spontaneous act led by clan leaders or other local powerful forces, the reclamation of river beaches and lakeshores on a large-scale in the 1960s–1970s was organized by administrative units, from production teams to county governments, and was no longer an individual action. Compared to the clans and local powerholders of earlier eras, these units had far more power to mobilize people to reclaim land. What was unchanged was that population pressure drove reclamation.

The Qing bureaucrats who opposed the enclosure of private polders thought these polders blocked waterways, a view shared by modern-day environmentalists. But not all imperial officials rigidly enforced the ban on private polders; some officials at the local levels in fact supported the enclosure of private polders. And in the Republic, some hydrological experts also strongly supported the reclamation of unoccupied land.[107] To reiterate, reclamation in the early PRC was often a state action led by local government. In the Qing and the Republic, bureaucrats had to take "national revenue and the people's livelihood" into account, whether they enforced the ban or covertly opposed it. The official rationale for reclamation in the early PRC also included the development of agriculture and the reduction of the threat of snail fever.

The official policy of the Qing since the mid-eighteenth century was to ban private polders entirely, but the Qing state constantly compromised, yielded, and gave in. Bureaucrats at the provincial level and above wanted to ban private polders, and prefectural officials usually supported the ban, but the evidence suggests that most county magistrates did not want to implement it. The local people, however, ignored the ban and constantly reclaimed and enclosed private polders. The policy of the early PRC also prohibited reclaiming river

107 Sun Fushi 1939: 72–73.

beaches and lakeshores, but many were reclaimed nonetheless, with the reclamation reaching a peak in the 1960s–1970s.

To be sure, the reclamation (particularly, the over-reclamation) of river beaches and lakes had side effects, including a decrease in the number and size of floodwater diversion zones, an increase of pressure on dikes, and the alteration of the local environment. However, China in the 1950s–1970s was still mostly a rural society, and the continual reclamation of wasteland was a primary way of increasing agricultural output. Without reclamation, who could feed a huge and fast-growing population, and how? In the early 1950s, the PRC state was able to ban such unhealthy social customs as gambling, prostitution, and opium smoking almost overnight, but once it came to the ban on the reclamation of river beaches and lakeshore, it had to face the facts and yield to population pressure. To maintain all the lakes and rivers in their original condition in the 1950s–1970s was simply unrealistic.

Once the economic situation changed, the peasants lost their enthusiasm for reclamation. The growth of China's population has largely been slowed due to the strict implement of the one-child policy from the early 1980s to the early 2010s—and even after the one-child policy was lifted in 2015, there had been no large population increase. At the same time, the steady growth of non-farming sectors siphoned a large portion of villagers to urban and/or coastal areas. Since these one-time villagers are no longer interested in farming, the original force driving reclamation disappeared. Meanwhile, the state has gradually withdrawn from rural governance, and it has become very difficult, if not impossible, for local governments to mobilize villagers to work on public projects, such as land reclamation, that require a great number of laborers. On the other hand, the focus of local governments and their cadres has also shifted from agriculture in the Mao era to non-farming sectors in the post-Mao era. Their struggle now is to draw in investment to boost GDP growth and thus they too are no longer interested in land reclamation.

At the same time, the rise of living standards has also boosted the people's demand for high-value items such as aquatic products, and aquaculture provides a higher income than farming. Therefore, to produce aquatic products, people in many places have returned farmland to lakes. Since the 1980s, the province of Hubei has returned millions of acres of farmland to lakes. After the 1998 flood, Hubei abandoned many beach polders and returned a great deal of low-lying land to the waters. The result is clear: both the number and water surface area of lakes in the Jianghan Plain are on the rise. We can optimistically conjecture that the reclamation of lakes and river beaches on the Jianghan Plain, a phenomenon that emerged in the Song dynasty, spread in the Ming-Qing era, and peaked in the 1960s–1970s, has now ended.

CHAPTER 8

Armed Forces, Military Affairs, and Dikes

The use of dikes has not been limited to water control. Historically, dikes have also served as defensive works. Thus, armed forces were widely involved in dike affairs, and war had a profound impact on dikes. Past dike-related studies, however, focused on water control, and few discussed the relationship between dikes and military affairs, or only briefly mentioned the subject but did not explore it at length. Even in a recent study of the "ecology of war," which analyzed the background and effects of the 1938 Yellow River flood—deliberately unleashed by the GMD by destroying the dike at Huayuankou to slow the advance of the Japanese army, the so-called Huayuankou Incident—dikes were not the focus.[1] In other words, the military function of dikes and their relationship with military affairs have long been overlooked. This is also true of studies of the Jianghan Plain. But the dike systems of the Jianghan Plain over the past three centuries in fact have served very important military functions and have had a close relationship with military affairs, particularly in the twentieth century.

1 The Military Function of the Dike Systems

Throughout Chinese history dikes have been frequently used in various ways in war. In the year 23, a subject of Liu Xiu, the founder of the Eastern Han dynasty, suggested that the Yellow River dike be broken to flood the rebellious Chimei peasant army, which "could turn a million people into fish," but Liu refused.[2] It is likely that he did not want to take such a brutal step to attack the enemy. Over sixteen hundred years later, in 1642 some government officials again planned to unleash the Yellow River to decimate rebellious peasants who had besieged the city of Kaifeng for a long time. Forewarned, the rebels moved to a high location, and then they themselves broke the Yellow River dike and let floodwater pour into Kaifeng, drowning hundreds of thousands of people inside the city.[3] But one of the most infamous and callous cases was the 1938 Huayuankou Incident. Not only did this man-made flood fail to stop the

1 Muscolino 2015.
2 Zhang Fang 2007: 43.
3 Zhou Kuiyi et al. 1990: 384–86.

 | DOI:10.1163/9789004739482_010

Japanese army but also caused the deaths of an estimated 890,000 innocent Chinese civilians. It also brought disaster to the inundated areas for years.[4]

Although all these cases had to do with the Yellow River dikes, the Jianghan dikes had also been used for the same purpose. In 1129, the magistrate of Jiangling broke the Yangzi River dike in an effort to resist the invading Jurchen army.[5] About eight hundred years later, in 1921 a warlord opened the Yangzi River dike at the town of Paizhou in Jiayu, drowning not only many enemy soldiers but local civilians as well.[6] In summer 1931, the Yangzi River high-water rose suddenly due to torrential rains, putting the segment of the Yangzi River dike at Shangchewan, in Jianli, in imminent danger. To flood the CCP base area in Honghu (in Mianyang) that was protected by this dike, the GMD forces intentionally opened the dike on July 27. The ensuing flood engulfed 95 percent of Mianyang. About thirty thousand people were drowned, up to ten thousand died of the cholera that followed, and four hundred thousand or more had to take refuge on dikes and platforms. In total, more than a million people were victims of this man-made flood.[7] Yet the flood failed to destroy the Honghu Base Area, which continued to exist for several years and became one of the most famous Communist base areas, and was even lauded by Mao.[8]

After the invading Japanese occupied most of the Jianghan Plain, the Bureau of Engineering Work on the Yangzi and Han Rivers was relocated to the province of Sichuan, as were many other government organs. A technician in the bureau argued that it should stay at Jingzhou to take care of the Great Jingjiang Dike, as the area was not entirely lost to the Japanese. His superior in the bureau responded by saying that that was true, but many areas protected by the dike had been lost to the Japanese, and he even wished to break the dike to thwart the invaders.[9] That is to say, he may have wished to destroy the dike, but he did not do so. About a decade later, some other people also wanted to blow up the dike. In 1949, when the defeated GMD troops had withdrawn from the Jianghan Plain to the upper reaches of the Yangzi River, they buried dynamite in the Great Jingjiang Dike and planned to explode it to stop the pursuing PLA troops. Luckily, the commander of the GMD troops in Jiangzhou—who had contacted the PLA earlier—refused to carry out the order and rebelled, and thus the dike was saved.[10]

---

4 Muscolino 2015; Edgerton-Tarpley 2016.

5 *Xiantao shuili zhi* 2008: 62.

6 *Wuchang xian zhi* 1989: 13.

7 *Mianyang xian zhi* 1989: 13; *Xiliuhe qu zhi* 1986: 9; *Jingzhou diqu shuiyun zhi* 1989: 229.

8 *Honghu xian zhi* 1992: 14.

9 Chen Zhangguan 1982.

10 Zhou Shangfan 1984; Li Guozhang 1986.

Dikes were built to protect the people from flooding, but they also could be used to protect the people from the enemy. Before the twentieth century, however, the use of cold weapons and the reliance on horses for transportation largely limited the defensive potential of Jianghan's dikes. That changed greatly in the 1930s and 1940s, particularly during the Anti-Japanese War. In the struggle with the invading Japanese, the GMD troops began to dig trenches on dikes. In 1938 in Hanyang, they dug six foxholes (2 × 2 × 1.8 meters) on a dike and two trenches (15 × 0.8 × 1.5 meters) nearby a dike in one place, and 32 foxholes (1.5 × 1.5 × 1 meters) on a dike and one trench (15 × 0.6 × 1.5 meters) nearby a dike in another place, but they were soon ordered to fill them up.[11] Around 1940, with the advance of the Japanese, the Jianghan area became a war zone, and 5,099 defensive works were built on the Great Jingjiang Dike—causing the dike to break in many places—and on one segment of the dike there were 77 machine gun nests and 330 foxholes.[12] In Shishou, to hinder the enemy's movement, the Nationalist troops built many mounds (one every 90 zhang) on the dike along the southern bank of the Yangzi River in 1939, then in 1943 they dug several hundred S-shaped air-raid shelters on the dikes along the Ouchi River and the Tuanshan River.[13] In Honghu, the Nationalists' 128th Division, which was largely independent, also dug trenches and built many blockhouses on dikes in the area it controlled.[14] In the fight against the Japanese, some polder dikes in Songzi were almost completely destroyed.[15]

After 1938, most areas on the Jianghan Plain protected by dikes either fell to the Japanese or became the front line. Furthermore, the Japanese also built many defensive works on dikes. Along the Han River, they not only bombed some dikes but also built blockhouses and trenches on the dikes.[16] They also built a dense network of defensive works, including trenches and blockhouses, on the Dongjing River dikes.[17] In 1943 they built numerous defensive bunkers (each 3 × 2.33 × 1.5 meters) on the Yangzi River dikes in Shishou, and in one place they even dug out the whole body of a dike to build a bunker that could hold thirty people.[18] From 1943 to 1945, they also damaged 23 dikes and destroyed 285 kilometers of dikes to build defensive works in Songzi.[19]

---

11 LS36-1-75.

12 *Jianli difang zhi* 1991: 74; *Jianli xian zhi* 1994: 151.

13 *Shishou xian zhi* 1990: 476.

14 *Honghu xian zhi* 1963: 121.

15 100-2-364: 41.

16 *Shayang xian zhi* 2013: 321.

17 *Dongjinghe difang zhi* 1994: 271.

18 *Shishou xian zhi* 1990: 476–77.

19 *Songzi xian minzheng zhi* 1985: 72.

In 1945, before their surrender, the Japanese dug trenches on a segment of the Yangzi River dike in Gong'an, which damaged the dike and caused it and some polder dikes to rupture in twenty-four places after the high-water rose. This led to 400,000 or more mu of land being flooded, and around 40,000 deaths caused by the floods and the ensuing diseases.[20] Although this death toll is smaller than that of the Huayuankou Incident, this time it was the Japanese who destroyed dikes and caused an inundation. A rupture of the Han River dike in 1944 was also caused by the fortifications the Japanese built on the dike.[21] The Japanese and the GMD troops built so many defensive works on Jianghan's dikes that on some segments of the Great Jingjiang Dike there was one such installation every 10 to 20 meters.[22] After the war, the Bureau of Engineering Work on the Yangzi and Han Rivers undertook many projects to fill up those defensive works. In Shishou on the dike of Luocheng polder alone, it filled up eleven heavy machine-gun dugouts, two light machine-gun dugouts, 156 foxholes, and many transportation trenches.[23] There were so many defensive works on Jianghan's dikes that, after the war, it took two years for the local government to remove them.[24]

Even so, before the GMD forces lost Wuhan to the CCP in 1949, they also built numerous defensive works on river dikes, and in some places they even broke the dikes.[25] A blockhouse group built in 1948 by the GMD troops on a Wuhan dike included a major blockhouse, three sub-blockhouses, and a lounge. The major blockhouse was able to hold more than thirty people, the lounge accommodated more than twenty people, and each sub-blockhouse could hold about ten people.[26] As late as 1952, dikes still ruptured because of hidden damage left by the Japanese, who once built blockhouses and dug trenches on them.[27] And up to 1952 the new government had used a huge amount of dirt to fill up the military works left on Jianghan's dikes.[28]

The defensive function of the Jianghan dikes continued in the PRC. A 1966 report suggested that, as a measure for war preparedness, the central government should take direct responsibility for the reinforcement of the Great Jingjiang Dike. This was because, the report claimed, the dike was too weak

20 *Gong'an xian zhi* 1990: 445.
21 40-1-195.
22 *Shashi shuili difang zhi* 1994: 57.
23 LS23-3-0866-001.
24 *Hubei shuili zhi* 2000: 530.
25 *Wuhan difang zhi* 1986: 63.
26 XX000084-WS01-0089-0004.
27 *Huchang qu zhi* 1987: 6.
28 SZ34-2-414.

to survive an attack by even small bombs (500 lbs. or less). If the dike collapsed, the Yangzi River could change its course, which would cause a huge loss of life and property.[29] For purposes of war preparedness and high-water control, some went further in 1973 by suggesting the construction of another dike behind the important segments of the Great Jingjiang Dike.[30] This suggestion was rejected, but the reinforcement of the dike was included in the national plan/budget of capital construction in 1974. Hence, the central government allocated funds for its annual repairs. This was a milestone in the management of the Great Jingjiang Dike, which became a national dike afterward. In the same year, the management of dikes in downtown Wuhan was also transferred to the central government, thus the dikes were dubbed "national embankments" (*guodi*).[31] At the county level, the local government also worried about the safety of dikes if they were attacked by the enemy and hence proposed strengthening them.[32]

At the same time, however, war preparedness carried to a fanatic level also damaged Jianghan's dikes. In 1969, each household was required to dig an air-raid shelter or dugout, but in Wuhan many air-raid shelters were built along dikes, which was very dangerous, and some were ordered to be quickly filled up.[33] On April 28, 1970, it was found in Wuchang that 229 air-raid shelters, each about two meters deep and encompassing an area of 5–10 square meters, had already been filled up. The rest were ordered to be filled up within three days.[34] Similarly, in 1970 in Jianli the more than two hundred air-raid shelters along five kilometers of the Great Jingjiang Dike were ordered to be filled up.[35] In total there were 1,608 air-raid shelters built along the Great Jingjiang Dike. One in Shashi (6 × 1.8 × 2 meters) endangered the dike; another and larger (14 × 1.5 × 5 meters) air-raid shelter in Shishou was even built on the body of the dike, creating a far greater danger.[36] According to incomplete statistics, in 1970 in Jingzhou alone, 6,890 air-raid shelters had been dug along the dikes of the Yangzi, the Han, and the Dongjing Rivers.[37]

When the historian Pierre-Etienne Will toured the Han River dikes around 1986, he was told that the units responsible for the management of the Han

29 SZ34-5-708.

30 SZ113-6-150.

31 *Jianli xian zhi* 1994: 157; *Hongshan qu zhi* 2009: 184.

32 39-1-95-2.

33 XX000065-WS01-1691-0021.

34 Wuchang 49-2-44a.

35 *Jianli difang zhi* 1991: 339.

36 Wuchang 49-2-44b.

37 40-1-10-1.

River dikes in Shayang had remained stable and independent during the preceding thirty years of political chaos.[38] While that may have been true in Shayang, it did not reflect the situation in other areas, particularly during the Cultural Revolution. A Jianli gazetteer claimed that water conservancy construction in the county was not influenced much by the Cultural Revolution; but another gazetteer—focusing on the dike systems—recorded many negative effects of the Cultural Revolution on dike management in Jianli, including the paralysis of the unit responsible for the dikes.[39] Moreover, archival materials show that people in Jianli were required to demonstrate their loyalty to Mao and to use "Mao Zedong Thought" in the fight against the annual high water.[40] Early in the Cultural Revolution, the Jianghan dikes also suffered from clashes among different Red Guard groups. Because of the chaos, the Hubei government in June 1968 underlined the crucial role played by the PLA in high-water control and called for the harsh punishment of "class enemies" who buried mines on dikes, dug trenches on dikes, farmed dikes, chopped down breakwater trees, and stole telephone poles.[41] On July 8, 1968, some Red Guards in Wuchang were blamed for beating dike keepers, taking high-water control materials, and occupying high-water control material warehouses and using them as their base camps. They were ordered to move out of the warehouses without delay.[42] It has been argued that the rupture of the Yangzi River dike at Tianjiakou in Honghu in 1969 was also caused by the "anarchistic thought" during the Cultural Revolution, or that the rupture was a "tragedy caused by the domestic turmoil of the Cultural Revolution."[43]

2 Dike Affairs as Military Affairs

Since dikes were so important and had been frequently used for military purposes, it is not surprising that they would be managed in a military fashion. In fact, in the Qing the harnessing of rivers was treated in the same way as commanding an army. First, in both cases good preparations were essential; otherwise, one would be at a loss as to what to do once something unexpected occurred. Second, some dikes functioned as city walls, and those who watched

---

38 Will 2006: 801.

39 *Jianli xian zhi* 1994: 154; *Jianli shuili zhi* 2005: 388.

40 20-1-053-005.

41 SZ139-6-0028-002.

42 Wuchang 49-2-42.

43 *Hubei shuili zhi* 2000: 15; *Jingjiang difang zhi* 2012: 237.

over those dikes performed some of the same functions as soldiers guarding walled cities. Such men should be capable, since simply grabbing people off street and impressing them would lead to defeat in battle. Third, as in commanding a military campaign, in harnessing rivers and managing dikes there should be clear, strict, and precise regulations.[44]

Since the situation in on-site emergency repairs during high-water control is fast changing and anything can happen within seconds, those in charge should be capable of making quick decisions and should have the authority to allocate any available personnel and to requisition any needed materials, just like military commanders in battle. This authority was guaranteed by law in the Republic,[45] and it was not something that merely existed on paper. In the 1931 flood, for example, in two cases of emergency repairs of the Great Jingjiang Dike, the cargo of a passing caravan and boats was instantly requisitioned to block openings.[46] In another example, during the 1947 high-water control, the governor of Hubei issued an order making it clear that, in high-water control, military law was to be applied—the on-the-site leaders were even authorized to immediately execute anyone who slacked off.[47] This is probably the harshest order regarding high-water control in modern times, but it sprang from a long tradition. In the Song dynasty, a high-ranking official in charging of high-water control angered by his clerks' tardiness and hesitation to press forward, warned them he would kill anyone who dared not follow him. Trembling, they all went to the worksite.[48]

This kind of thinking continued in the PRC. In Qianjiang in 1950, dikes were regarded as lines of defense, and if they were not solid, just like military bulwarks, could be breached by the enemy.[49] In Hanchuan in 1954, dikes were regarded as positions, the masses as soldiers, and materials as ammunition.[50] The same was true in 1957 in Qianjiang, where high-water control materials were considered as essential as ammunition to an army.[51]

The militarization of dike management thus continued in the PRC. The influence on dike management of the two decades of violent struggle for power between the CCP and the GMD is still evident today. In 1951, to collect information for high-water control, the Hubei government declared that "the control

44 Yu Changlie 1999 [1840]: 126; Ni Wenwi 2002 [1885]: 135; Shu Hui 2002 [1896]: 443.

45 100-2-52: 101.

46 Hubei sheng shuiliting and Hubei sheng dang'anju 1999: 398–402.

47 *Xiantao shuili zhi* 2008: 91.

48 Zhang Fang 2007: 240.

49 40-1-10b.

50 50-1-18.

51 40-1-148b.

of high-water is the same as fighting a war; it is difficult to defeat high-water without a sufficient information network," and thus the government considered it necessary to build many stations to record and report the high-water situation.[52] In the same year, laborers who participated in high-water control in Jingzhou were called "high-water control troops," though they were just normal peasants.[53] And high-water control in Jiangling was regarded as "a glorious political task, a sharp class struggle," since class enemies were supposedly still trying to sabotage dike work.[54] A 1952 report on Zhongxiang declared that there must be enough high-water control "weapons" (materials) and people must be vigilant in order to beat "the most vicious enemy—floods."[55] Two years later, the same tone appeared in a technological summary of the 1954 high-water control in Hubei, which stated that, for three months, a million "high-water control troops" fought, as if they were an army, "the most vicious enemy—floods."[56]

In 1952, the state launched a massive flood-control project—dubbed the Jingjiang flood diversion project—on the southern bank of the Jingzhou section of the Yangzi River. Among the three hundred thousand people who worked on this project were about a hundred thousand soldiers—an entire corps deployed by the central government. The balance consisted of forty thousand workers and a hundred and sixty thousand peasants. The soldiers were ordered to work on this project as if they were in a battle—that is, no different from building fortifications. They completed 48 percent of the earthwork in the first phase of this project.[57] The PLA also participated in high-water control (as discussed below).

Some of the peasants who worked on the Jingjiang flood diversion project were residents of Songzi and were led by the county head, Rao Mintai. When Rao mobilized the peasants on the worksite, he told them that he had received an "order" to do so, just as if he were leading soldiers into battle. By this he implied that he was acting under military orders, which could not be contested.[58] The management of laborers on dike worksites was also conducted in a military fashion. On the 1952 annual repairs worksite in Zhongxiang, for example, laborers were mustered to work and released for a rest break by the

52 SZ34-2-220.
53 40-1-14.
54 48-1-8.
55 54-1-5b.
56 Hubei sheng shuiliting and Hubei sheng dang'anju 1998: 4.
57 *Hubei sheng zhi junshi* 1996: 723; *Jingjiang fenhong gongcheng zhi* 2000: 97.
58 Jingzhou shi shuiliju et al. 2008: 269.

call of a bugle.[59] In high-water control, paramilitary management of laborers was even more common across the region. Indeed, laborers were widely organized through the militia system, and some were even inducted into militia high-water control and rescue commandos.[60]

Until recent decades in the PPC, laborers could be easily organized and managed in a military fashion because most of them were also militia members. Records show that in Hubei more than half a million participated in high-water control in 1954, 1 million in 1980, and 1.9 million in 1983. Militia members not only accounted for about 70 percent of all high-water control laborers in Hubei, but they usually also "took the most difficult, most dangerous, and most urgent tasks."[61] Rural militia were under the command of the Department of the People's Armed Forces (*Renwubu*, usually at the county level and below), and its leader always served as one of the major local leaders of high-water control, sometime as the most important leader. During the Cultural Revolution, the director of the Renwubu was made the top leader in high-water control since that job was too important to be swept up in the chaos of that time.[62] And the militia reportedly took water conservancy worksites to be as crucial as a battlefield.[63] Up to 1978, there were so many laborers who had to work away from home for months (in winter and spring) on water conservancy worksites in Hubei, that they were organized into military-like units: militia divisions, regiments, battalions, companies, platoons, and squads.[64]

Since 1979, and decades of emphasis on economic development, the influence of the military in the Chinese society has gradually faded, but not when it comes to dike affairs. Up to 1981, warnings were still being sounded that the public must be vigilant lest the "enemy" destroy the dikes.[65] At the turn of the twentieth-first century, in some places, such as Jingmen, high-water control laborers are still organized in a military fashion.[66] And to this day, the organizations responsible for high-water control are still called *zihuibu* (headquarters), a military term, as in "X zihuibu" ("such and such headquarters").

Another good example illustrating the paramilitary approach to dike affairs is the treatment of those who lost their lives in high-water control. In the 1788

59 54-1-5c.

60 *Hanchuan xian zhi* 1992: 162; *Zhongxiang shuili zhi* 1998: 189; *Tuoshi zhen zhi* 2008: 500; *Jianli xian zhi* 2013: 491.

61 *Hubei sheng zhi junshi* 1996: 632.

62 *Jianli difang zhi* 1991: 251.

63 42-1-158a.

64 *Hubei shuili zhi* 2000: 1208.

65 45-1-310b.

66 *Jingmen shi shuili zhi* 2012: 176.

flood, the emperor ordered that Lou Yeyao, the assistant prefect (*zhoutong*) of Jingzhou, who was swept away by floodwater as he was supervising the closing of the western gate of Jingzhou's city wall, be honored as if he had lost his life in battle.[67] This tradition continued into the PRC. A 1952 government report on Zhongxiang stated that people who lost their lives in high-water control would be treated as martyrs, and after the 1954 flood, the title of "high-water control martyr" was conferred on eighty people in Wuhan.[68] In other counties of the Jianghan Plain, people who lost their lives in high-water control were also honored with the title "revolutionary martyr" or "martyr," and were recorded in gazetteers, such as a man who jumped into a broken sluice gate to block it, and was engulfed by floodwater in Xiantao in 1991.[69] This is the highest honor for a soldier who has sacrificed his life, but these people did not die on the battlefield. This tradition has not changed after centuries, though the Qing and the PRC had very different ideologies and political systems.

3 The Impact of War on Dike Affairs

Throughout its long history, China frequently suffered from brutal civil wars and foreign invasions, most recently in "the century of humiliation" (stretching from the Opium War in 1839 to the Communist victory in 1949). War invariably affected local society and the economy, including the water conservancy system. Historically, the dikes on the Jianghan Plain were particularly subject to the ravages of war. The repercussions have been multifaceted, such as the use of dikes for defense discussed earlier in this chapter. Here I delve deeper into this subject.

In the long run, the most profound impact of war on dike affairs in the Jianghan area has been the opening of outlets. There were once many outlets along the Jing River to discharge floodwater to either side, but most of them had been gradually silted up by the mid-nineteenth century, with the exception of the Tiaoxian outlet and the Hudu outlet. The former was first opened by a northern general in his campaign to conquer the South in the year 280, and the latter was reopened by a general in the early Qing, also for the purpose of conquering the South. When the Qing state was reluctant or unable to block the breaches on the dikes in Ouchi and Songzi, partially due to its weak finances caused by frequent rebellions and foreign invasions, two additional

67 Ni Wenwei 2002 [1885]: 11.
68 54-1-5b; *Wuchang qu zhi* 2008: 172.
69 *Xiantao shi zhi* 2011: 976.

outlets were created: the Ouchi outlet in the 1850s and the Songzi outlet in the 1870s. The formation of all four of these outlets, which discharged floodwater from the Jing River into Lake Dongting, was related to war, directly or indirectly. By the end of the nineteenth century, these four southward outlets greatly changed the relationship between the Jing River and Lake Dongting, as well as almost the entire dike system in the Jing River area.[70]

The influence of war on high-water control and annual repairs is temporary and has been less profound than, for example, the formation of the four southward outlets. At the end of the Ming dynasty, as warfare swirled around the country, people chopped down trees along the dike of the Upper Baili Islet for fuel and the dike tithing system collapsed.[71] The Qing Empire enjoyed a period of relative domestic peace in the eighteenth century, but this was ended by the White Lotus Rebellion (1794–1804), and then half a century later, the massive and bloody Taiping Rebellion (1850–1864), which nearly overthrew the dynasty. During the Taiping Rebellion, some important repair projects of the Wancheng Dike were forced to a halt. More seriously, the Han River dike burst in Zhongxiang in 1852, and for several years thereafter the local people failed to block the rupture because of frequent rebel attacks. The ensuing floods caused widespread destitution and homelessness.[72] This in turn led to even more hardships for the people living along the lower reaches of the Han River. It also made it increasingly difficult for the locals to pay their national taxes and dike fees. For this reason, a special boat tax was levied in Zhongxiang County in 1858 to cover the maintenance of river dikes there. Later, even the likin (*lijin*, a commercial transit tax collected for military purposes) was used for annual repairs, a turning point in the source of dike funds in the region.[73]

The threat of the Taiping rebels also led to the locals organizing themselves for self-defense. Answering the call of the Qing state, the people of the mid-Yangzi River valley organized various militia units (*tuanlian*). In Hunan, they developed into the famous Xiang Army under the leadership of Zeng Guofan. On the Jianghan Plain in Hubei, tuanlian were usually organized by polders, some of which had several hundred militiamen (*tuanyong*). In total there were around three hundred thousand militiamen in Jingzhou and neighboring prefectures. They not only defended against the rebels but also maintained local social order by warding off bandits and thieves.[74] The tuanlian

70 *Tiaoguan zhen zhi* 2001: 10; *Hubei shuili zhi* 2000: 373, 382; Jiayan Zhang 2014: 23–27.

71 *Zhijiang xian zhi* 1866, vol. 3: 2a; *Jianli xian zhi* 1872, vol. 3: 10a.

72 *Zaixu xingshui jinjian* 2004: 349, 352.

73 *Zhongxiang shuili zhi* 1998: 4; Peng Yuxin and Zhang Jianmin1993: 218.

74 *Jingzhou fu Jiangling xian tuanlian qingce* n.d.

system continued even after the defeat of the Taiping rebels. In Tianmen, villagers were selected to undergo military training as members of the tuanlian. In 1877, tired of a dike fee surcharge levied by the clerks of the Bureau of Yuekou Dike Work, militiamen destroyed the bureau. A militia leader was captured by the magistrate, but the locals, weapons in hand, forced the magistrate to release him.[75]

The Republic was one of the most chaotic eras in modern Chinese history, with all manner of armed forces—warlord armies, the Nationalists, the Communists, the Japanese invaders and their puppet army, and others—fighting with each other as well as with numerous bandit gangs, all of which had an impact on dike affairs. In this respect, the Japanese invasion was particularly damaging. Aside from the threat to dikes caused by the construction of defensive works, as mentioned earlier, dikes also suffered from neglect as the locals fled the invaders or were drafted into the GMD army. People in inland China suffered from the invading Japanese even before they reached their area. In the 1931 flood, relief grain purchased in Manchuria could not be transported out because of the Japanese invasion. The invasion also disturbed the Nationalist government's relief efforts.[76] And the ten-month-long battle to defend Wuhan, the one-time de facto capital of wartime China, in 1938, displaced millions of civilians.[77]

After the fall of Wuhan in October 1938, most of the Jianghan area was also captured by the Japanese, with devastating repercussions. Annual dike repairs were impeded to the point that they were almost halted. In Xiaogan, for instance, many polders collapsed, and relief efforts were disrupted.[78] In Jiangling as well, almost no annual repairs were carried out during the Japanese occupation (1940–1945).[79] On the front lines, the Japanese destroyed not only sluice gates on some dikes but even the records of dike agencies. Many dike maintainers and local people were rendered destitute and homeless. Yet many also resisted as best they could. For example, in 1944 alone, Songzi contributed 1,393,039 units of labor (each unit consisting of one laborer working one time) to transport military supplies for GMD troops.[80] Yet, this had an undesirable side effect: it largely prevented these laborers from repairing their dikes. Even the attempt to maintain and repair the dikes could be fraught with danger.

75 *Tianmen shuili zhi* 1999: 331.
76 National Flood Relief Commission 1933: 18, 87.
77 On the symbolic significance and long-lasting influence of this battle, see MacKinnon 2008.
78 *Xiaogan shi zhi 1* 1992: 210.
79 *Jiangling difang zhi* 1984: 146.
80 100-1-182: 1, 11; 100-8-87: 4; 100-8-93: 30; 100-8-97: 18; *Songzi xian minzheng zhi* 1985: 21.

In one case, en route to a dike worksite, the director of the annual repairs and high-water control agency of Datong polder, a dike manager, and a dike security guard encountered more than thirty Japanese soldiers and were captured. Once the enemy realized that they served in a dike agency, they were threatened with execution if they did not pay a ransom of 300,000 yuan. Through negotiations via an interpreter, the amount was reduced to 120,000 yuan, and the captives were released six days later after they paid 150,000 yuan.[81] They were lucky they survived. In other cases, the Communist armed forces built dikes together with villagers to protect them from the Japanese and their puppet armies, and some sacrificed their lives.[82] Even under the Japanese occupation, there were people who struggled to protect the dikes, such as a certain Huang, the head of a local administrative unit in Shishou, who repaired dikes destroyed by military engineering projects and distributed relief to the locals. After the war, Huang was accused of collaborating with the Japanese, but the local people, out of gratitude, pleaded that he be forgiven.[83]

Before the Japanese invaded central China, the CCP and the GMD had already been fighting for years. Here we turn to their complicated conflicts on the Jianghan Plain, particularly in the Honghu Base Area. In summer 1931, the Yangzi River reached a record high level in the Jianghan area, and the local Soviet government in the Honghu Base Area did its best to prevent flooding, but the GMD troops opened the Yangzi River dike in Jianli and flooded most of the base area. Many other dikes also ruptured. To overcome this hardship and to reinforce the base area, but also to help the locals recover from the inundation, CCP guerrillas attacked Paizhou, seized grain there, and distributed it to flood refugees. They also helped the locals develop fishing to tide them over the famine. The local Soviet government even issued water conservancy bonds to raise money and organized the locals to repair the dikes along the Dongjing River. In Qianjiang, under the leadership of He Long, a legendary PLA marshal, the Red Army and twenty thousand local laborers together blocked an outlet of the Dongjing River, which eventually ended the long history of inundation caused by this outlet.[84]

Early in 1932, the local Soviet government considered dike repairs to be its first priority. It called on the people living in the GMD-controlled area to repair their dikes but at the same time resist the GMD's embezzling of dike

81 100-1-182: 53.

82 *Xiaogan diqu shuili zhi* 1996: 45.

83 LS7-3-00003675-069.

84 *Jingzhou diqu shuiyun zhi* 1989: 229; *Yanglinwei qu zhi* 1989: 8; *Xiliuhe qu zhi* 1986: 10; *Qianjiang shuili zhi* 1997: 3.

fees. It even suggested that people living along the CCP-GMD border call on the GMD troops for help in building dikes. The GMD authorities in fact cooperated with the local Soviet government in the Honghu Base Area in repairing its dikes, one of the few mutually beneficial cooperative efforts in their long history of enmity. In February, the CCP and the GMD agreed that the Communist forces would rebuild an important segment of the Yangzi River dike inside the Communist base area, and the GMD authorities would check (and accept) the project and then provide relief grain. The project was completed in May. This unusual instance of cooperation was even reported by *Shenbao* (Shanghai News), a Shanghai-based and very influential national newspaper, which noted that in the Communist area the leadership was unified, the action was swift, the relief wheat was fairly distributed, and there was no corruption in dike work.[85] That seems to imply that dike safety was so important that neither side could take it lightly. But ironically, it was the GMD forces that had intentionally broken the Yangzi River dike to flood the Honghu Base Area. After that the local GMD authorities cooperated with the Red Army to rebuild the dike.

The Red Army also led the locals to repair other dikes, such as a segment along the Han River in Tianmen. In 1932, it also organized the opening of what was locally known as the Red Army Channel, a sluiceway that benefited about ten thousand mu of nearby land. In 1943, the local Communist government built dikes for thirteen polders, which could protect 23,000 mu of land.[86] In Shishou in 1944, the local Communist Party committee led people to build dikes and enclose polders, which restored several polders that had been inundated for thirteen successive years.[87] In Xiaogan in the same year, the Communists also mobilized people who lived near the area controlled by the Japanese and their puppet troops to build dikes. To protect these peasants, armed Communist personnel were sent to work together with them.[88]

In all these cases, the sources portray the Communist forces positively. This is no surprise since all the sources (except for *Shenbao*) are from newly compiled gazetteers, which naturally favored the Communists. The Nationalist sources, however, paint a very different picture. These materials usually labeled Communists "bandits" and condemned them for harming the dike systems in many ways. The Nationalist sources frequently mention that dike fees were difficult to collect because of the activities of the "communist bandits"; that their military actions also made it difficult for the locals to carry out annual

---

85 *Jianli difang zhi* 1991: 238; *Honghu xian zhi* 1992: 12, 119.

86 *Tianmen xian zhi* 1989: 20, 24.

87 *Shishou xian zhi* 1990: 22.

88 *Xiaogan diqu shuili zhi* 1996: 45.

repairs and high-water control, thus leading to dike ruptures; that they burned account books or records of dike fees (this is likely an excuse to conceal corruption or theft). In fact, it was not unusual for the GMD to embezzle dike fees to fund military actions against the Communists.[89] Furthermore, the most powerful organization in Hubei in the early 1930s was the General Headquarters for Bandit Suppression in Hubei, Henan, and Anhui (E-Yu-Wan sansheng jiaofei zong silingbu). Both the government of Hubei and the Bureau of Engineering Work on the Yangzi and the Han Rivers were under its command.[90] The "bandit" in its name referred to the Communists.

However, not all PRC literature portrays the Nationalists negatively. A new gazetteer of a township of Shayang, for example, recorded a rare case in which the commander of a locally stationed GMD regiment forced local powerful people who had never personally done any dike work to go to the dike worksites to labor, and even doubled their originally assigned workload, while the poor people received a payment worth two liters of husked rice, funded by donations from the regiment. The project in the case was completed within twenty days, even though the area was at the risk of Japanese air raids.[91]

After the defeat of the Japanese in 1945, the CCP and the GMD continued to battle each other for another four years, which also impaired high-water control of the Jianghan dikes. In 1948, after an emergency on a segment of the Han River dike in Hanchuan, many people were on the site conducting repairs, presenting a great opportunity for the local GMD authorities to grab *zhuangding* (able-bodied men, subject to conscription). The youth on the site fled, but the older men alone were unable to affect the repair, and the dike collapsed,[92] a typical man-made disaster. On October 2 of the same year, as the water almost reached the top of the Han River dike in Shayang, a company of PLA soldiers, who were battling the GMD forces in downstream Tianmen and Qianjiang, were sent to Shayang to assist. But the GMD troops blocked the road, preventing the company from reaching the site. Since the local GMD authorities had collected dike fees but did not do much more than that, the dike broke in the morning of October 5, washing away a third of Shayang's houses and killing three to four hundred people.[93]

In the Republic, China suffered from continuous chaos: first, the tangled warfare among warlords, then the armed conflict between the CCP and the

89 *Jingjiang di zhi* 1937, vol. 2: 71–72; LS1-3-0236-005; LS19-4-6552.

90 LS1-5-614.

91 *Maliang zhen zhi* 2013: 287.

92 *Hanchuan shuili zhi* 2013: 191.

93 *Shayang xian zhi* 2013: 309–10.

GMD, followed by the Anti-Japanese War, and finally the Civil War between the GMD and the CCP. During all this time, most administrative positions were held by military personnel. Their main objective was to occupy as much territory as possible and to collect ever more tax revenue. Although they also recognized the importance of dikes and carried out some annual repairs and high-water control when that was possible, they clearly assigned a much higher priority to winning and holding territory and to tax collection than to protecting the dikes. This is obvious in the fact the warring parties threatened the integrity of the dikes by building defensive works on them and even went so far as to intentionally breach them. Furthermore, unremitting warfare, as we have noted, also made it impossible for the local people to carry out many dike projects. In short, the Jianghan dikes suffered more from warfare during these years than at any other time in history.

4 Armed Forces and Dike Affairs

The involvement of armed forces in dike affairs has a long history in China, stretching from around two millennia ago to today. For example, troops had frequently participated in the annual repairs of the Yellow River dikes.[94] In the tenth century, a special contingent of about ten thousand soldiers was assigned to dredge the rivers and build dikes in the Yangzi delta.[95] In the Qing, special river troops were assigned to protect the Yellow River dikes and take care of the seawall along the coast of Zhejiang.[96]

Armed forces also had been involved in dike affairs on the Jianghan Plain in various ways. First, they were engaged in dike management. Following the example of the management of the Yellow River dikes, after the 1788 flood on the Jianghan Plain, the Jingzhou Marine Battalion, numbering fifty-five soldiers distributed among twenty-six guardhouses, was assigned to manage some important segments of the Wancheng Dike. In 1869 this unit was renamed the Yangzi River Marine Battalion, but in any case was dissolved in 1876. A new unit, the Dike Protection Battalion of Jingzhou, was established in

94 Zhou Kuiyi et al. 1990: 47.

95 Liang Jiamian 1989: 333.

96 Zhang Fang 2007: 453; Chen Zhenhan et al. 1989, vol. 3a: 76; Zheng Zhaojin 1987: 219. In 1678, Jin Fu, director-general of waterways, suggested abandoning the use of hired laborers for annual repairs and high-water control of the Yellow River dikes and replacing them with "river soldiers" (*hebing*). These soldiers, Jin argued, would not flee. Moreover, using soldiers eliminated the worry over the malpractice of hiring old and weak people (Zhou Kuiyi 2002: 452).

that year, but its members were no longer military personnel, but instead peasants recruited from nearby villages.[97] Although they were not true soldiers, the name of the unit suggested it was still militarized.

In the Ming-Qing era, the Jianghan area had many grain transport stations (*wei*), whose main purpose was to facilitate transportation of so-called tribute grain (grain collected in the South and transported to the imperial capital). These stations, usually located along rivers, were essentially military posts. However, they were also like administrative units in that they were required to manage some dikes, dubbed "military dikes." The transport stations were required to manage the military dikes just as other dikes were managed or supposed to be managed: laborers patrolled and protected dikes during the period of high-water control and checked and repaired them after the high water receded. Autumn, however, was the busy season for transferring tribute grain, leaving the transport stations with very little time to supervise dike work. In response, dike maintainers would then postpone dike projects at will and collaborate with clerks to decrease the required amount of dirt, all of which lowered the quality of the dike projects. Since these military dikes were usually important, the Qing state decided in 1828 that they should be managed by the local civilian governments.[98]

Sometimes soldiers were directly sent to dike worksites as laborers, such as in 1887 when, in a step to save money, two battalions of soldiers were dispatched to participate in a dike project.[99] The same thing also happened in the Republic when troops were sent to build the Distance Dike in Zhongxiang in 1936–1937,[100] and in the PRC when PLA soldiers frequently worked in dike-related water conservancy projects.

Armed forces were also used to maintain order on dikes. Past scholarship has discussed how soldiers were frequently deployed to put down violent conflicts over the control of waterways, many of which were related to dikes.[101] These clashes, however, did not occur daily, yet armed forces (troops and police) were deployed to maintain order on dikes all year round. A department of dike police was first established on the Wancheng Dike in 1915. Each dike-work sub-bureau had two dike policemen, who had the right to detain and punish anyone who violated regulations, engaged in malpractices, and caused trouble.[102] The 1937

97 Ni Wenwei 2002 [1885]: 92–93; Ts'ui-jung Liu 1970.

98 *Qing huidian shili* 1991 [1899]: 674.

99 Shu Hui 2002 [1896]: 487.

100 *Zhongxiang shuili zhi* 1998: 66.

101 Jiayan Zhang 2013b.

102 *Jingjiang dadi zhi* 1989: 256.

Gazetteer of the Jingjiang Dike noted that the Wancheng Dike was two hundred or more li long, and during any project there were thousands of laborers on the dike every day. The gazetteer went on to explain that it was expected that some of the laborers would break the rules and cause trouble, but it would be a headache to go through all the formal procedures if they were sent to the local authorities for punishment. Moreover, this would hinder the project.[103] In other words, the dike policemen were in a position to discipline laborers and quickly handle any trouble. In the Republic, army units stationed in the area might lead and supervise high-water control of local dikes as well.[104] And, if needed, soldiers would be sent to dikes to maintain order during high-water control. In these instances, they were under the leadership of hydraulic technicians on the site.[105] The regular local police (not dike security guards) were also deployed to supervise laborers working on emergency repairs.[106] In Songzi it was common for annual repairs and high-water control agencies of people's polders to ask local police to prod those who refused to go to dike worksites to complete their assigned work. In the CCP discourse this sort of thing has been described as the GMD "oppressing" peasants, but in reality, it was the peasants who shirked their duty—they were not victims at all (see Chapter 4).

With the founding of the PRC, the relationship between the armed forces and dike affairs became even closer. In 1949, the armed forces stationed on dikes frequently offered their help to protect the dikes.[107] More commonly, militia were used to guard dike projects. In one case, in Jianli in 1951, 105 reed mats for high-water control were suddenly engulfed in flames and burned to ashes; a week later, in another case, two homes were burned, which almost ignited a huge pile of high-water control materials stored nearby. Although the source of these fires is unclear, it was widely believed they were likely set by "bandits, spies, and counterrevolutionaries." The government in fact issued an announcement that the public should learn a lesson from these two events and be more vigilant, and it sent the militia to watch over high-water control materials.[108] In the same year in Hanyang, landlords were not allowed to patrol dikes or participate in emergency repairs.[109] In 1952 in Gong'an, it was reported that during the high-water control period, landlords working on the dikes were carefully watched over by militia members, and it was even suggested that

103 *Jingjiang di zhi* 1937, vol. 4: 5.
104 LS1-5-784; LS1-5-785.
105 *Jianghe xiufang jiyao* 1971 [1936]: 194.
106 LS36-1-122.
107 SZ68-2-227.
108 Wuchang 49-2-1.
109 30-1-5.

some kind of special marks be made on the landlords so they could be easily identified.[110]

In the early 1950s, guarding against bandits and spies (or "special agents" [*tewu*]) was still a crucial part of maintaining the integrity of the Great Jingjiang Dike. Since the dike was so important, there was fear that it would be an ideal target for the enemy. Thus, people were mobilized to watch over it, and if someone was found sabotaging the dike but could not be caught, it is alright to shoot to kill. In the Mashan section, sentinels were posted at all intersections across this dike. And the militiamen in each sentry shack were charged with ferreting out any suspicious people.[111] These were still wartime policies. Although at first glance it may seem like the PRC authorities were overly sensitive, the civil war between the CCP and the GMD was not yet over, and the Great Jingjiang Dike was a perfect target for the enemy. In short, the new state had good reasons to be vigilant. Considering that in 1949 the GMD forces had tried to blow up this dike, the precautions taken by the new state should not be viewed as an overreaction.

Given these concerns, the PRC deployed troops to protect important dike projects. For instance, there was a company of regular PLA troops stationed at two sluice gates of the Jingjiang flood diversion zone to maintain order. This company was later replaced by the local police and then dissolved in 1966. This in effect replicated the history of the marine battalion in Jingzhou in the Qing. Ironically, when the order came down that a dike or sluice gate was be blown up to discharge floodwater, it was usually the PLA's engineering companies that were assigned the task. In 1954, the central government ordered that the Baili Islet polder in Zhijiang be used as a floodwater diversion zone. To minimize resistance from the residents, troops were ordered to dynamite the dike without notifying them. Soldiers who participated in this act could sense the mixed emotions of the local people after they realized that their dike had been blown open by the troops: they respected the soldiers, but they couldn't deny the fact that it was soldiers destroyed their dike.[112] During the Cultural Revolution, the militarization of dike management reached its zenith. At that time, all of the Great Jingjiang Dike was controlled by the military, as were all the other dikes in the Jianghan area. The PLA sent a representative to act as the director of dike management units and oversee their day-to-day operations in the various counties.[113]

110 D35-1-12.

111 48-1-5b; 48-1-8.

112 Zheng Longchang 2004.

113 *Jingjiang dadi zhi* 1989: 257.

After the Cultural Revolution, the military gradually withdrew from the day-to-day management of the dike systems, but the police became increasingly involved. Many government units managing dikes have a police department to maintain order on a regular basis. Under some special circumstances, such as a huge high water, the police from local administrative units have also been used. In 1980, for example, when the dike of a polder in Mianyang had already been breached, the upper-level government ordered that the dike be dynamited to discharge the high water of the Dongjing River. In a rush to harvest their crops, some women—carrying poles, sickles, and bags loaded with bricks, tiles, and dirt—guarded the spots that were to be blown up and pushed some cadres into water or beat them up. The incident ended after seven of the women were arrested by the local police.[114] Again, in 1998 in Shishou, the local county government decided to break open the dike of a small polder to protect a larger area. Residents of the polder, however, resisted and either stopped cadres from breaching their dikes or immediately filled up the openings. Eventually, the police were sent in to forcibly break open the dikes.[115]

The best-known military involvement in dike affairs on the Jianghan Plain in the PRC, in fact, happened in the high-water control of the Yangzi River in summer 1998. Numerous moving scenes of the heroic self-sacrifice of soldiers were reported in newspapers and broadcast to an audience of millions via national television. The following section is devoted to this historic event.

5 Troops and the 1998 High-Water Control of the Yangzi River

PLA soldiers participated not only in the building of many water control conservancy projects, but also in high-water control. In 1954, locally stationed PLA troops and police were required to dispatch groups to participate in that year's high-water control and to protect the dikes. In total 26,655 soldiers took part in the 1954 high-water control in Hubei. These soldiers, it was reported, considered "worksites as battlefields, and floods as enemies," and were ready to sacrifice themselves to safeguard the people. The Central Military Commission, China's highest national defense organization, even ordered that high-water control materials be air-dropped and that divers be sent to carry out emergency rescues.[116]

114 Xiong Fanlin 2008.

115 *'98 Jingzhou kanghong zhi* 1999: 250.

116 *Hubei sheng zhi junshi* 1996: 736.

In 1980, after some polder dikes in Jianli ruptured, military aircraft, warships, and military vehicles were used to bring in relief materials and to evacuate flood victims.[117] Since the 1980s, because of the increasingly tenuous situation of high-water control of the Jing River, PLA troops have been required to directly participate in high-water control on a regular basis. Every year, the military units charged with high-water control have sent representatives to inspect their assigned dikes and work out a plan in advance, and if needed, deploy soldiers to participate in high-water control as planned.[118] In 1981, 3,400 PLA soldiers took part in high-water control of the Jing River in Gong'an alone.[119]

In 1991, the State Council issued Flood Control Regulations of the People's Republic of China, the first document of its kind ever in Chinese history. Article 5 reads: "All units and individuals are obligated to participate in flood control. The People's Liberation Army and the Armed Police Force are important flood control forces." Article 43 of Flood Control Law of the People's Republic of China, issued by the National People's Congress in 1997, reads in part: "The People's Liberation Army, the People's Armed Police, and militia should carry out the task of flood control and emergency repairs entrusted by the state." In short, both the regulations and the law charge the armed forces with the task of participating in flood control (i.e., high-water control).

In 1998, when the entire Yangzi River valley suffered from a historic high water, the crisis was so severe that it prompted action from China's very highest leaders. On August 7, the Politburo of the Communist Party discussed the situation, and on August 16, President Jiang Zemin, who was also the chairman of Communist Party of China's Central Military Commission (i.e., the highest military commander of China), ordered that all troops stationed along the Yangzi River be deployed to its dikes. As a result, 130,000 troops (from the navy, air force, army, and armed police) were deployed to high-water control sites in Hunan and Hubei. Of this number, 95,000 were in Hubei (54,100 were in Jingzhou, the site under the gravest threat). This has been widely believed to have been the largest PLA deployment in the Yangzi River valley after 1949.[120]

The significance of the military's participation in this high-water control crisis can be understood in several different ways. First, it was a display of the strongest support from the highest level. China's top leaders frequented the high-water control front lines in Hubei and rallied local cadres and the public

117 *Jianli xian zhi* 2013: 2.

118 *Jingjiang difang zhi* 2012: 541.

119 *Gong'an xian zhi* 2010: 106.

120 *Hubei shuili zhi* 2000: 289–90, 317; *Jingjiang difang zhi* 2012: 585.

to fight against flood, and the government of Hubei promised to supply labor, materials, and funds, and anything else that might be needed.[121] The top leaders' speeches were repeatedly broadcasted on national television, delivering strong emotional support to people working on high-water control sites. On a practical level, the central government in fact met all that the local governments required, even going so far to deploy navy and air force units to Jingzhou. On August 20, the Navy Engineering Academy sent a team of fourteen members to an emergency site. A sergeant dived down 7.04 meters, found a crack in a sluice gate and blocked it with cotton slivers.[122]

All deployed soldiers were young, fit, and strong—just what was needed in high-water control, a laborious and exhausting job, particularly when it lasts for months. Usually, the Yangzi River high water remains at a critical level for just a few days. But in 1998 it lasted for several months. A huge number of laborers were needed, but most civilians, of course, also had other obligations, making it extremely difficult for them to stay on dikes for months. Moreover, after decades of economic reform, most young villagers—the backbone of traditional high-water control forces—no longer remained in their home villages, but worked off the farm in towns, cities, and coastal areas, in many cases hundreds or even thousands of miles away. But soldiers were readily available.

Furthermore, the peasants who remained were not always reliable, whereas soldiers were well disciplined. This is not to deny that over the years there were many model peasant laborers, but it is also true that the dike work done by soldiers in 1998 was of inestimable value. There is a history of peasant laborers fleeing dike worksites in the face of danger. To cite a few examples, in 1950 in Hanchuan, after hearing the sound of the emergency gong, laborers ran to rescue their own belongings, not the endangered dike; in 1954 in Honghu, all laborers fled when a dike split open; in Jianli, a large number of laborers ran off after some polder dikes collapsed; and in Yingcheng, after hearing the news that a river dike had ruptured, laborers on other dikes immediately ran back to their own villages.[123] In the 1998 high-water control, thanks to the soldiers, the story was different. In Hanchuan, where a sluice gate was in danger—if it broke, a large area would be flooded—all the dangerous and heavy tasks involved in emergency repairs were undertaken by soldiers.[124] In my field survey, I also heard a story that on an endangered dike segment in the 1998 flood, the situation was quickly deteriorating, and the dike was on the verge of

121 *Hubei shuili zhi* 2000: 289–90.

122 *Honghu shi zhi* 2014: 199.

123 50-1-1b; SZ1-02-0216-007; D35-1-41C; Wang Yingrui 2006.

124 *Xiaogan shi shuili zhi* 2011: 107–10.

collapse. Some of the on-site peasant laborers, who were average villagers, were frightened and fled. Soldiers were called, they rushed to the site and worked day and night, and the dike was saved. It is believed that that dike would have failed if it were not for the help from the army.

Here we can see some key differences between regular laborers (usually villagers or peasants) and soldiers. Regular laborers may not act quickly and not many of them are willing to sacrifice their lives for the sake of emergency repairs. Besides, they usually face no punishment if a dike breaks since the responsible parties are cadres. Soldiers, however, are young and vigorous, and more importantly, they are disciplined. They must obey their commander's orders and cannot simply run from dangerous situations. On the contrary, they rush to dangerous sites and immediately begin to act. Furthermore, in peacetime they may well welcome the opportunity to win honor and be rewarded with a promotion. In other words, they are the most capable and effective laborers for high-water control and emergency repairs.

Of course, the peasants had good reason to be fearful. Loss of life among dike laborers is not unusual. As an army commander said in 1998, "high-water control is like a war, fighting requires sacrifice."[125] On August 19, 1998, twenty-one-year-old Li Xiangqun, a PLA soldier, lost his life on a dike in Gong'an. Li, it was reported, "had participated in emergency repairs while he was ill and died due to overwork." Next year Jiang Zemin bestowed on him the title of "Heroic Soldier of the New Era," and Li was posthumously awarded "the China Youth May Fourth Medal" (the highest honor for Chinese youth). Although Li was a native of Hainan Province, his biography was recorded in the new gazetteer of Gong'an, the place where he lost his life.[126] This is unusual, since gazetteers are local histories and usually only record the biographies of important native figures. This treatment and the title of national hero made Li into one of the most celebrated heroes in modern China.

Li was also referred to as a "revolutionary martyr," an honor normally reserved for those who lost their lives fighting class enemies or in battle. In the 1998 high-water control, Li was certainly not the only person to die nor the only person to be recognized as a revolutionary martyr. Both an officer (whose troops were stationed in Hanyang) and a soldier (originally from Hanyang) lost their lives in the high-water control in Paizhou were also honored as revolutionary martyrs.[127] So too was a soldier who was washed away by floodwater in

125 *Shishou shi zhi* 2014: 182.

126 *Gong'an xian zhi* 2010: 13, 554–55.

127 *Hanyang qu shuiwu zhi* 2015: 235.

Hanchuan after giving his lifebuoy to another soldier.[128] Other than these soldiers, many civilians who lost their lives in high-water control were also honored as revolutionary martyrs, such as two people in Zhijiang in 1998—both of whom were experienced hydraulic technicians—who were so overworked that they suffered a cerebral hemorrhage, fell to the ground, and died later.[129]

Jingzhou Prefecture had the largest number of such revolutionary martyrs: twenty-one in total. Among them, nineteen (including one woman) lost their lives in high-water control in 1998, and nineteen were natives of Jingzhou (the only two non-natives were Li Xiangqun and Yuan Youfa—the former from Hainan and the latter from Shanxi), and eighteen were villagers. Most of them died of exhaustion or illness, and two lost their lives while rescuing drowning people. The sacrifices of all these revolutionary martyrs were recorded in local gazetteers.[130] Eight of them were residents of Shishou (nine if Yuan Youfa is included), including the only female martyr, a village cadre, who reportedly worked extra hours, and eventually fell to the ground and died. Another martyr of Shishou died because he worked for eleven days nonstop without a break. Their biographies were recorded in gazetteers, just the same as generals who had joined the Communist revolutionary movement before 1949.[131] The high number of high-water control martyrs in Shishou is indicative of the great significance of high-water control in that county, home to the most meandering section of the Jing River. That many of them died of exhaustion and/or ensuing illness also reflects the fact that the task of high-water control and emergency repairs is both physically and emotionally demanding.

---

128 *Hanchuan shi zhi* 2010: 847.

129 *Zhijiang shi zhi* 2012: 1083–84.

130 *Jingjiang difang zhi* 2012: 788–90; *Jianli xian zhi* 2013: 662; *Songzi shi zhi* 2012: 24, 695, 698, 700.

131 *Shishou shi zhi* 2014: 726–30.

CHAPTER 9

The Three Gorges Project and High-Water Control on the Jianghan Plain

The great number of dikes the Jianghan people built to protect their homes and land also blocked waterways and contributed to more water calamities. The construction of flood diversion zones was in some ways a compromise: villagers could still harvest one season of crops in these zones if that year's high water was normal or small, otherwise the zones would be used to discharge floodwater.

This idea is not new. In 1765 the governor of Hubei had suggested that the Jianghan region would be better off if there was no annual battle between land and floodwater; instead, he believed floodwater should be allowed to flow naturally over the land.[1] Wang Fengsheng, a famous Qing expert on water control, also argued that the discharge of floodwater is more important than diking.[2] In the early Republic, the director of the Bureau of the Wancheng Dike Work suggested as well that some open areas be set aside to temporarily store floodwater.[3] What he advocated was in fact the establishment of flood diversion zones. Later, in the Republic, some experts clearly proposed the construction of such areas.[4] It is a pity that these proposals were ignored, or perhaps they just could not have been carried out in those chaotic times.

At the very beginning of the PRC, the idea was again brought to the fore. Those in charge of high-water control of the Yangzi River argued that, based on the limited available manpower and the low level of technology, the construction of some flood diversion zones was the only effective way to reduce the flooding of the Jing River. The new state immediately took several steps to do so, including nationalizing river beaches, transforming some beach polders into flood-diversion zones, and constructing the Jingjiang flood diversion project. After using this project in 1954, however, it was found that there were still many unresolved problems. In particular, flood-diversion zones still could not protect the Jianghan Plain from an extreme high-water event (i.e., a once-in-a-thousand-years high-water event). For an event of that scale, the

1 Chen Zhenhan et al. 1989, vol. 2: 408.
2 Wang Fengsheng 1832.
3 *Jingjiang di zhi* 1937, vol. 4: 11.
4 Sun Fushi 1939: 74–77.

 | DOI:10.1163/9789004739482_011

government proposed relying on the Three Gorges Dam. Whether this dam should have been built or not is debatable, but it is simply impossible to have no dikes in the Jianghan area. One of the major concerns of the critics has been that the construction of the Three Gorges Dam required the relocation of more than a million people. However, compared to electricity generation, high-water control in the Jianghan area, and some unknown environmental consequences, population resettlement has been a relatively manageable challenge.

1 Impact of the Four Southward Outlets

In ancient times, there were many outlets along the Jing River, collectively known as the "nine openings and thirteen outlets" (*jiuxue shisankou*),[5] to discharge floodwater in different directions. Some of them were later blocked, then reopened, then blocked again. By the end of the nineteenth century, there were but four—the Tiaoxian, Hudu, Ouchi, and Songzi outlets—all of which discharged floodwater of the Jing River to the south to Lake Dongting.

The Tiaoxian outlet has an especially long history. According to a well-known legend from the late Zhou dynasty, one day when Yu Boya, a famous player of the qin (heptachord) moored his boat at a spot on the Jing River and to play his qin, he was surprised that a nearby woodchopper stopped and listened, enthralled by his music. The two became good friends. The place they met thus was called Tiaoxian ("adjusting the strings"). In the year 280, a Northern general ordered the opening of the mouth of a branch river at that location to defeat the South—that river was the Tiaoxian River. Over the centuries, the mouth, the Tiaoxian outlet, had been repeatedly silted up and reopened. For example, it was silted up and reopened in 1303, and was last reopened in 1855.[6]

The Hudu outlet was probably first formed in the Song dynasty, and then silted up later. In the early Qing, Wu Sangui, the notorious Han general who helped the Manchus conquered China, reopened this almost silted up outlet, to transport arms and ammunition to conquer the South.[7]

The Ouchi outlet was formed in the 1850s when an attempt to block an opening in the Yangzi River dike failed, leading gradually to a new branch river. From the opening at Ouchi to the formation of the Ouchi River took almost a decade. At that time, there were heated debates over the blockage of the opening and the discharge of floodwater. The general view favored the discharge of

5 Lin Yuan n.d.

6 *Tiaoguan zhen zhi* 2001: 4, 10–11.

7 *Hubei shuili zhi* 2000: 373, 380–81.

floodwater southward to Lake Dongting, and the opening at Ouchi did exactly that, hence neither the state nor the local people wanted to block it. Besides, China at that time was suffering from both internal rebellions and external invasion, and the state lacked funds to block the opening.[8]

The Songzi outlet was formed in the 1870s after the 1870 flood. Before that, the local people were responsible for 70 li of the Yangzi River dike, but after the outlet was created, they had to build and care for more and more dikes (both river and polder dikes), totaling 400 or more li in 1937. Although the dikes became longer, the local economy was not as prosperous as before. The people of Songzi thought that the opening on the Yangzi River dike in the 1870s should be blocked, but according to the compilers of a Songzi county gazetteer, because in imperial times the government was of preeminent importance, and the yamen of the upper-level government (i.e., the Prefecture of Jingzhou) was located in the city of Jingzhou, on the northern bank of the Jing River, the state favored the northern dike over the southern dike, and the south (Songzi) had to give in.[9]

The final formation of these four southward outlets, particularly at Ouchi and Songzi, had a profound impact on the relationship between the Jing River and Lake Dongting. After that, each year a huge amount of floodwater was discharged from the Jing River into Lake Dongting, which largely reduced the pressure of high-water control on the dikes on the northern bank of the Jing River, but by the 1990s the high-water level of the Jing River stood ten or more meters higher than the land on its northern side.[10] Most of the silt carried by the floodwater of the Jing River was deposited in Lake Dongting. Statistics from 1931–1936 and 1951–1964 show that nearly 90 percent of silt deposited in Lake Dongting was carried into the lake by floodwater from these four outlets. This silt created a wide expanse of land for the locals to reclaim, but reclamation also increased the frequency of water calamities in the region.[11]

At the same time, after these four southward outlets were formed, river beaches and islets in the Jing River were progressively reclaimed as polders.[12] This inevitably blocked even more waterways and contributed to increasingly frequent dike ruptures and inundations. In 1987, Sun Fushi, a respected hydraulic engineer, argued that allowing the formation of the Ouchi outlet and the Songzi outlet had been a huge strategic mistake: it not only overlooked the

8 *Hubei shuili zhi* 2000: 382.

9 *Songzi xian zhi* 1982 [1937]: 91.

10 *Jingzhou diqu zhi* 1996: 2.

11 *Hubei sheng zhi shuili* 1995: 29–30; Jiayan Zhang 2014: 25–27.

12 Peng Yuxin and Zhang Jianmin 1993: 195.

historical fact that the Jing River discharged its high-water to both the north and the south, but it also disregarded the relationship between the upper and lower reaches of the Yangzi River as well as the long-term consequences. If these factors had been considered, the situation along the middle reaches of the Yangzi River (i.e., the Jianghan area) would have been far better.[13]

The Tiaoxian River once was an important transportation route for the Hunanese who lived on the southern bank of the Jing River. These people had no access to other waterways that were connected the Jing River. In winter, the water level was so low that boats could not sail from the Jing River to the Tiaoxian River. Thus, cargo had to be offloaded, transported overland, and then reloaded, driving up the cost of transportation. The Hunanese wanted to deepen and widen the mouth of the Tiaoxian, but the locals resisted since they would lose their reloading business.[14] In the early PRC, however, the Hunanese wanted to block this outlet to reduce water calamities. In 1958 and 1959, the governments of Hubei and Hunan agreed to build a sluice gate to block the outlet, but concurred that the gate should be opened for flood diversion if the high water of the Jing River reached 36 meters (and was forecast to rise over 36.57 meters) in Jianli. This agreement, however, has been rendered moot, mainly because the outlet has been silted up over the past several decades, raising the elevation of the mouth of the Tiaoxian River from 24 meters in 1959 to 31.4 meters in 1989, making it almost impossible to discharge flood from the Jing River into the Tiaoxian River. The Tiaoxian outlet thus is effectively blocked and the Tiaoxian River is literally dead.[15] At the same time, since the collapse of the Yangzi River dike in Tiaoguan, a town at the mouth of the Tiaoxian River, would be a disaster for the Hunanese, when the situation in Tiaoguan becomes very threatening, Hunan will send laborers to participate in high-water control of the dike there.[16]

After the blockage of the Tiaoxian outlet, the Songzi outlet became the largest of the remaining three outlets, but its capacity to discharge floodwater has also been declining.[17] What's worse is that the mouths of the three remaining southward outlets have also been silted up over the years. Consequently, the floodwater discharged by these three outlets has been greatly reduced. In 1954, 40.8 percent of the Jing River's floodwater was diverted via four outlets into Lake Dongting, but that dropped to about 25 percent in 1998, which again

---

13 Sun Fushi 1987.
14 Ni Wenwei 2002 [1885]: 343.
15 *Jingjiang dadi zhi* 1989: 39; *Hubei shuili zhi* 2000: 571; *Jingjiang difang zhi* 2012: 67.
16 *Tiaoguan zhen zhi* 2001: 39; *Shishou shi zhi* 2014: 12.
17 *Songzi shuili zhi* 2008: 12–13.

increased the pressure of high-water control on the dikes along the Jing River, particularly in its lower reaches.[18]

2 The Construction and Usage of Flood Diversion Zones

In 1871, as the Wancheng Dike was nearing collapse, people rushed to some low-lying spots to build small dikes atop the Wancheng Dike to prevent the water from overflooding. Some called on the prefect to breach the dike on the southern bank of the Jing River to lower the high water and reduce the pressure on the dike on the northern bank. The prefect refused to do so, declaring that people who lived on the southern side of the Jing River were also his subjects (or "his people," as he put it)—how could he sacrifice one to protect the other? The high-water level soon dropped because a dike on the southern bank collapsed. The Wancheng Dike thus held, but the prefect became literally sick with worry.[19] This case suggests that under some extreme circumstances, reducing the pressure of high water on dikes might require a rupture, intentional or not. In other words, building flood diversion zones is important for protecting the Jianghan dikes. As discussed earlier, this idea had been frequently raised in the Qing and the Republic but had never been put into effect. It was the PRC that first set up flood diversion zones to protect a far larger area.

The construction of flood diversion zones, however, had been discussed in detail in the Republic. Sun Fushi once argued that some lakes should be used to store peak-time floodwater. But he also pointed out that lakes could only be used occasionally for this purpose, and therefore the local government should encourage the local people to farm them.[20] Both his explanation and methods are workable in theory only, since the social chaos at that time made building any flood diversion zones very difficult if not impossible.

The PRC state, however, was in a position to take immediate action. Lin Yishan, the director of the Yangzi River Conservancy Commission and widely regarded as "the King of the Yangzi River," saw the dangerous condition of the Great Jingjiang Dike in 1949 in person and realized that China's economic situation at that time made it impossible for the state to reinforce the dike significantly within a short time. In his view, the only and the best way to protect the

18 Hubei sheng shuiliting et al. 2000: 8.

19 Ni Wenwei 2002 [1885]: 303–04.

20 Sun Fushi 1939: 74–77.

Jianghan Plain from frequent flooding was to raise the dike as high as possible and, at the same time, build flood diversion zones.[21]

The central government acted on his suggestion and in 1952 decided to construct the Jingjiang flood diversion zone. Although this project was intended to benefit both Hunan and Hubei, the government of Hunan still worried it might increase the chance of water calamities in Hunan. Zhou Enlai required the two provinces to cooperate and compromise, and concluded that the project was Mao's idea (Mao was a native of Hunan), and hence it must be built. The government of Hubei also regarded it as a must-do political task.[22] The first phase of the project was completed within seventy-five days, starting on April 5 and finishing on June 20, 1952; the second phase was started on November 14, 1952, and completed on April 5, 1953. At the same time, to relocate the evacuated residents, the local government also built many safety zones (surrounded by dikes) and safety platforms (built along dikes or on high locations) inside the flood diversion zone.[23]

In 1954, the high-water of the Yangzi River reached a record level. This threatened the Great Jingjiang Dike, which in turn prompted the State Council to decide to use the Jingjiang flood diversion zone. Before the sluice gates were opened, the Gong'an County government was ordered to evacuate everyone from the zone before July 10. It immediately sent cadres down to all the villages to supervise the evacuation. Many older people refused to leave, but cadres picked them up and carried them out, with some old folks striking and biting them. Cattle and grain also had to be moved out. The result was "no one was drowned" and "no one was starved to death,"[24] a result that would have been difficult to reach in earlier eras. In 1954, the Jingjiang flood diversion zone was actually used three times, greatly reducing the damage from flooding that year.[25]

The Jingjiang flood diversion project, however, was hardly problem-free. The safety zones/platforms were overcrowded, vulnerable to fire, and distant from the land where the relocated people still farmed. This led many to move back into the diversion zone. It took four to five years for the people who returned to the diversion zone to restore production and rebuild their homes and water conservancy infrastructure. Before the Jingjiang flood diversion zone was first used in 1954, about 164,300 people lived inside the zone, but by 1995, the

---

21 *Jingjiang fenhong gongcheng zhi* 2000: Lin's introduction.

22 Gan Xinghong 2013.

23 *Jingjiang fenhong gongcheng zhi* 2000: 95, 120–46.

24 Li Shouhe 2001.

25 *Hubei shuili zhi* 2000: 9.

population had jumped to 498,000, of whom 331,700 lived in the area designated to receive discharged floodwater.[26] This not only greatly exceeded the number of residents it could reasonably accommodate, but it also made the job of evacuation much more difficult. It is no wonder the state has become increasingly reluctant to use the flood diversion zone.

Using these flood diversion zones is costly, but the cost would be even higher if they were not used. It is widely believed that if there had been no flood diversion zones in 1954, the Great Jingjiang Dike would have failed, flooding most of the Jianghan Plain, and even causing the Yangzi River to change its course. Since this would be a horrendous disaster, the conclusion can only be that using flood diversion zones, and thereby sacrificing a small area to protect a larger area, is reasonable, and perhaps even unavoidable.

Since the Jingjiang flood diversion zone is one of the nation's most important, its use requires an order from the State Flood Control and Drought Relief Headquarters of the central government. But the order will be carried out by the local government.[27] As noted earlier, this zone is in Gong'an, a county with no less than eighteen rivers, including three southward branch rivers that discharge water from the Jing River into Lake Dongting. For this reason, the local people built a great many dikes, including those for the safety zone inside the Jingjiang flood diversion zone. Gong'an's dikes stretch over 800 kilometers; no other county in the nation rivals Gong'an in the length of its dikes. Maintaining these dikes has inevitably imposed a heavy burden on the people of Gong'an.[28]

The situation along the Han River was much the same. At the end of 1949 experts agreed that the emphasis at that time should be on high-water control, and the best way to protect the Han River dikes from flooding was to construct one or several flood diversion zones in the middle of the Han River valley to reduce the pressure of high-water on the dikes along river's lower reaches. Although the idea was not novel, the new government immediately took action by sending experts to investigate the situation in the Han River valley in 1950.[29] Before any major flood diversion zones were built along the Han River, the region was hit by a historic flood in 1954. To protect the Beijing-Hankou Railway and Wuhan Municipality, the government of Hubei province decided in July to open the Han River dike at Yuwanggong in Mianyang to discharge floodwater. The Mianyang County government was ordered to relocate more

26 *Jingjiang fenhong gongcheng zhi* 2000: 119, 240–41.

27 *Jingjiang fenhong gongcheng zhi* 2000: 239.

28 *Wuyin Gong'an kanghong zhi* 1999: 1.

29 SZ34-2-116.

than six hundred thousand people to higher ground overnight—a daunting task and one that cost Mianyang dearly.[30]

After this flood diversion, the state sped up the pace of constructing flood diversion zones along the Han River. One of the major zones is the Dujiatai flood diversion zone in Mianyang, constructed between November 1955 and April 1956 by 150,000 laborers who came from several counties. It was used twice in 1956, preventing damage that would have amounted to more than the cost of building the project. From 1956 to 1984 it was used nineteen times and twice from 1984 to 2014. If other polder land in Mianyang were used to discharge floodwater, that would inundate more than 3 million mu, and the immediate loss would be equivalent to two times the value of Mianyang's agricultural output.[31]

Other than these large flood diversion zones, many small polders were also used, or designated, to receive discharged floodwater if needed. For example, in 1954 when Tianmen was threatened by both floodwater and logged water, and many polder dikes collapsed, the county government intentionally breached the dikes of some small polders to protect larger ones.[32] In another example, in Jingzhou Prefecture in 1980, when 142 polders were overflowed or collapsed, seven flood diversion zones in three counties were used; and in 1983, when 103 polders were overflowed or collapsed, eight flood diversion zones were used.[33] In a word, the use of flood diversion zones was common.

By 1995 the province of Hubei had forty-two large flood diversion zones, mostly on the Jianghan Plain.[34] In addition, the province has many smaller ones. In Jingmen, for example, there are twenty flood diversion zones,[35] and seven in Xiaogan, where the smallest has only 6 square kilometers of flood storage area.[36]

Naturally, residents of these places were not eager to see their land covered with discharged floodwater. In 1980, the residents of one polder were ordered to break the dike to discharge floodwater of the Dongjing River. To thwart the breaching, some female residents stood on the dike where it was to be broken, armed with shoulder poles, sickles, and bags of bricks, tiles, and dirt—they even pushed some cadres into the water or beat those who tried to reason with

30 Zhou Xiancai 2008.
31 *Xiliuhe qu zhi* 1986: 23; *Hanjiang difang zhi* 2015: 129; *Hubei shuili zhi* 2000: 614–15.
32 *Tianmen xian zhi* 1989: 145.
33 *Jingzhou diqu zhi* 1996: 588.
34 *Hubei shuili zhi* 2000: 589.
35 *Jingmen shi shuili zhi* 2012: 73.
36 *Xiaogan shi shuili zhi* 2011: 50–54.

them.[37] In 1984, residents in a flood diversion zone on the outskirts of Wuhan felt crushed because they repeatedly suffered serious losses when floodwater was discharged onto their land.[38]

The residents of flood diversion zones wanted what every other villager in China wanted: to develop the economy and live a better life. According to a 2007 survey, the living standard of people in the Jingjiang flood diversion zone was relatively low. The respondents complained that each year they had to spend three months on high-water control and another three months on annual repairs, taking time away from their farmwork. And farmwork was all they had since there was no industry in their villages.[39] To them, all this was incredibly unfair. But up to the 1990s, most people in the Jianghan area also had to spend months on high-water control and annual repairs. Besides, flood diversion zones were designed for sacrifice, not for protection. If a flood diversion zone also becomes an economically important area and the state decides it should be protected, then it is no longer a flood diversion zone. That is exactly what has happened in recent decades.

According to the central government, however, all the designated flood diversion zones in the Jianghan area, except for the Jingjiang flood diversion zone, were formerly lakes or beaches that functioned as natural flood diversion zones. Moreover, the government contended that regulations required that they be used to discharge floodwater, and their residents were not "sacrificed" for the sake of others.[40]

The problems caused by flood diversion zones have accumulated over time, just as with water calamities in the Ming-Qing era. At first, there were many outlets to discharge floods, and water calamities were relatively infrequent. Later, however, an increasing number of outlets were blocked, and if dikes were mismanaged, the result was an increase in water calamities caused by frequent dike ruptures. Although the reopening of former outlets would have been the best way to reduce water calamities, doing so was impractical since former flood diversion areas had already become densely populated villages and fertile farmland. Nowadays the dilemma is even worse since there are far more people living in these zones than before and there is a great deal of industry as well. In short, it has become increasingly difficult to use these flood diversion zones. A case in point occurred in 1983 when Hubei's leaders repeatedly ordered the Mianyang County government to operate the Dujiatai flood

37 Xiong Fanlin 2008.

38 XX000084-WS03-0137-0004.

39 Jiang Tanyu 2007: 389.

40 Li Jiansheng 1999: preface.

diversion zone to its full capacity. A leader of Mianyang risked his political career by refusing to carry out the order, and yet he managed to convince the provincial leaders to change their minds by arguing that the cost to Mianyang was unreasonably disproportionate.[41] His success signaled that the state recognized that using flood diversion zones had become increasingly problematic.

In the 1998 flood, China's State Council decided to use the Jingjiang flood diversion zone again, and the Gong'an County government was ordered to evacuate all residents and their livestock within twenty-four hours. The county's 1,654 cadres were sent to every village to ensure that all residents and livestock were moved out and sent to the assigned places, to find shelter for them, and to guarantee their safety. Within sixteen hours, 332,003 people and 18,000 cattle were evacuated, at a cost of six billion RMB (roughly 723 million USD) and 99 deaths (from accidents, suicides, and illness).[42] However, even though all the residents of the Jingjiang flood diversion zone were evacuated, at the last moment the order to use the zone to receive floodwater was cancelled. Again, the main concern was the high cost. The local leaders admitted that, "emotionally," they really did not want to see the zone inundated with floodwater, and as far they were concerned, all areas under their jurisdiction were equally important.[43] With the rapid development of the rural economy in recent decades, it is likely the state will be even more reluctant to avail itself of these flood diversion zones and will have little choice but to increasingly tie the safety of the Jianghan Plain to the Great Jingjiang Dike and the Three Gorges Dam.

3 The Three Gorges Dam

The Three Gorges Dam is one of modern China's most controversial engineering projects. It was first proposed by Sun Yat-sen in 1917. However, there was no possibility of building it in the chaotic Republican era. The issue was brought up again in the early PRC and sparked lengthy and heated debates.[44] In the 1950s, Lin Yishan argued in favor of a dam for high-water control of the Jing River, but opponents argued it would cost too much. In 1969, Zhang Tixue wanted to see such a dam constructed to generate electricity, but Mao opposed it out of fear that it might be attacked by the enemy, resulting in a disastrous

41 Zhou Xiangsheng 2008.

42 *Wuyin Gong'an kanghong zhi* 1999: 2–4; *Gong'an xian zhi* 2010: 111.

43 Zou Dongting and Zheng Zehua 1998: 151; *Jingjiang difang zhi* 2012: 421; Liu Keyi 2000: 20.

44 Su Xiangrong 2007.

flood. And in the 1980s, Deng Xiaoping supported it to stimulate economic growth, but some party leaders, for various reasons, were still opposed.[45]

The main criticism of the opponents of the Three Gorges project was that it would entail the relocation of more than a million people, the destruction of landscapes, and the degradation of the local ecosystem. Yet, for many technicians, particularly hydraulic engineers who were involved in the design and analysis of the project, and scientists who participated in the discussions and studies of the project, the destruction of some landscapes and other adverse environmental impacts were certainly not a major concern. They recognized that the Three Gorges project had to be the centerpiece of the solution to the floods in the Jingjiang area. A representative from the Chinese Academy of Sciences worried about the negative environmental consequences at first, but after hearing that many people would die due to floods, he changed his mind and supported the project.[46] This story seems to imply that he considered life to be more important than the environment. It is hard for anyone who has never suffered from a flood to imagine how devastating it can be, particularly in the Jianghan region.

According to Wang Renzhong, a long-term provincial leader of Hubei, both the Great Jingjiang Dike and the Distance Dike were must-be-protected dikes, not because they were intended to protect the farmland behind them, but because they protected the lives of millions of people elsewhere. In fact, in 1954, farmland behind these dikes had already been inundated, yet they protected about three million people. If the Great Jingjiang Dike broke, at least a hundred thousand people would drown, even in the best-case scenario.[47] To explain why the Distance Dike had to be protected, the head of Zhongxiang County in 1955 said it was because the dike protected hundreds of thousands of people and the city of Wuhan, an "industrial and transportation center." If the dike failed, the result would be an unmitigated disaster for both the people and the nation. Thus, ensuring its integrity was a political and economic task of monumental proportions.[48] That was the situation in the 1950s. At the end of the twentieth century, about 27 million people and 30 million mu of farmland were protected by Jianghan's dikes.[49] If these dikes broke today, far more people would lose their lives—Wuhan Muncipiality alone has 11 million residents.

45 Hong Qingyu 1999.
46 Cao Le'an 1990.
47 Li Rui 1996: 147.
48 54-1-15.
49 *Hubei shuili zhi* 2000: 226.

Beginning in the 1950s, hundreds of thousands of engineers and technicians studied the possibility of damming the Yangzi River and experimented for four decades and amassed an enormous amount of data.[50] After lengthy debates and discussion, the fate of the Three Gorges project was sent to the National People's Congress (NPC) for a decision—the only engineering project ever to go through this process in the PRC. On April 3, 1992, representatives of the NPC made their decision: 1,767 voted in favor of the project, 177 voted against it, 25 did not vote, and 664 abstained.[51] That is to say, even within the NPC, the PRC's national legislature, a so-called rubber-stamp organization, not every member favored the project—in fact, almost a third did not support it.

Among the public both at home and abroad, the Three Gorges project is famous for being a hydroelectric plant, the world's largest. The project has certainly been widely recognized as an energy project. But in the early discussion of the proposal to construct the project, the emphasis was on its major function: high-water control of the middle Yangzi valley, the site of the Jianghan Plain. According to hydraulic technicians, the Three Gorges project is crucial to high-water control of the Jianghan Plain.

First, until the 1980s, high water from the rivers on the Jianghan Plain was mainly discharged to flood diversion zones. But these zones had many shortcomings. More than a few of those established in the 1950s had been gradually reclaimed, which greatly reduced their discharge capacity. For example, almost three-quarters of the Dujiatai flood diversion zone had been reclaimed as numerous polders. In another example, the sluice gates of the Jingjiang flood diversion zone were last used in 1954 and eventually became almost inoperative. Furthermore, using flood diversion zones, as noted earlier, is very expensive. According to a 1991 survey, using the Jingjiang flood diversion zone would cost 2.8 billion RMB, and the direct cost of using all the flood diversion zones would add up to 21.2 billion RMB (2.56 billion USD). The amount would be far larger if indirect costs were also included. This was the cost in 1991. After that, costs steeply escalated. Furthermore, evacuating hundreds of thousands of people and their livestock in an extremely short time is, needless to say, challenging. The task would be formidable even for troops, let alone villagers. And sadly, many people will lose their lives during an evacuation. In short, using flood diversion zones has become increasingly problematic. Moreover, there is simply no way to contain an inundation from a once-in-a-thousand-years high water such as that in 1870, when 105,000 cubic meters per second were discharged from the Jing River. The maximum safe flood discharge volume of

50 Lin Yishan 2011: 546–87.

51 *Hubei shuili zhi* 2000: 466.

the Jing River is 50,000 cubic meters per second, but from the year 1153 to the end of the twentieth century the volume of the Jing River exceeded 80,000 cubic meters per second eight times. At the turn of the twentieth-first century it was believed that there was still no reliable method to handle such a huge discharge of high water, and hence the state felt it had to rely on the construction of the Three Gorges project.[52]

Second, most dikes along the Jing River are built on alluvial plains, with their bases usually consisting of sandy soil, which is susceptible to collapse. During the high-water season, the water level of the Jing River is typically several to ten or more meters higher than the surrounding ground—an extremely dangerous situation if a dike breaks. Although the Great Jingjiang Dike is the most important dike on the Jianghan Plain, it is more than 1,600 years old, and even after decades of reinforcement since 1949, it is still too weak to cope with historic high-water levels.[53]

Third, from a purely hydrological perspective, in 1931, 1935, and 1954, more than 95 percent of the Jing River's high water came from the upper reaches of the Yangzi River, stretching from Yichang—the site of the Three Gorges Dam—westward. Almost all historic super floods such as those of 1788, 1860, and 1870 were mainly caused by the huge volume of high water from the upper reaches of the Yangzi River. Hydrologists have argued that only the Three Gorges project can effectively control such huge floods and reduce the pressure of high-water control along the Jing River segment.[54]

As various sources show, hydraulic technicians usually favored the construction of the Three Gorges project because it could reduce the pressure of the annual high water of the Jing River on its dikes and lessen reliance on flood diversion zones. These are exactly the sort of benefits brought by the Danjiangkou Reservoir, which was built in the middle reaches of the Han River. After the reservoir was constructed in 1968, the safety of the Han River dikes, particularly the Distance Dike, was ensured and the frequency of the use of the Dujiatai flood diversion zone was greatly reduced—from 1968 to 1985, it was used only five times in four years. If the Danjiangkou Reservoir did not exist, the Dujiatai flood diversion zone would have had to be used twenty-five times.[55]

Once the Three Gorges Dam was built, it formed a gigantic reservoir—the Three Gorges Reservoir—which, like the Danjiangkou Reservoir, is used to stagger floodwater. Reservoirs have been widely used for this purpose all over

52 Wang Shengfu 1985; *Jingjiang fenhong gongcheng zhi* 2000: 2, 347, 372.

53 *Hubei sheng zhi shuili* 1995: 204–05, 228–31.

54 Xu Qianqing 1992: 80–89.

55 *Hubei shuili zhi* 2000: 426.

the nation. In 1998, 763 large- and middle-size reservoirs in five provinces stored or blocked 34 billion cubic meters of floodwater to stagger or reduce the high-water peaks on the Yangzi River. Nationwide, 1,335 such reservoirs stored or blocked floodwater to protect more than two hundred cities.[56]

The Three Gorges Dam was officially completed in 2009. After that, the burden of high-water control in the Jianghan area was drastically reduced. The Hubei Headquarters of High-Water Control has noted that in normal years, cadres and laborers are no longer needed to go to dikes to patrol during the high-water season,[57] a milestone in the history of high-water control on the Jianghan Plain. Thus, for example, in Yongjiu Village in Songzi, most years no or only a few laborers were needed to patrol dikes during the high-water season.[58] In another example, in Shishou the period of high-water control was substantially shortened.[59] Changes occurred even earlier in Jianli, where there were almost no floods after 2002 when the Yangzi River was dammed for the construction of the Three Gorges Dam.[60] The dam also reduced the frequency of the use of flood diversion zones to discharge floodwater, and, in some cases, entirely eliminated their use.[61]

The completion of the Three Gorges Dam, however, does not mean that high-water control on the Jianghan Plain can be relaxed. On the contrary, it requires even more care because a dike is safer if it has been immersed in water year after year, since this can reveal any hidden threats. Also, the Three Gorges Dam is designed to protect against super high water (i.e., a once-in-a-thousand-years high water), not regular high water. Thus, flood diversion zones are still needed. National leaders, including Li Peng, the premier from 1988 to 1998, who was trained as a hydraulic engineer and was widely believed to be the main patron of the Three Gorges project, warned in 1986 that "the high-water control of the Jing River ... is crucial to the national economy. ... In the future, even after the completion of the Three Gorges project, the construction of the Great Jingjiang Dike and its flood diversion zone cannot be relaxed."[62] Backing up Li's words, the state made a large investment in reinforcing the Great Jingjiang Dike after the 1998 flood.

At the same time, another major burden on local laborers, that is, the annual repair of dikes, has also gradually disappeared since the end of the twentieth

56 *Jingjiang difang zhi* 2012: 765.
57 *Jingzhou shuili zhi 1* 2016: 353.
58 *Yongjiu cun zhi* 2013: 88.
59 Yuan Li 2014: 178.
60 *Jianli xian zhi* 2013: 81.
61 *Jingjiang difang zhi* 2012: 8.
62 *Jingjiang fenhong gongcheng zhi* 2000: 5.

century. On the Jianghan Plain, one reason for this is the water now released from the Three Gorges Dam does not contain much silt and as it flows it slightly but gradually dredges the river course and reduces the high-water level.[63] The result is that the local people do not have to build their dikes higher and higher year after year, as they did for more than a millennium. Although this may seem to be great progress, hydraulic experts actually have become even more cautious and hesitant, since the water not only can dredge the river course but can also scour out the foundation of river dikes, particularly in sections with no river beaches. If a dike's foundation is scoured out so deep that the dike collapses, it would be too late to do anything to rescue the dike.[64]

The Three Gorges project will also reduce the floodwater discharged to Lake Dongting via the three remaining outlets, even though the diversion of floodwater from these outlets to Lake Dongting is vital to the integrity of the Great Jingjiang Dike. If there were no longer any outlets, relying on the Three Gorges Dam alone could not ensure the safety of the Great Jingjiang Dike. Thus, dredging branch rivers and increasing the volume of floodwater they discharge into Lake Dongting is crucial.[65]

4 Water Conservancy Relocatees

According to a 1993 report by the Yangzi River Conservancy Commission, the Three Gorges Reservoir would submerge 257,400 mu of farmland, 58,000 mu of river beaches, 110,200 mu of gardens, 49,100 mu of woods, and 4,700 mu of fishponds in twenty-two counties in Hubei and Sichuan (now Chongqing), and require the relocation of 846,200 people.[66] But according to the journalist Dai Qing, the most outspoken opponent of the project, 1.9 million people would be forced to relocate,[67] one million more than the official estimate. Later data show that more than 1.3 million people were eventually relocated.[68]

The difficulty of relocating more than a million people and the misery they would suffer were perhaps the main concerns of those who opposed the Three Gorges project. Even some of China's top leaders also regarded

63 *Jingjiang difang zhi* 2012: 906.
64 Personal communication with Xu Hailiang, a retired senior engineer, Ministry of Water Resources, October 2017.
65 *Jingjiang difang zhi* 2012: 906.
66 *Hubei shuili zhi* 2000: 471.
67 Dai Qing 1998.
68 Wang Yingzheng 2010: 233.

population resettlement as the key to the success of the project.[69] It is true that the people of China have preferred to remain in one place for generations and loathed moving, but it is also true that the history of China is a history of migration, with people moving from the North China Plain to the South or Manchuria (especially in the Qing), from the lower reaches to the middle and upper reaches of the Yangzi River valley, from plains to mountains and even abroad; in a word, Chinese have moved in every direction over millennia. Among the most powerful forces that propelled migration were war, famine, and population pressure. The Jianghan Plain itself was a famous destination of millions of immigrants (mostly from ten provinces, particularly Jiangxi) in the Ming-Qing era, but many Jianghan people also started to migrate out in the mid-nineteenth century due mainly to frequent water calamities.[70]

The picture in the PRC is different in two major respects. First, after private landownership was abolished early in the PRC, land was technically nationalized but in reality was managed by the villages. Second, most of the immigrants from Hubei were displaced by water conservancy projects. China has the largest number of people—about 26 million by the end of 2006—who were relocated due to the construction of large and middle-size reservoirs.[71] The majority of water conservancy relocatees in Hubei have also been forced to move because of the construction of reservoirs. From 1949 to 1995, 5,817 reservoirs were built in Hubei, 53 of which were large (the most in the nation).[72]

The situation on the Jianghan Plain is more complicated. There are three main categories of relocatees in the region: those who moved to state farms, mostly in the 1950s–1970s; those who were sent to the Jianghan area because of the construction of water conservancy projects in their homelands; and those who were relocated nearby due to the construction of water conservancy projects.

First, immigrants to state farms. From the 1950s to the 1970s, the government of Hubei organized the widespread reclamation of wasteland and established many state farms in the Jianghan area (see Chapter 6). Most of the people involved in reclamation stayed, and many more moved in. The latter included cadres, workers, peasants, sent-down urban youth (young people sent from the city to the countryside during the Cultural Revolution), former PLA soldiers, and fresh college graduates; and they came from all over the country. The employees of Taihugang Farm and Caobuhu Farm, for example, came

69 Wang Yingzheng 2010: 4–5.
70 Zhang Guoxiong 1995; Jiayan Zhang 2014: 33–35.
71 Wang Yingzheng 2010: 50.
72 *Hubei shuili zhi* 2000: 981.

from 87 counties of eighteen provinces and 165 counties of eighteen provinces, respectively.[73] In another example, by 1985, Zongkou Farm had a population of 36,000, hailing from 374 counties of twenty-eight provinces, and by 2005, Linjiaohu Farm's Nanhu Village had 986 residents from seventeen provinces.[74] Although the total number of people who moved to Hubei's state farms was far less than that of migrants in the Ming-Qing era, their places of origin were more diverse. In fact, as immigration was such a core element of state farms, the relevant upper-level government organ—the Committee of (State) Farm Management or the Bureau of (State) Farm Management in Jingzhou—was simply renamed the Bureau of Immigration in 2007.[75]

Second, water conservancy relocatees from outside of the Jianghan area. From 1958 to 1986, the central government constructed three large reservoirs in Hubei: Danjiangkou, Lushui, and Gezhouba. Lushui Reservoir, in Puqi County, was an experiment for the Three Gorges project. Most of the people dislocated by the reservoir were moved within Puqi. Construction of Gezhouba Dam, the first dam on the Yangzi River, started in 1970 and ended in 1988. The dam, 38 kilometers downstream from where the Three Gorges Dam would be built, was another experiment for the Three Gorges project. Most of the people who were dislocated were moved up to the mountains at the same location, except for 1,463 people who were relocated to Zhijiang. Finally, the residents dislocated by the construction of the Danjiangkou Reservoir, a large project in the middle reaches of the Han River, were either moved up to nearby mountains or moved out to other places, particularly the Jianghan Plain. For these three projects, 345,289 people (including 68,867 from Henan) were relocated from 1958 to 1980.[76]

In the 1960s, relocatees from smaller reservoir projects in Jingzhou were usually separated up and relocated to a few selected villages. In the receiving villages, typically every ten households received one household of relocatees.[77]

At the turn of the twenty-first century, the Jianghan Plain also received many people displaced by the Three Gorges project. These people were allocated to several different counties and state farms, and then further assigned, either separately or collectively, to administrative units at the next lower level. All relocatees from the Three Gorges project who were sent to Shishou were peasants. The majority—92 households with 319 people—were moved into

73 *Taihugang nongchang zhi* 1988: 1; *Caobuhu nongchang zhi* 1988: 12.

74 *Zongkou nongchang zhi* 1986: 1; *Jingzhou shi nongken zhi* 2010: 110.

75 *Jingzhou shi nongken zhi* 2010: 196.

76 *Hubei sheng zhi minzheng* 1994: 371–74.

77 D35-1-319.

Xinchang Township.[78] In Gong'an, 342 households and 1,509 people were allocated to ten townships.[79] In one township in Jingmen, in the year 2000, eighty-five people from twenty households were moved in.[80] The situation was similar on state farms. In Taihugang Farm, 2,061 people (from 449 households) were allocated to each branch farm, while Huanghu Farm built three new villages to collectively settle the 1,104 people (in 214 households) assigned to it.[81]

The Jianghan Plain also received people displaced by other water conservancy projects such as the Qing River Reservoir (the Qing River is a branch of the Yangzi River) and the South-to-North water diversion project (a scheme to transfer water from the Yangzi River [in South China] to North China).[82]

Third, the relocation of local residents due to water conservancy projects. This is the largest but also the most complicated category of water conservancy relocatees in the Jianghan area. To simplify the discussion, I further divide the category into five sub-categories.

First, peasants relocated to state farms. By 1985, the PRC state established forty-nine state farms in Hubei, mostly on the Jianghan Plain.[83] The residents of these farms came from all over the country. The state farms in Jingzhou alone had people hailing from 330 counties of twenty-eight provinces.[84] They also included people from the Jianghan Plain. In People's Dayuan Farm, for example, there more than 60,000 people had been relocated from Gong'an, where they lost their land to the Jingjiang flood diversion zone project.[85] These people constituted the majority of residents on this farm. The farm also took in people from nearby counties, such as Shishou. These Shishou peasants, however, preferred to stay in their home villages and did not want to move. The county government thus did its best to relocate entire villages or production teams and settled them together in their new homeland. These settlers, it was reported, were happy later since they produced more in their new villages than they had in their villages of origin.[86] In 1955, when the Dujiatai flood diversion project in Mianyang engulfed the land of 2,040 households (consisting of 8,710

78 *Shishou shi zhi* 2014: 673–74.
79 *Gong'an xian zhi* 2010: 388.
80 *Duodao qu zhi* 2011: 48.
81 *Jingzhou shi nongken zhi* 2010: 87–88.
82 *Jingzhou shi nongken zhi* 2010: 87; *Qianjiang shi zhi* 2011: 526.
83 *Hubei sheng zhi nongye 2* 1999: 379.
84 *Jingzhou shuili zhi 1* 2016: 1289.
85 *Jingjiang fenhong gongcheng zhi* 2000: 240.
86 Liu Wei 1987.

people), the residents were relocated to eight townships in the same county and were settled within 84 days.[87]

Second, peasants relocated to newly enclosed polders. As discussed in Chapter 7, local administrative units—counties and communes, and even brigades and production teams—enclosed many small polders from river beaches, islets, and lakeshore in the 1950s–1970s. As a rule, the people who were moved to a new polder were selected from among the residents under the jurisdiction of the administrative unit that enclosed the polder. According to my on-the-site survey of one such polder, most of the people who were moved to the new polder did not want to relocate. However, each brigade was assigned a quota. In one brigade, everyone who lived along the dike was mobilized to relocate together. Some moved willingly since they did not get along with their fellow villagers and hoped to begin a new life in a new land.

Third, relocatees from local small water conservancy projects. Since various kinds of relatively small water conservancy projects also occupied farmland, the local governments generally relocated those who lost their land to nearby villages. For example, in 1957 in Pengxing Township, when a tiny reservoir was built in a village, 16 households (with 72 people) were moved to other production teams in the same village.[88] In another example, in 1969 in Xiaxindian Township, two villages (consisting of about 280 households or around 1,500 people) were entirely moved from the southern bank of the Hanbei River to its northern bank due to the construction of the Hanbei River water conservancy project.[89]

Fourth, people who were moved from the water side of a dike to the land side after the 1998 flood. To make more room for floodwater, after the 1998 flood the central government ordered the local governments to abandon some river beachland and to relocate people to the land side of the dikes. Thus, in Shadaoguan Township in Songzi, for example, 1,023 households were relocated from river beaches to land on the other side of the river dike.[90] In Tiaoguan Township in Shishou, all residents of three polders on the river side of the dike were relocated to the town of Tiaoguan, which is the land side of the river dike.[91]

Fifth, people who freely and voluntarily migrated. This is the main category of immigration in China before 1949 when people could freely move for any

87 *Mianyang xian zhi* 1989: 431.

88 *Pengxing xiang zhi* 2011: 228.

89 *Xiaxindian zhen zhi* 2012: 530.

90 *Shadaoguan zhen zhi* 2009: 210.

91 *Tiaoguan zhen zhi* 2001: 175.

reason: war, famine, water calamities, financial considerations, and others. For instance, Mantianxing (Stars in the Sky) Village in Xiaogan was populated entirely by people who had migrated from many different places before 1949.[92] The same was true of Zaxing (Miscellaneous Surnames) Village in Hanchuan, where all the residents had migrated from other places.[93] Residents of Mabuguai, an "immigrant village" along the Great Jingjiang Dike and on the boundary of Jiangling, Shishou and Jianli, came from seventy-two counties of nine provinces.[94] This tradition continued in the early 1950s. In 1952, when 500 mu of farmland in two townships in Mianyang was ravaged by a flood, some people moved to nearby counties to join their relatives and friends.[95] This kind of free movement, however, became very difficult or almost impossible after the application of the household registration system in 1958. Four decades later, however, peasants impacted by the Three Gorges Dam, if they could find a place willing to accept them, were free to move. Moreover, the government provided them with some financial support. Some of these people relocated to the Jianghan area. Recorded instances include 53 households with 173 people who moved to Xiaochang, 76 households with 327 people who relocated to Shayang, and 842 households with 3,149 people who relocated to twelve townships in Songzi.[96]

Most of the people displaced by reservoirs in Hubei in the 1950s–1970s received both land and some monetary compensation from the government. In 1985, the central government started to promote a policy of developmental resettlement—that is, a program that went beyond simply giving money to relocatees and aimed to help provide a basis for development.[97] Because China faced financial difficulties in the 1950s–1970s, the compensation paid to relocatees was modest.[98] After 1979, as China's economy mushroomed, the state was able to provide greater compensation. In fact, the budget for the relocation of peasants displaced by the Three Gorges project, set up in 1993, was 40 billion RMB (about 44.4 percent of the entire budget of the project),[99] or roughly 6.9 billion USD. While the central government has provided some subsidies to people relocated because of large water conservancy projects, the local

92 *Xiaxindian zhen zhi* 2012: 87.
93 *Miaotou zhen zhi* 2010: 147.
94 Zhang Yunqian 2004.
95 *Mianyang xian zhi* 1989: 431.
96 *Xiaochang xian zhi* 2013: 781; *Shayang xian zhi* 2013: 1061; *Songzi shi zhi* 2012: 652.
97 *Hubei shuili zhi* 2000: 983, 1016.
98 *Zhanghe shuiku yimin zhi* 1996: 6.
99 *Hubei shuili zhi* 2000: 472.

governments have been responsible for people relocated because of small projects.[100] Generally speaking, as discussed above, in the past two or three decades, water conservancy relocatees, including people displaced by the Three Gorges project, have been much more generously compensated than relocatees of the 1950s–1970s.

5 Relocatees and Local Society

None of the abovementioned three large reservoirs—Danjiangkou, Lushui, and Gezhouba—is located on the Jianghan Plain. But it is likely because the Jianghan Plain benefited from both the Danjiangkou Reservoir and the Gezhouba Dam that some of the people displaced by these projects were relocated to the Jianghan area. In the Gezhouba Dam project, people who were moved higher up the mountains in the same general location were not very satisfied because land there was in short supply. Those who were relocated to Zhijiang, however, were very pleased. Most of them had resided in hilly areas and had grown corn and mandarin oranges, but after they were relocated to the flatland in Zhijiang they raised cotton, and their lives were far better than before, and thus no one returned.[101]

The fate of the Danjiangkou Reservoir relocatees has been mixed. Among those who were moved out, 43,989 people from 9,358 households in 41 brigades and 286 production teams were relocated to Zhongxiang in three batches from 1966 to 1968 to reclaim wasteland between the old Han River dike and the Distance Dike. This area was made into a township (Dachaihu Township), the sole relocatee township in the nation, but many of the relocatees were still very poor decades later.[102] In 1968, 160 households (751 people) were relocated to Xiliuhe Ward in Mianyang, but many of them either moved back or somewhere else since they could not get used to the local lifestyle. By 1985 only fifty households (255 people) remained.[103] In Jingmen, from 1965 to 1967, 3,933 people were relocated to Duodao District, where the first group of relocatees mostly lived on infertile high hills. Some of these people later were redistributed to various production brigades and teams.[104] The situation was similar in Dongbao District, where 25,955 relocatees (in 5,634 households) were assigned

100 *Jingmen shi shuili zhi* 1989: 252.

101 Huang Yonggui 1993.

102 Chen Tao 2007; *Zhongxiang shi zhi* 2013: 764; *Chaihu zhen zhi* 2013: 8–9, 34.

103 *Xiliuhe qu zhi* 1986: 390.

104 *Duodao qu zhi* 2011: 452.

land that was insufficient and infertile. Moreover, they lacked agricultural implements and cattle, making their lives very hard.[105] In Jingzhou District, in the mid-1980s some people displaced by the Danjiangkou Reservoir project were still poorer than the locals.[106]

The 147 people in thirty households who moved in 1967 to Tanglin Village, in Jingmen, however, were luckier. The village allocated 330 mu of land and built relocatee settlements for these people. The village also assigned some experienced older peasants to teach these mountain people how to grow rice. Most of relocatees, it was reported, quickly adapted to their new setting.[107] Another positive outcome was reported in Maozui Township in Mianyang, which took in ninety-two households displaced by the Danjiangkou Reservoir in 1969. These people were relocated to villages with ample land and, it was said, were treated equally with the local villagers.[108]

No matter what the outcome, however, relocation was challenging. In April 1966, the first group of Danjiangkou Reservoir relocatees, about 15,000 in total, were transported from Xichuan in Henan to Zhongxiang in Hubei, via several different means: most people, together with their baggage, were transported by motor vehicles; materials and elderly and ill people as well as pregnant women were transported by boats; large cattle were driven to Danjiangkou, then were transported to their destination by a combination of trains, ships, and trucks.[109] To move the second group of Danjiangkou Reservoir relocatees in 1967, the state assigned a large number of cadres including some PLA officers to furnish support and also provided the means of transportation to complete the task. Within fourteen days, 23,521 people, 10,468 tons of materials, and 2,380 cattle were relocated from Xichuan to Jingmen and Zhongxiang.[110] Today this is merely a 321 kilometer (or roughly 200 miles) trip on a modern highway, but since there were no highways back then, the trip had to be made by trucks, boats, and/or trains. In 1969, due to the difficulty and mounting burden of transportation, the Hubei government decided that the following items would not be relocated: stone implements, some wooden materials (doors, trees, and firewood), livestock, coffins and fragile items, and poultry.[111] Until 1976, however, there were still about 43,300 people who had not been relocated or resettled. These people did not have their own houses

105 *Dongbao qu zhi* 2013: 716.
106 *Jingzhou qu zhi* 2013: 499.
107 *Tanglin cun zhi* 2008: 154.
108 *Maozui zhen zhi* 1990: 111.
109 SZ67-01-0965-004.
110 SZ139-1-000007-018.
111 SZ139-1-000360-012.

or land and relied on government relief.[112] The Hubei government issued a regulation in 1963 that no future project would be allowed to start unless the relocatees would be resettled reasonably well.[113] However, governments at different levels certainly did not follow this regulation.

It appears it was not difficult for people who were relocated within the jurisdiction of their original administrative units, whether it was a county or a commune/township or a brigade, to quickly adjust to their new settings, as they did not have to change their lifestyle and way of making a living. But it was certainly a challenge for those who moved a long distance into a completely different environment. If their standard of living remained the same as the locals or was far better than in their original villages, all might have gone well. But if their circumstances were poorer or worse than that of the local people (this was, unfortunately, common), relocatees definitely were not happy.

Many other factors also contributed to an unhappy outcome. One, as mentioned earlier, was that the land allocated to relocatees was usually limited and of low fertility. Another was the relocatees' reluctance to adapt to the local society. This was particularly the case with Dachaihu Township. All the residents of Dachaihu, who had been relocated from the province of Henan, were reluctant to be moved out in the first place, and they did whatever they could to maintain their own identities after relocation. They were relocated with their entire villages, and they kept the name of these villages (38 in total), they continued to speak their Henan dialect (making communication with local people difficult), they ate wheaten food instead of rice, they watched Henan opera, and they married only within the Henan relocatee community. To alleviate their poverty, the government of Hubei relocated ten thousand of them again in 2004. Many of these people, particularly the younger generation, have gradually begun to integrate with local society.[114]

The story of the Weishui Reservoir relocatees in Songzi is different. Relocation began in 1959 and immediately ran into difficulties. The government of Songzi County required that solving these difficulties be taken as "a serious political task." Most of the relocatees were moved to nearby communes and brigades, and some were relocated to several state farms in other counties. Although the residents of Nanhu Village in Lingjiaohu Farm came from seventeen provinces, 90 percent of them had been displaced by the Weishui Reservoir. These people got along with relocatees from other provinces and

112 SZ67-04-0515-007.

113 SZ113-3-393-001.

114 Quan Runhua 2013.

intermarried with them, but they were so powerful that the other relocatees began to speak the Songzi dialect and to follow Songzi customs.[115]

It is no secret that water conservancy projects sparked frequent protests.[116] We certainly should not sugarcoat the conflicts between relocatees and local residents either. In 1966–1967, 17,345 people from 3,539 households were moved from Henan to Shayang because of the Danjiangkou Reservoir project. It was arranged to resettle them in one location. In 1969, a three-day-long fight broke out between these relocatees and local residents, causing the death of fourteen people and the serious wounding of sixty-six. A major impetus behind the mayhem was that the relocatees faced hardship in their new homeland: they were heavily in debt, the land allocated to them was poor, they lacked farm machinery and cattle as well as houses and sufficient building materials. They were relocated to one place and, because of their number, they were very powerful. To avoid a repeat of this kind of conflict, the local government divided the second batch of Henan relocatees into small groups and sent them to seventy-one villages. The same method was used in the following decades. The 1,835 relocatees of the Three Gorges project, however, were either settled separately or in groups at one location in Shayang, by the construction of the so-called Three Gorges villages.[117] In the twentieth-first century, people displaced by the South-to-North water diversion project refused to be relocated separately in Shayang, and instead wanted to be relocated in groups to one location, which would be very difficult because of the limited availability of land in Shayang. After great efforts by the local government, and with the agreement of delegates of the relocatees (who secretly visited their assigned locations) and of local villagers, all of them, 1,525 in total, were relocated in groups to nine locations. The relocatees were reportedly satisfied with this arrangement.[118]

The conflicts between relocatees from Henan and local residents in Mianyang were even worse and lasted far longer. In 1968, about twenty thousand people displaced by the Danjiangkou Reservoir project were relocated to Mianyang. Some of them began to make trouble in 1970 by asking that they be allowed to move back to their hometowns, and then in the late 1970s they started to refuse to work in the fields, paid no heed village cadres, and fought with local villagers. Misjudging the tolerance of the local government as weakness, in the late 1980s they even organized "illegal" meetings, assaulted the

115 *Weishui shuiku zhi* 1996: 187–89; *Weishui liuyu zhi* 2010: 230.

116 Jun Jing 2000; Ying Xing 2001.

117 *Shayang xian minzheng zhi* 2012: 174–75; *Shayang xian zhi* 2013: 1061.

118 Li Jialin 2014.

local government by breaking into its offices, beating local cadres, and destroying a police car. These activities clearly crossed a line. The local authorities acted and arrested and jailed two of the relocatees' leaders in 1990.[119]

Although the government at various levels invested a great deal of effort in relocating people affected by the Danjiangkou Reservoir project, there were, as we have noted, seemingly endless problems, such as people moving back without authorization, protesting, and fighting. Particularly unruly were people had been relocated several hundred kilometers away. The state learned a lesson from this, and later most people displaced by water conservancy projects in Hubei were relocated locally.[120]

Based on previous experiments, and with China's increasing financial strength, relocatees in the twentieth-first century have been treated much better than their predecessors. In Qujialing Administrative District (formerly Wusan Farm), 108 households and 443 people displaced by the Geheyan Reservoir (created by the construction of the Geheyan Dam on the Qing River) were relocated to this region in the years 2002 and 2003. To accommodate these relocatees, the local government built 103 houses and one office, a 1,500 meter street, and a town replete with electricity, water, TV, phone service, and other amenities.[121]

Among all the types of relocatees discussed above, the least numerous by far were those in the last sub-category, that is, people who moved to a new place on their own. In other words, the removal and relocation of the vast majority of people displaced by water conservancy projects were organized by the state, which is very different from the Ming-Qing era when virtually all immigrants came to the Jianghan Plain on their own. From the above discussion it can also be seen that the population movement due to water conservancy projects is not a phenomenon that first appeared only in recent decades. The PRC has dealt with millions of people displaced by water conservancy projects since 1949. From 1949 to 2005, more than half of China's relocatees moved, or were moved, because of the construction of reservoirs. And in the Yangzi River valley, more than five million people had been relocated due to the construction of reservoirs.[122] From 1958 to 1964—the heyday of water conservancy

119 Ma Daofu 2009.

120 *Hubei shuili zhi* 2000: 990–91.

121 *Qujialing guanliqu zhi* 2014: 429. This kind of improvement of resettlement housing also occurred in the province of Yunnan, where people relocated because of the construction of hydraulic power stations on the Lancang River usually approved of the resettlement housing (Tilt 2015: 76, 81).

122 Changjiang shuili weiyuanhui 2005: 1–2.

construction—in Hubei alone 222,000 people were relocated because of the construction of reservoirs. According to regulations, if the average farmland per capita in any place is below 0.6 mu of paddy fields or 0.4 mu of paddy fields and 0.4 mu of dry land, all "extra" people must be relocated to other places, and the hosting villages must treat them the same as their own villagers.[123] Unfortunately, not all hosting villages have followed this regulation.

Since the movement of the majority of relocatees in the PRC has been directed and organized by the state, a closer look at the role played by the government at different levels is needed. At the very beginning of the Three Gorges project, the Yichang County government had to immediately move 12,257 people to higher locations. The government approached this challenge as a political task that had to be completed.[124] In his case study of a small hydroelectric station in the Three Gorges area, the sociologist Ying Xing vividly describes how each level government—from the township to the central government—was involved in resettling the displaced population.[125] The state played a no less prominent role in similar projects in the Jianghan area. For example, in the 1950s–1960s, when many reservoirs were built in Jingzhou, the Jingzhou government considered the task of dealing with the displaced people as not simply an economic one but also a political one.[126] In another example, in the early twenty-first century the governor of Hubei indicated that the relocation of 77,000 people displaced by the South-to-North water diversion project was a political task that must be completed. Officials who failed in their duty, the governor added, would be fired.[127] No matter what the period or level, population resettlement was always a political task. In the political ecology of the CCP, if a task is labeled "political," it absolutely must be completed, and no excuses will be brooked.

Although the Chinese state has been gradually withdrawing from the governance of rural society since 1979, it is still powerful enough to remove and resettle people by "administrative order." It was by the exercise of this power that the population displaced by the Three Gorges project was distributed all over China, including the Jianghan Plain. The relocation of about 1.3 million people would be a formidable challenge or simply impossible in many other nations, where land is privately owned and decision-making is more democratic or

123 SZ113-1-268.
124 Yang Liming and Jian Xin'an 2000.
125 Ying Xing 2001.
126 D35-1-319.
127 Li Jialin 2014.

subject to public scrutiny, but under the unique party-state system of current China where land is owned by the state but managed by villages, the challenge was manageable, though there were many problems and difficulties. One cannot be as confident, however, that the long-term environmental consequences of the Three Gorges project, which remain largely unknown, will be so easily handled.[128]

128 For a recent discussion on the mixed environmental consequences of the Sanmenxia hydropower project in mid-twentieth-century China, see Xiangli Ding 2024: 204–230.

CHAPTER 10

The Evolution of Dike Systems and Socioeconomic Life

In the Qing and the Republic, as dike-building and polder-enclosure forged ahead, the mismanagement of the dike systems led to frequent flooding. The Great Jingjiang Dike, for instance, was breached in 39 of the 161 years from 1788 to 1949.[1] The Han River dikes also frequently broke, and almost every year some polders in the Jianghan area suffered from water calamities caused by dike breaches.

The collapse of a polder dike might affect only the residents of that polder, but the collapse of a river dike would cause suffering among far more people across a much larger area. The 1788 flood, for example, breached the Wancheng Dike in twenty-two places and flooded a large part of the Jianghan Plain, drowning about ten thousand people in Jingzhou city alone. The Jianghan area suffered another record-setting flood in 1870, which eventually resulted in the formation of the Songzi River. In the twentieth century, the most disastrous flood to hit the Yangzi River valley occurred in 1931. This flood drowned twenty-three thousand or more people in the northern part of the Jianghan Plain alone, and led to the deaths of still more through starvation and disease.[2] In fact, more flood victims died of disease than starvation.[3] The total number of deaths by drowning in the flood—which inundated 269 counties in the middle and lower reaches of the Yangzi River—is difficult to estimate. The GMD's National Flood Relief Commission claimed that about a hundred and forty thousand drowned, while Rewi Alley believed the toll might have been in the millions.[4] Four years later, the Han River dike in Zhongxiang broke at eighteen locations, and the ensuing inundation caused about eighty thousand deaths in eleven counties on the northern bank of the Han River; or, according to another source, the flood drowned ninety-six thousand people.[5] In Qianjiang only a hundred people were drowned, but fifteen thousand died of starvation. Reportedly, the dike was in disrepair and there was a lack of

1 *Jingjiang dadi zhi* 1989: 71–74.
2 *Hubei shuili zhi* 2000: 189–91; *Honghu xian zhi* 1963: 32.
3 Courtney 2018a: 76–82.
4 National Flood Relief Commission 1933: foreword; Alley 1956: 11.
5 *Zhongxiang shuili zhi* 1998: 4, 64; *Hubei sheng zhi shuili* 1995: 219.

 | DOI:10.1163/9789004739482_012

preparations. It was said that the dike collapsed even before laborers were sent to it.[6] Although there were no more huge floods after that, people still suffered from the breach of river dikes, such as in 1945 in Shishou, where the Ouchi River dike and the Yangzi River dike ruptured at thirteen locations and flooded twenty-eight polders. An epidemic of diseases—cholera, smallpox, measles, and others—ensued, killing about ten thousand people.[7]

With the coming of the PRC in 1949 dike ruptures drastically declined. In the early years of the PRC, the riverbed of the Yangzi River continued to rise, the volume of water grew, and the high-water season lasted longer, all of which increased the threat to the integrity of the dikes, but the dike systems became more secure than ever before. The Yangzi River dike in Jianli, for example, ruptured in thirty-seven years from 1608 to 1935, but from 1949 to 1985, even though in fourteen years the high-water level was higher than that in 1935, the dike held except in 1954, when a segment was intentionally opened to discharge floodwater.[8]

The 1954 flood was the greatest flood to hit the Jianghan area in the twentieth century. Extensive tracks of land were flooded, many polder dikes ruptured, and some river dikes also collapsed.[9] After that, however, all major river dikes remained secure with a few exceptions such as the Yangzi River dike rupture at Tianjiakou, Honghu, in 1969, which flooded 800,000 mu of land. Even in the 1998 flood—the second greatest flood in the twentieth century—all major river dikes held.[10] But the rupture of polder dikes, particularly dikes of river beach polders, constantly recurred. Some of them were over-flooded and others were opened to discharge floodwater and hence reduce pressure on more important river dikes. Occasionally some small branch river dikes also ruptured.

Generally speaking, the Jianghan people have built river dikes to protect their farmland from the annual high water for about two millennia, and they began to build polder dikes on a large scale about a millennium ago. This endless diking, however, naturally reduced the flood diversion area along the Yangzi and the Han Rivers and the water surface of lakes, which contributed to the increasing incidences of water calamities, particularly the frequency of dike ruptures. Both the state and local society therefore bore the very heavy burden of providing relief for flood refugees. And the local people also had to grow relief crops or collect aquatic products for survival. Furthermore, frequent dike

6 *Qianjiang shuili zhi* 1997: 93–94.

7 *Shishou xian zhi* 1990: 22.

8 *Jianli difang zhi* 1991: 51.

9 SZ113-2-60-001.

10 *Honghu xian zhi* 1992: 25; *Jingjiang difang zhi* 2012: 762.

ruptures led to conflicts over the control of waterways, particularly the opening/closing of river outlets/breaches. All these are general features of Jianghan society in the Qing and the Republic that underwent dramatic changes after 1949,[11] with the realignment of rivers, the improvement of water conservancy, and the stabilization of the dike systems. This chapter explores some of these changes from a historical perspective.

1 Disaster Relief Measures and Their Effect

Based on the Qing dynasty's regulations, it is believed that government relief early in the dynasty was effective, but gradually declined after 1800.[12] Although throughout its history China frequently suffered from disasters and famines, it also enjoyed a two-millennia-long tradition of public relief. The Qing government inherited this tradition and promulgated detailed regulations on relief. During the era often described as the "High Qing," that is, from the 1680s to around 1800, China enjoyed both peace and prosperity. In the nineteenth century, however, the Qing regime became financially weak and politically unstable, and concomitantly, relief became weaker and less effective, particularly in North China.[13]

The most common disasters in the Qing were water calamities, which accounted for nearly 60 percent of all catastrophes. And most water calamities were caused by waterlogging and flooding. According to the Qing regulations, the procedures for providing disaster relief included reporting the disaster, surveying the disaster, and distributing relief—tax exemptions, relief money and food, work relief, refugee settlement, and so on.[14]

Since the Jianghan area frequently suffered from water calamities caused by dike ruptures, the local governments had to report calamities to the court time and again and distribute relief to flood refugees. In 1788, 37,000 taels of silver were provided as relief to about three hundred thousand flood refugees in Jiangling, and 196,000 taels for repairing houses, city walls, prisons, and granaries in Jingzhou city. In this instance, relief was usually provided for one month. In other cases, relief money could also be used to set up shelters for flood refugees. The state was also responsible for sending refugees back to their hometowns. Depending on the extent of a disaster, the central government granted

11 Jiayan Zhang 2014.
12 Hsiao Kung-ch'uan 1960; Will 1990; Will and Wong 1991.
13 Lillilan Li 2007: 250–82; Edgerton-Tarpley 2008.
14 Li Xiangjun 1995: 15, 19, 23–47.

remissions of the agricultural tax and/or tribute grain in disaster-stricken areas or postponed the collection of tax for one to several years.[15]

The standard food relief in the early Qing was half a *sheng* (about half a liter) of husked grain per adult per day, which merely sustained a person at a subsistence level.[16] The Qing classified victims older than sixteen as "big mouths."[17] But the standard for a big mouth was much lower in the Republic. In 1946 the Hubei branch of the United Nations Relief and Rehabilitation Administration determined that 10 percent of emergency relief grain should be allotted to "the elderly, weak, women, children, and disabled people, with no more than 20 pounds for a big mouth and 12 pounds for a small mouth (under twelve years of age)."[18] In 1948 in Zhijiang, the targets of emergency relief were people who could not survive without relief. Individuals more than five years old were classified as "big mouths" and received two *dou* (about ten liters) of relief grain, while those under the age of five were "small mouths" and received one dou.[19] In terms of harvests, a loss of 60 to 100 percent of a harvest was regarded as a "disaster" in the Qing.[20] In the PRC, a loss of 70 or more percent of the harvest was regarded as a "heavy disaster," and 60 or less percent as a "light disaster."[21]

Once a flood was reported and confirmed, other than exempting taxes in the flood-stricken area for one or more years, depending on the seriousness of the disaster, the Qing state also released relief grain from state granaries known as ever-normal granaries (*changping cang*). But relief from these granaries was slow and relied on obtaining approval from the upper-level government to transfer grain. Communication also took time. Relief from community granaries was quicker and could be provided directly to those in need.[22] And the number of flood refugees could be huge. In 1823, 448,300 flood refugees in Jiangling and Jianli qualified for relief.[23] The sources of relief funds varied. In 1827, 50,000 taels of silver from that year's land tax (*diding yin*) were used to relieve flood refugees in these two counties.[24] In 1830, as a top official from Beijing en route to Yunnan passed through Jiangling, a degree-holder accompanied by about a hundred local residents submitted a petition to him asking

15 Shuili dianlibu 1991: 485, 514, 699, 833, 837; Chen Zhenhan et al. 1989, vol. 3b: 133–43.

16 Will 1990: 130–32.

17 Li Xiangjun 1995: 26.

18 100-8-38: 32.

19 *Zhijiang minzheng zhi* 1990: 41.

20 Li Xiangjun 1995: 24.

21 *Jingmen shi minzheng zhi* 2016: 175.

22 *Yunmeng xian zhi* 1883, vol. 1: 3a–13a.

23 Shuili dianlibu 1991: 658.

24 *Zaixu xingshui jinjian* 2004: 56.

the county government to investigate their situation. This top official found that half a month after the flood, the water still had not receded, and the people were suffering. This petition seems to imply that the Jiangling County government did a poor job of surveying and dealing with the disaster.[25]

Furthermore, aside from the state, there were several other sources of relief in imperial times, and official relief only covered those who actually worked the land (but excluded wage laborers, since the state expected their employers to provide relief) and a third of poor degree-holders. Hence relief provided by private organizations and charities could help to cover those who were not eligible for state relief.[26] Yet, even these sources of relief were subject to peculation. There is evidence that in Tianmen in 1832, for example, half of the donations collected from the local people might have been embezzled. In addition, the donations received by the Tianmen County government were to purchase cotton clothing, but flood refugees only received straw clothing.[27]

In traditional China, a common form of relief was to provide porridge to flood victims, a practice influenced by both Confucianism and Buddhism.[28] Although the state also set up soup kitchens, most were run by non-state organizations or individuals. In the mid-nineteenth century, because of frequent dike ruptures, many Jianghan people fled to Hankou, a major entrepôt absorbed into the newly created municipality of Wuhan in 1949, and to house themselves erected shacks along the banks of the Yangzi and the Han Rivers. The presence of these people became a common sight every year in autumn and winter.[29] In autumn 1849, for instance, about 230,000 refugees fled to Hankou, and in 1909, flood refugees again escaped to Hankou, where the local government, in cooperation with local gentry-merchants, set up soup kitchens in thirty places and fed more than a hundred thousand victims daily.[30]

This tradition continued in the Republic. In 1931, after Hankou was flooded, the Hubei provincial government established a Hubei Emergency Relief Commission, which set up no less than thirty refugee camps, each housing from one thousand to more than a hundred thousand refugees.[31] Soup kitchens were set up and dished out a big spoonful of porridge for adults and a small

25 *Zaixu xingshui jinjian* 2004: 107–08.
26 Will 1990: 135–37.
27 *Zaixu xingshui jinjian* 2004: 139.
28 Lillian Li 2007: 228.
29 Wang Baoxin 2002 [1933]: 52.
30 Shuili dianlibu 1991: 881, 1196–97.
31 LS1-5-644.

spoonful for children—excluding women wearing makeup and silver jewelry.[32] Some local wealthy households also provided porridge for flood refugees.[33]

Since broken dikes, particularly major dikes, had to be repaired quickly, a task that required a great deal of labor and money, the Chinese state from the early Qing all the way to the early PRC chose to kill two birds with one stone by providing work relief (*yigong daizhen* or *gongzhen*), that is, hiring flood victims to build or repair dikes, particularly after the rupture of a major dike when the rebuilding project would be large and costly. Funds for work relief came directly from the state, or through state loans, fines, donations, or even the likin. And there could still be other sources. For example, after the 1788 flood, the state confiscated the property of the Xiao family (who occupied an islet that was blamed for causing the flood) and ordered that it be used in the future for work relief.[34] In the mid-nineteenth century, the cost of one person's monthly grain ration was only 0.15 taels of silver, or 1.8 taels per year, and thus if a work-relief dike project cost a hundred thousand taels and all the money was paid as wages to the laborers, it could support 55,555 people for one year. Flood refugees who were hired as workers were usually able-bodied men. The preferred form of relief for the elderly and weak was to provide them with porridge. Work-relief projects also could stabilize disaster-stricken areas. If stability was not achieved, flood refugees, out of desperation, could turn to banditry.[35]

Many of these relief measures—particularly work relief and soup kitchens—continued in the Republic. After the 1931 flood, the GMD government established five work-relief bureaus in Hubei. These bureaus hired 38,200 laborers in total, handed out 93.5 million kilograms of relief grain, and rebuilt 1,295.4 kilometers of dikes along the Yangzi and the Han Rivers.[36] The relief also included flour, tools, and money to build shelters.[37] The Bureau of Engineering Work on the Yangzi and Han Rivers was ordered to repair dikes via work relief, and flood refugees often directly requested funds from the bureau to repair their dikes, including polder dikes, which were usually maintained by the locals themselves.[38] In 1946 the bureau made a detailed proposal to improve work relief on the Jianghan dikes, such as setting the workday at seven hours

32 Hubei sheng shuiliting et al. 1999: 66–67.
33 *Jianli xian zhi* 1994: 836.
34 Ni Wenwei 2002 [1885]: 26.
35 *Zaixu xingshui jinjian* 2004: 247, 388–89, 785.
36 Yu Tao 2005.
37 LS1-5-785.
38 LS1-3-0214-007; LS1-3-0236-005.

and providing 2 kilograms of relief grain per laborer per day as well as a reward for catching badgers.[39]

But even this kind of relief suffered from corruption. In one case, the official in charge of a project appointed his relatives as overseers and falsely reported the number of laborers and the volume of dirt and pocketed the difference. In other instances, officials simply embezzled relief funds.[40] Rewi Alley, who participated in the relief work after the 1931 flood, found that the relief organized by the GMD government was insufficient and ineffective. Moreover, Alley asserted, the GMD was uncooperative, leaving real relief work up to "a few enthusiasts." Alley went on to charge that the GMD was only interested lining its pockets and attacking the Communist base area in Honghu.[41] Although Alley was unquestionably an ardent Communist sympathizer, archival materials confirm that relief was more commonly carried out by individuals and non-governmental organizations. The public generously provided porridge, medicine, and money to flood refugees—although some people had warned that it was better to give refugees materials rather than money since the latter might easily be embezzled.[42]

Archival materials also show that the GMD was not as inept and callous as Alley and others would have us believe. After the 1931 flood, to relieve flood refugees, the government went so far as to issue relief bonds and to cut the pay of government employees by 3 to 20 percent for three months to raise relief funds. It also acted to stabilize the market price of rice and ordered county governments to safeguard cattle. Flood victims were not to be allowed to kill or sell their cattle. If this created undue hardship, county governments were to provide loans to the affected flood victims.[43] In Songzi in 1945, targets of work relief included anyone—male or female—who voluntarily went to the dike worksite to repair the broken dike. "Big mouths [were paid] 400 yuan and small mouths 300 yuan," again, with no gender differences.[44]

Work relief was also used in the early PRC. In fact, it was the most common form of relief. The new state had learned from past experience that the restoring of dikes was the best way to relieve a huge number of flood refugees.[45] County governments thus usually organized hundreds of thousands of flood

39 LS30-1-0889-004.

40 Wang Zhengben 1999; LS1-5-785.

41 Alley 1956: 11–14.

42 LS1-5-644.

43 Hubei sheng shuiliting and Hubei sheng dang'anju 1999: 34–37.

44 100-5-280: 4.

45 SZ1-5-65.

refugees to repair broken dikes.[46] The standard for determining the pay for work relief was that each laborer should be compensated enough to support the laborer and another 1.5 to 2 people.[47] In Mianyang the principle behind relief work was slightly different: if the laborers were poor peasants, they contributed labor and received grain; if they were middle peasants, they contributed labor but did not receive grain; if they were landlords and rich peasants, they contributed both labor and grain.[48] In other words, these kinds of undertakings were not simply engineering or water conservancy projects—they were also political projects.

To better relieve flood refugees, in Jianli in 1950 even women and children were encouraged to go to the diking sites, where they could be useful by gathering firewood and cooking meals.[49] In 1951 in Zhijiang, each laborer could earn 2.5 jin unhusked rice per day. Since a laborer's daily consumption was around 1.5 jin, that left a surplus that could help feed other members of the laborer's family. Indeed, many people who came to help repair dikes did so to ride out a lean year.[50] For example, in spring 1955, a rural household in Zhijiang was able buy desperately needed rice seed by using money it had earned from work relief on that year's annual repairs.[51] In the same year in Mianyang, the money laborers earned from work relief could purchase enough grain to feed the population of the whole county for nineteen days, a huge source of support for the locals, who "enthusiastically endorsed" the policy.[52] At the turn of 1953, Hubei conducted work relief in 68,000 or more water conservancy projects, with peasant laborers using the money they earned from these projects to purchase grain, run sidelines, and invest in agriculture. The sources claim they thanked the Communist Party and regarded the government as their savior, boosting the prestige of the party and government.[53] It has been argued that, in the early PRC, the campaign against snail fever was aimed at legitimizing the party's rule,[54] just as with the construction of water conservancy projects.

Work relief in the early PRC, however, should not be understood purely as public welfare. Examples of this truism abound. After the 1954 flood, many

46 *Xiaxindian zhen zhi* 2012: 20; *Qianjiang xian zhi* 1990: 286.

47 40-1-55a; *Dongjinghe difang zhi* 1994: 307; *Hanyang xian shuili zhi* 1990: 160; *Jianli difang zhi* 1991: 366.

48 SZ1-5-59A.

49 1-1-010-004.

50 45-1-1.

51 45-1-11a.

52 Wuchang 49-1-2a.

53 SZ113-2-022-003.

54 Gross 2016.

small bundles of willow were needed for the rebuilding of a dike in Tianmen. The job was assigned to the women of Yuekou Township, but since there were many women and each day different ones were employed, the result was that the willow bundles were of poor quality. A document in the Jingzhou Municipal Archives argues that the quality of assigned projects should not be sacrificed.[55] In Shishou, people were told that they could get almost anything if they went to work on the dike, and thus all kinds of people went, including a blind person led by a boy. And among the 839 people in a township in Mianyang who sought work relief, 350 were children, the elderly, the ill, the disabled, and pregnant women—all were considered unqualified.[56] In Jianli in 1954, some cadres also simply understood work relief to be purely public welfare, and as a result, many laborers went to the dike without bringing their tools, and even old women with bound feet, children, the ill, and pregnant women went.[57] Similarly, in Jiangling in 1955 almost everyone went to the annual repairs worksite. That included 1,577 people (550 elderly and weak, 320 children, 26 pregnant women, 249 disabled people, and 432 others) who were sent home. One old man claimed that if he stayed home he would starve to death, so it was better to go to the dike.[58]

In the PRC, even urban residents benefited from work-relief projects. In 1963, 15,475 jobless urban laborers were recruited to participate in forty-six work-relief projects, fourteen of which were dike projects. The income a person earned from these projects could support three people, which not only improved the lives of many but also significantly reduced the relief burden on the local governments.[59]

At the end of the Qing and in the Republic, foreigners became increasingly involved in relief in China. In winter 1931, a Christian church in Hankou provided porridge and medicine to flood refugees, reportedly inducing many residents of Mianyang to convert to Christianity. But some of them later left the church since they refused to give up their tradition of ancestor worship.[60] In the following year, when flood refugees from Changtangkou Ward of Mianyang returned home, some Christian clergy followed them and convinced them to be baptized by giving them cash. Many of these converts later also left the church, since they "could not bear to discard the memorial tablets of their

---

55 D35-1-50b.
56 40-1-55b.
57 1-1-092-008; 1-1-092-009.
58 48-1-32.
59 XX000065-WS01-1946-0014.
60 *Xiliuhe qu zhi* 1986: 10.

ancestors."[61] After World War II, materials from the United Nations Relief and Rehabilitation Administration were also distributed in the Jianghan area.[62]

Before 1949, the outbreak of epidemic diseases such as dysentery and cholera after dike ruptures was common. In Shishou, for example, the rupture of many polder dikes in 1918 caused an outbreak of cholera and the death of about a thousand people; dysentery broke out after the 1935 flood and caused around three thousand deaths; the rupture of the dikes of fourteen polders in 1945 drove many refugees to live on dikes, amid trash and night soil, with the result that cholera soon broke out and claimed about ten thousand lives.[63] After 1949, however, these epidemic diseases were largely brought under control, as the PRC state sent medical personnel to flood-stricken areas to treat the ill and prevent disease from spreading. In 1954 in Jingzhou, 2,800 medical workers (800 of whom came from outside of Jingzhou) were sent to flood-stricken zones for this purpose.[64] The state even relocated some flood refugees to safer places in neighboring or more distant counties and provided them with relief grain monthly.[65] Madian Village, for example, received some flood refugees and cattle in the 1954 flood.[66]

After the 1954 flood, because of the state's great effort to prevent epidemics, there was no outbreak of influenza, hepatitis, or dysentery in Qianjiang.[67] But in Changtangkou Ward, after the diversion of the 1954 flood, 70 to 90 percent of the villagers contracted dysentery and malaria.[68] Although the state in both the Qing and the Republic was well aware of the danger of the spread of epidemic diseases following an inundation, the Qing state did not send or was unable to send medical personnel to provide care. It was not until the Republic that the state started to dispatch medical personnel to treat the ill and prevent diseases and it was not until the PRC that this became routine.

In another change brought by the PRC, support was elevated socially and extended nationwide—all of society was mobilized to help flood refugees, and relief materials poured in from all over the country. In 1954, people fighting the flood in Wuhan received machinery from Harbin (in Northeast China), materials from Sichuan (in the upper reaches of the Yangzi River), and help from

61 *Changtangkou qu zhi* 1990: 7, 432.
62 *Jianli shuili zhi* 2005: 417.
63 Cheng Gongneng 2004.
64 *Jingzhou weisheng zhi* 1990: 8.
65 *Qujiawan zhi* 1986: 14; *Xiliuhe qu zhi* 1986: 22.
66 *Madian cun zhi* 2004: 2.
67 Yue Xinfa 2006.
68 *Changtangkou qu zhi* 1990: 14.

medical personnel from Shanghai (in the lower reaches of the Yangzi River).[69] Some people believe this flood "did not destroy the foundations of the PRC, but instead enhanced the prestige of the CCP."[70]

As a result, in striking contrast to the Qing and Republican eras, when deaths from disasters were all too common, in the PRC, as official propaganda repeatedly proclaimed, the new state would not let a single person starve to death. Although this is a fundamental task for any state, it was hardly possible before 1949. After the 1954 flood, the state indeed did its best to prevent deaths, and many official documents boasted that "not a single person had starved to death" after the flood. This was far from being a hollow boast. In Xiliuhe Ward, for example, about a thousand people were killed in the 1931 flood (it is unclear how many of them died of starvation), but not a single person died of starvation in the 1954 flood.[71] While it is probably true that no one died of starvation in Xiliuhe, death from starvation occurred elsewhere and people still drowned in floods. And even more people died of flood-related diseases. In Jiangling alone, the 1954 flood drowned 101 people and 181 cattle, and 1,494 people died of pestilence, and some even committed suicide, though no one died of starvation in Haoxue.[72] And in Yingcheng, the 1954 flood killed 33 people and injured 340 in that year's high-water control and emergency repairs.[73] According to a survey, of the almost ten million people affected by the 1954 flood in Hubei, 149,507 died, including 69 who froze or starved to death, 2,010 who committed suicide, 5,001 who died from "drowning, or were crushed to death by collapsed houses, or otherwise suffered abnormal deaths." All the other deaths—142,427 (or 95.26 percent)—were caused by disease (Table 10.1).

As Table 10.1 shows, about 3.87 million residents of Jingzhou—a key area of the Jianghan Plain—were flood victims and 101,637 of them died of disease. As the original notes indicated, these are rough statistics. An official Hubei gazetteer on health—its numbers should be more accurate—stated that there were about 3.7 million flood victims in Jingzhou, of whom 95,000, mostly the elderly and children, died of disease, principally enteritis and dysentery.[74] Although these figures are obviously contradictory, and far smaller than the death toll before 1949 (particularly from drowning), in absolute terms they were still high.

---

69 Alley 1956: 45–49. On this matter, the best-known example would be the relief effort after the Wenchuan earthquake of 2008; in the Jianghan Plain, it would be the relief campaign during and after the 1998 flood.

70 Zeng Fanrong 1997: 60.

71 *Xiliuhe qu zhi* 1986: 297–98.

72 *Haoxue zhen zhi* 1986: 13, 59.

73 Yingcheng 49-1-26.

74 *Jingzhou weisheng zhi* 1990: 8.

TABLE 10.1 Deaths in flood-affected areas of Hubei, 1954

Prefecture	Flood-affected population	Mortality				
		Total	Deaths from freezing and starvation	Suicide	Deaths from diseases	Deaths from other causes
Total	9,904,801	149,507	69	2,010	142,427	5,001
Huanggang	2,606,130	11,260	5	109	10,152	994
Xiaogan	2,435,999	31,699	9	516	29,692	1,482
Jingzhou	3,873,335	104,598	54	1,349	101,637	1,558
Yichang	248,957	829		25	683	121
Xiangyang	598,814	889		1	163	725
Enshi	108,629	107		7		100
Huangshi	32,937	125	1	3	100	21

Notes from the original source:
1. Data are from reports of affected counties.
2. Data are dated from the beginning of the flood (1954) to around February 1955. Since each county reported at different times, the accuracy of the data was affected.
3. "Deaths from other causes" includes "deaths due to drowning, being crushed by collapsing houses, and other abnormal deaths."

SOURCE: SZ039-001-0012-0002

The death toll was drastically reduced in the following decades. In Jingzhou in 1980, dikes in 142 polders collapsed and seven dikes were opened to discharge floodwater, and in 1983, dikes in 103 polders collapsed and eight were opened to discharge floodwater, but the death toll had been reduced to 119 and 207 people, respectively,[75] though it is unclear from these data how many of victims were drowned or died of other causes.

The PRC state obviously has a greater capability of delivering relief goods than its predecessors. In 1980, after some polder dikes collapsed in Jianli, the state and the PLA even used airplanes, warships, and military vehicles to transport relief materials to the flood-stricken area and to transfer flood refugees.[76] These were certainly fast and efficient and had never been used for the same purpose before.

75 *Jingzhou diqu zhi* 1996: 588.
76 *Jianli xian zhi* 2013: 2.

Even with all these advantages, it is still almost impossible to prevent deaths from floods. In July 1958, when a polder in Hanyang was over-flooded because of the use of the Dujiatai flood diversion zone to discharge floodwater, ten people died.[77] In Xiaogan in the 1998 flood, seventeen people died of drowning, sunstroke, accidents, and lightning.[78]

Since state relief before 1949 was insufficient, flooding forced many people to sell their land. In the 1831 flood, for example, dikes ruptured in more than seventy locations in the Jianghan area. The Qing state spent more than 290,000 taels of silver to block these breaks.[79] In the next year, the Xiong family in the town of Yuekou purchased hundreds of plots of land from peasants who moved out or faced financial hardship and had a poor harvest.[80] The same fate befell peasants in the Republic who sold their land because they could not pay the dike fees.[81] For peasants, it is hard to imagine any fate worse than losing their land. Thus, there was a long tradition, which continued into the Republic, of peasants resorting to witchcraft or sorcery to deal with the worst effects of disasters.[82]

After 1949, the state also took full responsibility for relief for flood refugees, eliminating the rationale for sorcery. With the decrease in the number of dike ruptures, relief in the PRC has gradually become a state endeavor. The state sometimes organizes individuals to donate money or materials, but over time the share of their contributions has been drastically reduced. For example, after Baili Islet's dikes were opened to discharge floodwater in 1954, the state provided emergency relief funds that could support all the residents for more than forty days. Not a single person was rendered homeless, and none died of starvation.[83] As noted earlier, even cattle were taken good care of by the state. Needless to say, cattle were extremely important to peasants and agriculture, thus in 1954 when many peasants suffered from floods and wanted to sell their cattle, they were persuaded by the state not to do so since the animals were "vital" to agriculture.[84] The state did more than exhort the peasants. For example, when peasants in Yunmeng found it extremely difficult to transport their livestock when they were transferred to safety areas, the state-run local supply and marketing cooperative purchased the pigs and cattle and found someone

77 *Hanyang xian zhi* 1989: 20.
78 *Xiaogan shi shuili zhi* 2011: 74.
79 *Zaixu xingshui jinjian* 2004: 123.
80 Zhang Jianmin and Tang Gangmao 2014: 122–343.
81 LS20-11-00001132-001; LS20-11-00001162-001.
82 Ao Wenwei and Li Xiang 2003.
83 *Zhijiang minzheng zhi* 1990: 160.
84 Yingcheng 49-1-26.

to raise them, then loaned the cattle to peasants after the flood receded. Peasants reportedly praised the CCP for this.[85]

Since 1954, no major river dike in the Jianghan area has ruptured (with a few exceptions). There have been breaks in secondary or polder dikes, but even then, many of them were intentionally opened to discharge floodwater. The serious breach of the Honghu dike at Tianjiakou in 1969 was mostly caused by a lack of vigilance, which the Honghu County government considered a great political and economic loss. After the rupture, the government at different levels immediately took action to rescue flood victims and to distribute relief, in stark contrast to the tardiness and ineffectiveness of the 1931 relief effort.[86]

2 Environmental Stability and Crop Choices

To cope with the frequent water calamities in the Qing and the Republic, the Jianghan people developed a sophisticated cropping system by carefully choosing water-resistant crops and a dominant single-cropping pattern of middle season rice that suited their precarious environment. They also planted various famine-relief crops such as buckwheat, millet, beans, kaoliang (Chinese sorghum), and even vegetables. Taking advantage of an environment dominated by rivers and lakes, they also grew or collected aquatic products as substitute foods. These practices greatly contributed to their odds of survival in famines or other disasters.[87]

This tradition continued in the early PRC, particularly in the GLF. The Jianghan people had generations of experience in dealing with water calamities and coping with famines, which was of great value in confronting the GLF famine. In a state farm, for example, after a flood in 1958, the residents planted cabbage and radishes. They were also mobilized to chop reeds, catch fish, and engage in sidelines to overcome the disaster.[88] But in general, because this famine was not caused by floodwater, the locals did not grow water-resistant crops, but instead planted famine relief crops, especially buckwheat and sweet potatoes, and collected readily available aquatic products such as lotus roots, water chestnuts, and wild rice. These kinds of substitutes, however, were nutritionally

85 Xu Linguan 2006.

86 SZ113-2-342-001; SZ113-2-342-003. In his critical reassessment of the 1954 high-water control, Chris Courtney also notes that the Maoist state was "relatively effective" or did "relatively well" in relief work (Courtney 2018b: 1810, 1835).

87 Jiayan Zhang 2014: 103–15, 185–90.

88 *Caobuhu nongchang zhi* 1988: 17.

inferior, and malnutrition led to widespread diseases. But these practices still saved many lives and substantially reduced the death toll from starvation. The situation on the Jianghan Plain was therefore generally better than in many other parts of rural China.[89] In some cases, however, the Jianghan people certainly suffered greatly. In winter 1959 and spring 1960, despite severe food shortages, Yangling Commune in Yingcheng overtaxed the endurance of the peasants by carrying out an undue number of water conservancy projects. The result was 4,599 people suffered from edema, 1,577 moved out, and 1,695 died of starvation, illness, and violence.[90]

The food shortages around the year 1960 (the depth of the GLF), the large area of dike surfaces, and the relaxation of vigilance over dike safety led to some residents growing crops (principally barley, wheat, soybean, peas, and particularly sweet potatoes, and vegetables such as pumpkins and radishes) in dike exclusion zones or on dikes themselves. This was extremely dangerous, and the state immediately took action to uproot these crops. Although the dangerous consequences were clear, it still took some time for local governments to restore the dikes that had been damaged (see Chapter 2).

This tradition of dike governance continued after the GLF. In summer 1969, when a river dike ruptured in Yunmeng the government immediately began to provide relief grain—26 jin of husked rice per person per month for two months—to flood refugees. The county government also quickly allocated buckwheat, *sagu* (a kind of rice), and radish seeds to every production team without charge.[91]

With the constant construction of ever more water conservancy projects and the increasing stability of the dike systems, the Jianghan peasants gradually stopped raising relief crops. In Jiangling, for example, the annual acreage under millet decreased from about 70,000 mu before 1949 to less than 1,000 mu in 1985, and the acreage under soybeans fell from about 100,000 mu to 8,200 mu. Similarly, around 20,000 mu of kaoliang were planted in 1949, but in 1985 only a few households planted it sporadically.[92] In many other counties of the Jianghan Plain as well, by the 1980s, buckwheat and sweet potatoes, important traditional relief crops, had either almost disappeared or were only grown on piecemeal plots or planted sporadically.[93] In Miaotou Township, Hanchuan,

89 Jiayan Zhang 2013a.
90 Xiong Chuanxian 2006.
91 Xu Xinzhou 2006.
92 *Jiangling xian zhi* 1990: 274.
93 *Qianjiang xian zhi* 1990: 233; *Jianli xian zhi* 1994: 135; *Hanchuan xian zhi* 1992: 126–27.

potatoes and sweet potatoes were no longer categorized as food grains after 2000 and were recategorized as vegetables.[94]

The increasing safety of Jianghan's dikes also influenced peasants' choice of regular crops. For centuries, in deciding which crops to raise, the Jianghan people had to give first priority to survival. They favored crops or varieties that produced a relatively reliable harvest or were water-resistant over crops or varieties that had high yields or high economic value.[95] But in the PRC peasants gradually assigned a high priority to boosting their cash income and selected crops with high commercial value. In Hanchuan, people grew more barley than wheat before 1949, because barley matured earlier and could be harvested earlier to avoid being flooded. They also devoted more than half of their grain fields to so-called miscellaneous food-grain crops, such as kaoliang, corn, sweet potatoes, millet, buckwheat, and various beans as famine relief crops. After 1949, however, wheat gradually became the second most important food-grain crop after rice, and many miscellaneous food-grain crops were only planted sporadically or not at all by 1985.[96] In Xiaxindian Township, for instance, barley, buckwheat, kaoliang, and millet had entirely disappeared by 2007.[97]

Cotton, the most important non-grain cash crop in the Jianghan area, traditionally could only thrive on land that was not prone to flooding. Mainly because of this, although cotton had been planted in the Jianghan area for centuries, it was not widespread until the end of the nineteenth century and the early twentieth century, when outside market demand and the relative security of the Great Jingjiang Dike, which suffered no major break from 1871 to 1931, combined to stimulate the expansion of cotton cultivation in the region, particularly along the Yangzi River.[98] For the same reason, the acreage under cotton in both Tianmen and Mianyang—important cotton producers in the region—was either unstable (Tianmen) or limited (Mianyang) before 1949, since these two counties famously suffered from frequent water calamities at that time, mainly caused by breaks of the Han River dikes. But the acreage under cotton in both counties not only increased but also was more stable in the PRC. From 1920 to 1947, the acreage under cotton in Tianmen varied from 16,000 mu to 528,600 mu, and the yearly output of ginned cotton varied from 4,100 dan to 198,400 dan. But from 1949 to 1985, in twenty-five years more than

94 *Miaotou zhen zhi* 2010: 65.

95 Jiayan Zhang 2014: 103–10.

96 *Hanchuan xian zhi* 1992: 126–27.

97 *Xiaxindian zhen zhi* 2012: 97.

98 Jiayan Zhang 2014: 121–25.

700,000 mu of land were planted in cotton, and in 1976, 1977, 1979, and 1984 the yearly output of ginned cotton exceeded a million dan. One reason for this expansion was that water conservancy had been improved.[99] In Mianyang, the acreage under cotton was only 232,400 mu in 1920, but by 1976 it had increased to 625,000 mu, even though part of Mianyang had been split off in 1950 to create the county of Honghu.[100] Again, the expansion was due mainly to the improvement of water conservancy. At the same time, the Han River dikes became far more secure and stable.

Since the 1990s, views on some aquatic products have also changed. These are no longer considered relief products or substitute foods but instead "green" or "organic" foods.[101] For example, lotus seeds, lotus roots, and several other aquatic products of Lake Diaocha have recently been certified as "green foods."[102] The annual output of aquatic vegetables (mainly lotus roots) in Hanchuan increased from 23.08 million kilograms in 1986 to 119 million kilograms in 2005.[103] And Caidian District (formerly Hanyang County), where land was returned to lakes to grow lotus after 1979, had 150,000 mu devoted to raising lotus at the beginning of the twentieth-first century. This, it was claimed, made Caidian the number one producer of lotus roots in China.[104] In many places, people turned their rice paddies into plots devoted to aquatic products, which had become more profitable than rice.[105] The logic behind this change is clear: the local people welcome the opportunity to compete in an expanding market.

3 Improving Water Conservancy and Reducing Dike Disputes

The frequent water calamities in the Qing and Republic contributed to the reputation of the Jianghan people, especially those who lived in the lower reaches of the Han River, as litigious and bellicose. Lawsuits around water calamities arose from disputes over such things as landownership (which could become confused due to changes of waterways and the collapse and silting up of riverbanks), the control of waterways (which could determine access to fishing and irrigation), and the control of dikes (which could decide, e.g., whether an outlet

99 *Tianmen xian zhi* 1989: 188–91.
100 *Mianyang xian zhi* 1989: 87–88.
101 *Xiaogan shi nongye zhi* 2011: 227, 237.
102 *Hanchuan nianjian* 2008: 305.
103 *Hanchuan shi zhi* 2010: 226.
104 Fang Bin 2012.
105 E.g., *Xiaxindian zhen zhi* 2012: 353–54.

should be closed or opened). Most of these lawsuits involved disputes between residents of different polders, and cases could drag on for decades or even centuries. In other words, lawsuits were not only time-consuming but also were an ineffective way of resolving real disputes. Thus, more commonly, the Jianghan people chose instead to find a quick solution themselves through physical confrontation, particularly when the control of a dike segment or the opening/closing of a dike breach was involved. Fighting, however, usually did not resolve disputes either. Some fights, especially when there were deaths, would lead to even more lawsuits, a common phenomenon in the lower reaches of the Han River where dike ruptures occurred almost every year. In the late Qing, residents of some neighboring polders even became bitter enemies because of endless lawsuits and fights.[106]

Most of these clashes were more than scuffles between two individuals. Since rivers snake through several counties and some lakes were shared by two or more administrative units, and some river dikes were managed by multiple administrative units, fights were usually large-scale violent conflicts between residents of the upper and lower reaches of a river or between residents on the opposite banks of the same river. The warring parties could number in the hundreds or even thousands, and they could come from different counties or prefectures. Moreover, these conflicts were planned and organized. The organizers were usually people who had military experience, or had a background as gangsters, or were polder leaders or clan leaders. And their methods of organizing and mobilizing combatants could be remarkably sophisticated, such as distributing flyers and establishing (illegal) dike bureaus. And some polders or clans had special regulations to compel their members to participate in fighting. The combatants wielded firearms, such as hunting rifles and blunderbusses, and bladed weapons, such as spears, knives, sickles, and even fishing tridents. These kinds of organized violent conflicts continued in the PRC, but their scope and frequency, the state's attitude toward them, and the ways the government has handled them all changed.[107]

Frequent water calamities, including dike ruptures, were a major cause of the violent conflicts over water control in the Jianghan area in the late Qing and the Republic. Hence, those conflicts could have been diminished if water calamities, or more precisely the frequency of dike ruptures, had been reduced. The PRC state indeed made great efforts to improve water conservancy and to maintain the integrity of dikes, such as the nationalization of rivers and lakes, the adjustment of administrative boundaries, and the strengthening of dike

---

106 Jiayan Zhang 2014: 197–201.

107 Jiayan Zhang 2013b.

management. As a result, the frequency of dike ruptures and ensuing conflicts were significantly diminished. And because the use of firearms was strictly controlled, there were far fewer fatal conflicts in the PRC.

All these efforts by the PRC state made good sense. First, it is difficult to manage all the rivers and lakes in the Jianghan Plain since some rivers run through different counties and prefectures, and many lakes are shared by several neighboring administrative units. And more importantly, they were privately owned by individuals or clans before 1949. While it was of course infeasible for anyone to really "own" rivers, different groups, such as clans, de facto managed them, such as the Yangzi River and the Ouchi River in Shishou.[108] This naturally led to conflicts. Therefore, after the PRC was founded, it immediately decided to nationalize both rivers and lakes, a keystone of the 1950 Agrarian Reform Law (Article 18). This nationalization significantly reduced the impetus behind conflicts.

Second, it was always difficult to manage dikes that snaked through several counties or prefectures. In the late Qing in the lower reaches of the Han River, the locals frequently refused to build "other people's" dikes, although they also benefited from these dikes.[109] Even worse, one county's dike might be in another county. For example, several segments of Zhongxiang's dike along the Han River were actually in Jingshan.[110] Enclave dikes of this sort were naturally subject to disputes.[111] To reduce conflicts over dikes that belonged to different administrative units, the PRC state adjusted the boundaries of many administrative units, from provincial boundaries all the way down to boundaries between villages or production teams, by using rivers as natural boundaries, then adjusted dike management by placing dikes under the jurisdiction of nearby administrative units, or adjusted the management of dikes according to their location.[112]

Another difference is that in the Qing and the Republic the state only managed some important segments of river dikes, leaving the local people responsible for other river dikes and all polder dikes. In Gong'an in the Republic, for example, disputes over polder dikes were regarded as an "unbreakable habit."[113] The PRC state made great efforts to reinforce the Jianghan dikes, including both river dikes and polder dikes, and also paid special attention to high-water

108 *Shishou xian zhi* 1990: 207.
109 *Xiangdi cheng'an* 1969, vol. 1: 56b.
110 *Zhongxiang xian zhi* 1867, vol. 3: 17b.
111 Wang Fengsheng 1832, vol. 2: 44a.
112 *Tianmen xian zhi* 1989: 137; *Yingcheng xian zhi* 1992: 74; *Yunmeng xian zhi* 1994: 218.
113 *Hubei xianzheng gaikuang* 1934: 1004.

control and annual repairs. As a result, the major dikes in the Jianghan area, which frequently were breached before 1949, have remained secure since 1954, with a few exceptions such as the breach of a river dike in Honghu in 1969. This impressive achievement naturally reduced disputes and violent clashes over dike ruptures.

Third, another major change was that the deliberate opening of other people's dikes, particularly in the Han River valley, virtually disappeared. The sources record many instances of this problem in the Qing era as well as in the Republic.[114] For example, in 1934 the dike of Penggong polder, on the boundary between Mianyang and Hanchuan, was deliberately breached, and in the following year a river dike in Mianyang was broken open.[115] A 1938 file indicates that this kind of deliberate opening of other people's dikes and ensuing fights were still common along the Han River.[116] But in the PRC, people lost the incentive to damage neighbors' dikes mainly because private landownership was abolished and rivers and lakes were nationalized, but also because once the state took over the management of almost all dikes, conflicts of this kind became virtually impossible.

Finally, one must keep in mind that the number of polders shrank in the PRC, and thus the source of many conflicts between neighboring polders disappeared.

4 Changes in Dike Systems and Daily Life on the Jianghan Plain

Although all of Jianghan society benefited from the protection afforded by the dike systems, throughout history diking was such an exhausting job that people would have been happy to avoid it, but they had no choice if they wanted to survive. This was still the case until the 1980s, after which manual labor was gradually replaced by machinery. In the Qing and the Republic, as noted earlier, the local people had to take care of both river dikes and polder dikes. With the increased construction of river dikes in the PRC, however, many polder dikes were abandoned or turned to farmland. The large scale of water

114 On the deliberate opening of dikes in the Qing, see Ni Wenwei 2002 [1885]: 238; Shuili dianlibu 1991: 796; *Mianyang zhou zhi* 1894, vol. 3: 26b.

115 The head of Hanchuan County reported that the breach of the Penggong polder dike was caused by a clash between laborers and security guards (LS1-3-0234-002); another report from five Hanchuan natives who lived in Wuhan, however, claimed that the dike was opened after someone was bribed (LS1-3-0234-001). On the opening of the river dike in Mianyang, see *Jianghe xiufang jiyao* 1971 [1936]: 258.

116 LS31-4-221.

conservancy construction and the increasing stability of the dike systems in the PRC, and particularly the completion of the Three Gorges Dam in 2009, brought dramatic changes to dike management and had a profound impact on the Jianghan people, whose daily lives have been interwoven with dikes for centuries.

First, dike work in the PRC gradually became less arduous. Before 1949, some laborers had to walk long distances to get to dike worksites,[117] carrying their rations and tools with them. Occasionally they had to stay outdoors all night, even in cold weather.[118] This situation did not change much in the early PRC. Some laborers still had to walk a long distance, such as those in Qianjiang in 1954, who walked 150 li to get to the worksite of the Great Jingjiang Dike annual repairs.[119] Laborers in Mianyang in 1959 even walked up to 1,000 li—a fifteen-day slog—to get to the Danjiangkou Reservoir construction site.[120] The big change, however, was that most of time, laborers did not have to worry about going hungry and did not have to sleep outside, as the state provided basic food (except on occasion during the GLF) and shelter, either arranging for them to stay at nearby villagers' houses or providing them with temporary sheds.

Second, the psychological and emotional toll of anxiety over the failure of dikes has been largely alleviated since the 1980s simply because the dikes have been more secure. Indeed, the people have gradually taken dike integrity for granted. In the past, whenever the high water rose, the local people worried about the rupture of their dikes, a perfectly realistic concern when dikes were poorly maintained and managed. In one place in Mianyang, for example, the quality of the Han River dike before 1949 was so poor that the local residents feared it would rupture during the annual high-water season.[121] In the early Republic in a spot of Jianli, after the dike had failed every year for five successive years, the local people had given up all hope of fighting the high water and instead simply fled when the water came again.[122]

This kind of anxiety lasted into the early PRC. In the 1954 flood in Yingcheng, for example, after the rupture of their dikes, some pessimistic local residents were ready to "flee from a calamity," and in Jianli, four flood refugees were

117 The same was the case with the imperial management of the Yellow River. For example, in the Song dynasty, some "hydraulic laborers" from the Huai River valley "traveled over a distance of hundreds of kilometers to serve in Hebei" (Ling Zhang 2016: 183).

118 *Jiangxia xian zhi* 1869, vol. 2: 41a.

119 D35-1-50a.

120 Fu Zhongjin 2006.

121 *Zhengchang qu zhi* 1985: 60.

122 *Jianli difang zhi* 1991: 52.

so despairing that they committed suicide by jumping into the water.[123] In the following year, worried about the use of the flood diversion zone again in Mianyang, some residents packed up their belongings overnight and were ready to flee from discharged floodwater.[124] To avoid panic among laborers, the authorities in Wuchang in 1956 decreed that beating the gong to sound the alarm must be a last resort.[125] In 1964 in Qianjiang, many people bundled their quilts and put them underneath their roofs before they went to the high-water control site, numerous households made rice cakes in anticipation of a food shortage after inundation, and one woman moved her belongings three times in a single day. And some people simply refused to participate in high-water control.[126] In this situation, a careless rumor could easily spark panic. In 1973, worried that their dikes would be opened to discharge floodwater, some people in Zhijiang slept at night without extinguishing their lamps so as to make a quick escape in case a dike ruptured, and many sent-down youths from Wuhan received telegrams asking them "return home immediately" since their "parents were terminally ill."[127]

But by the 1980s the situation had changed. In the early 1980s, when I first participated in high-water control of a local dike that had been built in the mid-1970s, en route to the assigned worksite, I was thinking how panic-stricken and anxious the locals would be. To my surprise, on the site I found the locals who lived not far from the inner side of the dike were calm and carried on with life as usual. This, I assumed, was because they had either become inured to the fear of flooding or they had become confident that the dikes would hold. I felt that the most nervous people were the cadres, because if anything untoward happened to the dike, they could lose their position or, if there was a disaster, they could be jailed. That is similar to the penalties meted out to governmental officials throughout China's history—and its positive aspect should be recognized.

Thus, in the 1998 flood, when floodwater reached the same level or even higher than the top of river dikes, governmental officials at various levels worked assiduously on their designated high-water control sites, while the common people continued with life as usual. A frequently used contrasting comparison is the image of the Wuhan Customs House (Wuhan Guan): in 1931, boats loaded with flood refugees floated in front of its gate; in 1998, however,

123 Yingcheng 49-1-26; Hubei sheng shuiliting and Hubei sheng dang'anju 1998: 249.

124 Wuchang 49-1-2b.

125 Wuchang 49-1-5.

126 40-1-305.

127 45-1-116.

the street in front of the building was full of buses, cars, and bicycles, and there was no sign of floodwater. Critics could regard this contrast as Communist propaganda, but, in any case, it was also a reality. In fact, the highest floodwater level at the Wuhan Guan in 1998 was 1.15 meters higher than that in 1931.[128] Since floodwater now has been constrained within dikes, a dike breach or an inundation has become alien to the locals. For this very reason, if the high water were to rise to a potentially threatening level, it would become national news. Nowadays, the loss caused by a breach of a major river dike indeed would be huge.

Although the water level of the Han River in 1954 was higher than that in 1931, and the Han River dikes in Hanchuan broke in many places in 1931, there were no breaches in 1954. This the locals attributed to the leadership of the CCP. In many cases they thought the dike was doomed to collapse, but once again, the government was able to overcome one threat after another and maintain the security of the Han River dikes. This of course enhanced the reputation of both the party and the government.[129] Although this assertion is made in an official report and could be dismissed as no more than empty boasting, that would be unwarranted. Residents of the Jingjiang flood diversion zone even "revered Mao Zedong as a god" since, according to them, he predicted the record high flood of 1954 and built the flood diversion zone ahead of time.[130] This, of course, was just a coincidence.

The changes in the dike systems and the fact that they have become relatively secure in the PRC also have had a profound impact on the local people's daily lives. This can be seen, for example, in the lower reaches of the Han River, where residents of areas that were not protected by dikes before 1949 constantly found it necessary to shift from farming and fishing. This changed in the PRC, and with it the quality of life for many people greatly improved. In one village, before dikes were constructed, many adult males remained unmarried due to poverty caused by water calamities; but it is said that after dikes were built in 1976, that even disabled men have been able to get a wife.[131]

In the whole Jianghan area, an obvious and universal change was the location of houses and the way they are built. Traditionally the Jianghan people build their houses on relatively high locations to avoid flooding. If no such high spots were available, they would build a platform and erect a house on it. And

---

128 The highest water level at the Wuhan Guan in the 1931 flood was 28.28 meters, while it was 29.43 meters in the 1998 flood (*Hanyang qu shuiwu zhi* 2015: 169, 189).

129 50-1-16b.

130 Zeng Fanrong 1997: 71.

131 *Pengxing xiang zhi* 2011: 21, 448.

some people simply built their houses on dikes.[132] This pattern has gradually changed since the 1950s. First, people were no longer allowed to build houses on dikes, and houses that had been on dikes for decades or even centuries were gradually removed (see Chapter 3). Second, with the protection afforded by the dike systems, the Jianghan people no longer worried about frequent inundations when they chose a location to build their houses or factories. For example, since the town of Shahu is the lowest place in Mianyang (and in all of the Jianghan Plain), at an elevation of 22 meters, while the water level of the Han River in the high-water season is 36 or so meters, if it were not for the protection provided by the Han River dikes, 14-meter high floodwater would wash away everything in its path. Shahu was indeed only a small town engaged in the fish trade before 1949, but it has gradually turned into a large, prosperous town in the PRC. When residents were asked in 1986 why they build new houses in very low locations, they responded that the area had not been hit by a flood for thirty-one years, and so they did not worry about flooding anymore.[133] The situation was similar in the village of Yangma in Shayang. Up until 1970, the villagers constructed their houses on platforms. But after 1970, with the decreasing threat from the Han River floodwater, they no longer took inundation into consideration and abandoned the use of platforms.[134] In reality, however, it is still possible for the Han River dikes to fail.

Even the materials the locals use to build their houses have also changed. Before 1949, houses were usually wood framed. Such houses could be relatively easily rebuilt after floodwater receded. But in the PRC, wood-framed construction gradually gave way to brick-walled and tile-roofed, and in some cases, reinforced concrete, constructions. Thatched cottages gradually disappeared.[135] At the same time, detached houses on separate homesteads were replaced by multi-unit housing in concentrated settlements.[136]

Transportation also underwent change. Both the netted waterways and the lengthy dikes contributed to the unique transportation system—particularly the use of boats—in the Jianghan area before 1949. This also determined that most important market towns were located along rivers or waterways. Until the 1950s, water transportation was dominant in Hubei, especially in the core

132 Jiayan Zhang 2014: 191–92.

133 Wang Yingming 1986. In summer 2018, I hired a taxi in Tianmen to survey dikes along the Han River and the Dongjing River and visited the town of Shahu in Xiantao (formerly Mianyang). The taxi driver, a native of Tianmen in his thirties, was surprised to hear that both Tianmen and Mianyang had been frequently flooded in the past.

134 Chen Tao 2007.

135 *Mianyang xian zhi* 1989: 598; *Pengchang zhen zhi* 1994: 313; *Gong'an xian zhi* 1990: 584.

136 *Madian cun zhi* 2004: 79–80; *Tanglin cun zhi* 2008: 3.

area of the Jianghan Plain, where water-borne transportation might have accounted for 80 to 100 percent of all transportation.[137] In some places, particularly in urban areas, dikes also served as local highways for land transportation, which is still the case today. This is immediately apparent if one drives over Jianghan's dikes. And in some instances, former dikes have been turned into highways, and dike sections also function as thoroughfares.[138] Since some dikes (particularly polder dikes) were usually very narrow on top, they were hardly ideal for transportation.[139] Dikes are still used as local highways, but the traffic on them is controlled, and vehicles not related to high-water control are not allowed to enter during the high-water season, and trucks are not allowed on many dikes all year around. Interestingly, with the rise in the number of private cars, I found that in recent decades some segments of river dikes close to urban areas have become popular routes for training new drivers, particularly after the 1998 flood when the tops of most river dikes were paved with concrete.

The large-scale realignment of waterways, however, has had a negative impact on water transportation. Due mainly to the building of numerous bridges and sluice gates, the navigable distance of rivers in the Jianghan area has been gradually shortened. Some rivers—which traditionally served as water highways—have recently even become completely unnavigable.[140] Also, as the highway system has expanded, towns that were located along rivers in the past are now growing toward highways.[141] Consequently, many riparian market towns have declined or disappeared. The major items transported via waterways have also undergone drastic changes. Since Hubei was an important agricultural center of late imperial China, the preponderance of goods transported by boats and ships before 1949 consisted of grain, textiles, tea, and salt. But these cargoes gradually diminished after 1949. By the 1980s and 1990s the main ship-borne cargoes had become raw materials such as stone, coal, and other minerals, as well as materials such as cement and chemical fertilizers.[142]

At the same time, land transportation gradually became dominant in the region. In the past, where locals relied on boats for travel, nowadays they can go anywhere via local highways. In Honghu, for example, boats remained the main means of transportation until the early PRC, but waterborne transportation drastically declined after the 1970s and yielded to land transportation.

---

137 *Hubei hangyun shi* 1995: 10.

138 *Jiangdi xiang zhi* 2006: 232; *Wuhan shi zhi 1* 2006: 349.

139 *Yanglinwei qu zhi* 1989: 224.

140 *Jingzhou hangyun shi* 1996: 255.

141 *Mianyang jiaotong zhi* 1985: 3.

142 *Hubei shuiyun zhi* 1996: 13–15.

By 1987, waterway transportation accounted for only 12.9 percent of all transportation in Honghu.[143] Hubei began to construct highways on a large scale in 1980, and ten years later, they reached 98 percent of townships and 82 percent of villages in the province. By 2005, all townships and 97.09 percent of administrative villages (former production brigades) in Hubei were connected to a highway.[144] And since the 1990s, more and more superhighways have been built in the Jianghan area. Finally, for about a century, the region had no railways except for the Beijing-Hankou Railway. That too has largely changed since the 2010s—now a high-speed train railway runs right across the center of the Jianghan Plain.

The increasing connection of river dikes, though good for high-water control, also has had negative consequences for fishing. When rivers and lakes were connected, fish freely swam from rivers into lakes to search for food and for reproductive migration. But when dikes were connected, lakes were separated from rivers, making it impossible for fish in the rivers to swim into lakes, and hence the catch of river fish thus dropped dramatically.[145] The Dongjing River simply lost its value as a fry fishing site since 1956 due to the construction of a new dike.[146] Even in Lake Hong, the largest lake and a major fishing ground in Hubei, catches also drastically dropped in the 1980s after decades of water conservancy construction, which severed the connection between the Yangzi River and Lake Hong, and the enclosure of lakes. Many fisherfolk had to give up their boats and settle down on land and become fishery farmers.[147]

The changes in the dike systems and the local economy are interconnected and multifaceted. Here I would add one hard-to-imagine change related to dikes. Recall that, as discussed in Chapter 3, in the 1950s in Zhongxiang pigs were regarded as harmful to dikes and were to be eradicated. According to Ni Wenwei, pigs raised by people who lived along dikes were more harmful than termites and badgers, since pigs turned over the soil and rooted up grass, which resulted in potentially dangerous pits after rain. Ni also explained that it was not enough to just post some signs to warn the local people, since that alone would not work; instead, he called for dike managers to keep an eye out for threats to the integrity of dikes.[148]

143 Ye Hengchang 2004.
144 *Hubei gonglu shi* 2013: 1–4.
145 *Zhijiang xian zhi* 1990: 155.
146 *Jianli xian zhi* 1994: 187.
147 *Wagouzi zhi* 2013: 85.
148 Ni Wenwei 2002 [1885]: 144.

For the same reason, it was absolutely forbidden to grow crops on dikes, as this would not only damage the dikes but also make it difficult to find hidden dangers such as leaks. However, it was fine to graze cattle and horses on dikes since they would compact grassroots and the soil.[149] But in the early PRC, even cattle were not allowed to graze on dikes. Hubei's 1950 regulations on dike maintenance made it clear that "pigs, goats, and other livestock" were not allowed to graze on dikes.[150] But it is not enough that cadres alone carried out the regulations; all residents who lived close to dikes had to be involved. For example, in Jingmen in the early 1950s, cadres tried to drive pigs and cattle away from dikes everyday but in vain. But after the establishment of a dike maintenance organization with the participation of local villagers in 1957, the number of pigs and cattle on dikes gradually was reduced.[151] In 1958, Jingzhou Prefecture drafted a new regulation on the maintenance of water conservancy projects, which stated that cattle could be grazed on dikes as long as there was someone to "watch over" them.[152] The Shishou County government also had a regulation that prohibited the grazing of pigs and goats on dikes, but allowed cattle to be grazed as long as that did not affect high-water control materials and the security of the dikes.[153]

In short, in the early PRC, convincing the local people not to graze their cattle on dikes remained a significant challenge. This has gradually changed in recent decades. As a child, I often went to the nearby river dike to collect cattle dung and horse manure as fertilizer. Even in the 1990s and 2000s, if one drove on any river dike in the Jianghan Plain, one could still see cattle, in twos and threes, grazing on dikes. While driving over river dikes in the Jianghan Plain in the past two decades, several times I saw herds of goats grazing on the dikes—in Jiangling (Mashan) and Gong'an (Doufudi and Beizha) along the Great Jingjiang Dike, and in Zhongxiang along the Han River dike. These goats were not supposed to be allowed on dikes. In my tour of the Jianghan dikes in 2018, I only saw a few cattle grazing on dikes. However, they were not draft buffalo, but yellow oxen (i.e., beef cattle).

Cattle were also a major disseminator of snail fever, as cattle usually excreted in water, and the number of schistosome (blood fluke) eggs in the excrement of one cow equals that of a hundred or a hundred and fifty people. But before the 1990s, it was very difficult to persuade peasants to give up their

149 Hu Zuhe 1999 [1838]: 187.

150 22-1-006-012.

151 D35-1-148b.

152 22-1-266-027.

153 20-1-394-24.

infected cattle, since they were crucial to farming, particularly plowing.[154] In recent decades, however, this has changed. The government has banned keeping cattle along rivers, and peasants have increasingly turned to machinery for farming. In some places, many villages have become "cattle-free farming villages," where machines have completely replaced draft cattle.[155] The result has been a huge reduction in the number of cattle grazing on dikes, a change unimaginable just two decades ago.

---

154 *Songzi xumu shouyi zhi* 2013: 215–16.

155 *Dongxihu qu jianzhi* 2012: 152.

Conclusion

Based primarily on the rich source materials in the archives, this study challenges many conclusions in past scholarship that have a weak empirical foundation. For example, past scholarship emphasized such negative aspects of dike management in the Qing and the Republic as corruption and incompetence among officials. While these phenomena certainly did exist, not all governmental officials were corrupt and incompetent: some were dedicated public servants. That fact deserves more recognition. Orthodox Marxist scholars in China particularly highlight the exploitation of the peasants in the Republican era, but that often turns out to be an exaggeration. In the West, scholars condemn the CCP's water conservancy projects by claiming they were imposed in a top-down fashion by a despotic state and ignored the peasants' interests and wishes. Moreover, such scholars point to the supposedly disastrous environmental consequences of these projects. While there can be no doubt that there is some truth in these charges, the reality is more complicated. In the early PRC, many water conservancy projects actually complied with the people's wishes. At the same time, the local people did not always readily yield to the state. This can be seen in, for example, the continuing enclosure of land from riversides and lakeshores despite frequent government bans. When scholars and observers criticize the Three Gorges project, they mostly concentrate on the difficulties involved in population resettlement and the misery suffered by the relocatees, as well as the environmental damage wrought by the project. Rarely do they mention high-water control in the middle Yangzi River valley, one of the major purposes of this project.

1 Dikes and Society

Since it would be unrealistic to have no dike protection on the Jianghan Plain, any discussion of Jianghan's dikes and society should concentrate not on the financial cost of diking but on its social, environmental, and political significance. From a political perspective, once the PRC state decided that the safety of some dikes had to be guaranteed no matter what, it had to accept the costs involved, no matter how high. In fact, in the PRC in recent decades, unlike in the Qing and the Republic, with few exceptions the Jianghan dikes have been secure. This change, however, had a profound impact on local society and the local people's daily lives. The largest impact was that peasants gradually

 DOI:10.1163/9789004739482_013

no longer took dike ruptures into consideration when selecting what crops to raise, and in the process gave up famine relief crops. In addition, building houses and factories on low-lying locations became acceptable. Even residents living in designed flood diversion zones have gradually become reluctant to see their land used to discharge floodwater. Since flooding farmland could be extremely damaging and costly, many people assume "the state will prevent it from happening"—that is, do its best to avoid using flood diversion zones to discharge floodwater. This has completely altered the state's original intention, and in this case, it seems the state has been hijacked by the local people.

To solve frequent water calamities, as soon as the PRC state was established it took action. On the Jianghan Plain that included realigning the Jing River, and, in 1952, constructing the Jingjiang flood diversion project, the first large water conservancy project in the PRC. This demonstrates that the PRC state inherited the imperial tradition of rebuilding water conservancy projects and restoring agricultural production after the establishment of a new state to showcase that it was better than the old regime and had the right to rule. But both the ideology and power of this new state were far greater than those of any of its predecessors. Furthermore, both the number of water conservancy projects and their scale have been incomparable larger than those in the past. Although peasants suffered a great deal when they built these projects, they also benefited tremendously from them and continue to benefit to this day.

Generally speaking, in the Qing and the Republic the locals concentrated more on building dikes than on maintaining them, while in the PRC the state has stressed dike maintenance as well. Jianghan's dikes frequently ruptured in the Qing and the Republic, but the major river dikes on the Jianghan Plain, except for a few incidences such as the one in 1969, have remained secure since 1954. This can be attributed to improvements in both dike management as well as the quality of dikes and their year-round maintenance. All of this again attests to the importance of dike management in preventing ruptures and dikes rarely ruptured simply because floods were too huge to be contained. The rupture of a river dike in Honghu in 1969 and the collapse of the dike of Mengxi polder in Gong'an in 1998, however, should be a warning to people that if dikes are mismanaged, they can still fail.

Another dramatic change has been the structure of the workforce. Before 1949, laborers were mainly men. Women usually did not work outside the family compound because of both social customs and physical barriers—many women had bound feet, making working at dike worksites difficult. After Land Reform in the early PRC, however, so-called feudal social customs were pushed aside, and women officially were accorded the same political status as men. The revolutionary ideology of gender equality also stimulated women to ignore their physical and physiological differences from men and enthusiastically participate in "socialist construction," including diking and high-water control.

Technology has always played an important role in dike management. It is generally true that dike-related technology did not change much from the early Qing to the late twentieth century, but that does not mean that technology was not important. Although manpower provided the sole power for diking until the 1950s, since then machines have been increasingly used, and by the turn of the twentieth-first century, diking no longer relied mainly on human physical power. Other changes include materials (e.g., steel and concrete in place of dirt and wood), transportation (with motor vehicles now quickly moving people and materials in emergencies), and communications (the internet, computers, and cellphones can help alert people to the coming of floodwater and to the need for adjustments in flood storage in various reservoirs, and can make it possible for them to immediately communicate with one another). The development of modern hydraulic technology and the improvement of communications have enabled the responsible people to instantly understand the water regimen of the whole Yangzi River valley, which can help them to make informed and immediate decisions.

This, however, does not diminish the importance of effective management in ensuring the integrity of dikes, since deciding how to use available technology is a part of effective management. Many technologies employed in the PRC are traditional but have been used more effectively than in the past. This includes such things as the growth of breakwater forests, the clearance of badger dens and termite nests, and the removal of houses on dikes—all these have contributed to the safety of the Jianghan dikes in the PRC. By the same token, the availability of new technologies does not mean that people can now sit back and relax when it comes to dike management. On the contrary, governments at all levels still have to take annual repairs and particularly high-water control seriously. The new materials and technologies would be useless without good management, and dikes would still rupture. Once a dike ruptures, the blame will not be attributed to insufficient technologies, but instead to the failure of the responsible government officials. And those officials will almost certainly be punished accordingly.

It should also be noted that there were almost no modern hydraulic engineers in China before the twentieth century. In the early twentieth century, the Republican government started to train modern hydraulic engineers, though their numbers were very limited. Since 1949, however, a growing number of such engineers have been trained, and nowadays the dike systems are basically managed by professionals.

It is often argued that Chinese society has undergone greater change in recent years than at any other time in past centuries or even millennia. We should add that dike management reflects one small slice of this epoch-making transformation.

2 State and Society

Unlike the North China Plain, which suffers from a shortage of water, the Jianghan Plain has too much water. Hence the importance of the dike systems in the Jianghan region. From the Qing to the PRC, the state's involvement in dike management increasingly deepened and intensified, and followed a trend of regularization and specialization or professionalization: the dike tithing system (or, the system of dike headmen) was common in the Qing, annual repairs and high-water control were separated in the 1930s, and the PRC relied on administrative orders and established permanent organizations to deal with annual repairs and high-water control. The task of environmental management, or more precisely dike management in the case of the Jianghan Plain, is impossible to fulfill without the involvement of organized institutions and effective governance. The fact that water calamities, particularly dike ruptures, frequently occurred on the Jianghan Plain in the Qing and the Republic was not because the state was unaware of the problems. On the contrary, many government officials clearly understood the serious consequences of over-reclamation of upstream mountains and riverside and lakeside wastelands as well as the widespread mismanagement of the dike systems on the plain. But the Qing state intentionally left the management of local dikes up to local government and local people. In the Republic, the government assumed a much bigger role, but government departments failed to cooperate, and the state in both the late Qing and the Republic suffered from constant chaos caused by frequent rebellions and foreign invasions, which greatly weakened its ability to properly manage the dike systems.

Thus, the local people before 1949 played a leading role in dike building and polder enclosure, as well as in the transformation of the local environment, with the state only playing a passive and distant secondary role. After 1949, however, the state initiated and dominated almost all water conservancy projects, except for the enclosure of some small polders from lakesides, river beaches, and islets in the collective era that most likely were initiated by village leaders who disobeyed government regulations. Nevertheless, these polders were enclosed for the benefit of the public (a village or a production team), not, as in the Qing and the Republic, for a clan or individual households.

For centuries, ensuring the integrity of dikes has been the most important official duty of the chief administrators on the Jianghan Plain and their top priority in governance. No one dared to take this task lightly. But dikes in the Qing and the Republic nonetheless frequently burst. It is widely believed that this was largely caused by the corruption of officials and mismanagement. But it would be unfair to assume that all responsible officials in the Qing and the Republic were corrupt, incompetent, and negligent. On the contrary, many were not only capable and experienced, but also diligent in their management of dikes; some were even willing to sacrifice their lives for the sake of protecting the dikes, a tradition still seen in the PRC.

The construction and management of dikes relies on collective efforts. In the Qing, it was usually the rural gentry and/or clan leaders who led the local people to build and repair dikes and to control high-water, under the supervision of local officials on some dikes. In the Republic, the gentry and/or clan leaders still played an important role in both annual repairs and high-water control, but the state also gradually increased its involvement in dike affairs, particularly for river dikes. After 1949, the gentry and clan leaders as a social group quickly disappeared and were replaced by newly emerged village leaders who were usually poor peasants and represented the state at the village level. The state was able to deeply penetrate rural society, down to the village and even household level in the collective era. Since the 1980s, the state has been gradually withdrawing from rural governance, making it increasingly difficult to mobilize peasants to work on water control projects. Since 2005 it has been in fact illegal for local governments to force peasants to work without compensation. Instead, since the new millennium, the government has increasingly relied on the use of machinery at water control sites.

In the Qing and the Republic, there was more conflict than cooperation in dike affairs, but it is obvious that the reverse has been true in the PRC. People in the Qing and the Republic often debated whether frequent water calamities, particularly dike ruptures, were caused by natural phenomena (e.g., too much rain or too huge a flood) or were man-made (i.e., the result of mismanagement). In some cases, the question was whether the officials in charge of the management of the ruptured dikes were accountable. In the PRC, however, that has no longer been an issue. After the state decided in the early 1950s that the integrity of some important dikes must be guaranteed (i.e., there must be no ruptures under any circumstances), any rupture would be automatically attributed to dike mismanagement, and someone would be punished. For this reason, the PRC state has made great efforts to improve and strengthen dike management by creating a permanent top-down hierarchical system of organizations. This stands in contrast to dike management in the Qing and

Republic, when organizations responsible for high-water control were mostly temporary—after the high water receded, they were disbanded. In the PRC, more governmental departments were created to manage the dike systems, more staff were added to these departments, and most importantly, both annual repairs and high-water control gradually became a year-round task, not just a seasonal job, conducted by full-time professional employees instead of informal peasant dike keepers as in the past.

In their struggle for the control of waterways, the Jianghan people in the Qing and the Republic were well-known as litigious and bellicose. Some even dared to challenge the state's authority. The local governments handled this kind of confrontation neither actively nor effectively, and as a result, it was not uncommon for residents from hundreds of polders to be involved in fatal fights. Some feuds lasted for decades or even centuries. After 1949, however, the nationalization of rivers and lakes and the adjustment of administrative boundaries so that they correspond with waterways whenever possible largely reduced disputes over control. In addition, the disappearance of rural gentry and clan leaders made multi-clan or multi-village alliances impossible. Furthermore, the improvements in water conservancy and the destruction of most polder dikes eliminated the raison d'être of residents of hundreds of polders struggling over the control of waterways. The strict control of firearms has also minimized the potential for fatal disputes, and the powerful party-state system has brought state power directly into local society—no community or individual is able to successfully oppose the state, making it easier to solve disputes at the local level.

As a part of the state apparatus, the military has always been involved in dike affairs. After the 1788 flood, for example, a small army was deployed to protect the most endangered segments of the Wancheng Dike. Soldiers in both the Qing and the Republic were also assigned to build dikes, to participate in high-water control, to maintain order on high-water control and diking sites, and to suppress dike-related disputes. The PRC state went a step further by deploying more troops to build dike projects and to participate in high-water control—particularly in emergencies. The work of the Chinese People's Armed Police Flood Control and Rescue Force, particularly after the 1998 flood, has become well-known.

3 Changes and Continuities

The Jianghan Plain has experienced many continuities and changes over the past three centuries. The continued importance of the dike systems and society-wide participation in high-water control—until the 2000s—are perhaps most notable. The most significant changes have been in dike management, with regulations governing rewards and punishments, responsible positions made official and institutionalized, an exponential expansion of state funding, and so on. Other important changes have been the decline and disappearance of organized violence over water control and the disappearance of relief crops. Unlike most relevant research, which focuses either on the Qing and the Republic or the PRC era, this book covers all three: the Qing, the Republic, and the PRC. Although China's political system underwent many changes during this time, the dike systems remain as important as—or even more important than—before. Looking closely at the building and management of the dike systems, it becomes clear that trying to rigidly separate the Qing, the Republic, and the PRC—as PRC scholarship usually does—is a fool's errand. The reality is that the threat of floods, the importance of dikes, some methods of high-water control, and so on, have not changed much over time. On the other hand, one cannot ignore the great differences after 1911 and especially 1949, such as the methods of mobilizing people for annual repairs and high-water control, the effectiveness of annual repairs and high-water control, and some materials and procedures used in annual repairs and high-water control—all of these changed dramatically.

There has been a new trend in recent scholarship toward dividing up the history of the PRC into two thirty-year periods. Some scholars go even further and claim that all the twentieth century should be divided into three thirty-year periods: the Republic, the Mao era, and the reform era. These scholars have tried to trace the changes in China after the 1979 reform back to the first thirty years of the PRC or even to the Republic. This periodization, however, does not fit well with the history of the construction of water conservancy projects on the Jianghan Plain. Some major changes—such as selecting the unit in charge of a dike project through public bidding, with the winner perhaps coming from thousands of miles away, and shifting from compulsory labor to hired labor, from manpower to machines, and from dike fees collected from villagers to funds from the state or bank loans—only occurred at the end of the twentieth century.

Up to the 1990s, both annual repairs and high-water control of dikes still largely relied on compulsory and uncompensated labor provided by people from nearby villages, a method that was very much out of step with the trend toward marketization in the post-Mao era. With economic development, rapid urbanization, and the constant outflow of young villagers, mobilizing villagers as compulsory laborers via administrative order became unworkable, and after 2005 it became illegal. After the completion of the Three Gorges Dam in

2009, however, the locals have rarely needed to go to dikes for high-water control during the high-water season. Since the disappearance of the locals from regular high-water control is a recent phenomenon, what might happen if the region is hit by another record high flood is an open question.

Scholarship in the PRC once portrayed "Old China" and "New China" in black and white. The resultant image in many respects was faithful, particularly regarding the political systems in the two eras. But in other respects, it was distortion. This is clearly the case with dike management on the Jianghan Plain. Although the classic CCP discourse treated the GMD negatively, sources from the Republic, as we have seen, show that the chief leaders went to dikes to supervise high-water control, that troops participated in diking and high-water control, and high-water control was supported by society—all this was very similar to what happened in the PRC. It is clear that the Republic was a transitional period during which new materials in diking and high-water control were introduced, as was the idea of a nationwide coordination and unification of water control, the promotion of merging polders, the requirement to plant breakwater forests along dikes, and the construction of flood diversion zones. All these provided a foundation for similar policies in the PRC. Other elements of the transition include the influence of nation-building, which was reflected in the organization and level of the units responsible for water conservancy, and the campaign against superstition—both the GMD and the CCP, after all, were revolutionary parties.

The establishment of annual repairs and high-water control agencies in people's polders in the Republic also represented a transition, since in the Qing, the local people themselves, not the state, usually managed polder dikes. In the Republic, the state began to be involved in their management. The director of the annual repairs and high-water control agency in each people's polder was elected by the residents of the polder, but the election was monitored by a governmental official and the elected director had to be officially appointed by the county government. In the PRC, the state became further involved in dike management as all dikes, including those in polders, were now managed by the government at different levels. Even during the collective era, when local peasants were hired as informal (*yigong yinong*) dike keepers, the units they worked for were still government agencies. In the post-Mao era, the state has gradually withdrawn from rural governance but at the same time it has reinforced its management of the dike systems, and replaced informal peasant dike keepers with full-time professionals.

Until the early PRC, the dike systems were an integral part of the daily lives of the Jianghan people. They had to provide money, materials, labor, and time for annual repairs and high-water control. And they constantly faced the grim prospect that if the dikes ruptured, they would suffer grievously. In the past three decades, particularly since the beginning of the twentieth-first century, however, the dike systems have become remarkable secure, and perhaps because of this they occupy a less central place in the local people's daily lives. Moreover, the local people are no longer required to provide labor, not to speak of money and materials, for annual repairs and high-water control. This lifting of responsibility from the shoulders of ordinary citizens has not been without drawbacks. Before the PRC, it was not unusual for peasants and leaders to shirk their duty to protect the dikes. This can be seen in a case from the Republic discussed earlier in the book, where many directors of annual repairs and high-water control of people's polders in Songzi resigned on the excuse of illness and unavoidable family commitments. This kind of evasion of responsibility became virtually impossible in the early PRC, when moral incentives were more important than material incentives. Hence, in the early PRC many rural cadres continued to work on water conservancy worksites even after the death of a family member. But today such selflessness and dedication are unthinkable. What did not change but remained consistent from the Qing to the PRC is the honoring of those who lost their lives in high-water control as martyrs.

In the picture of changes and continuities, tourism stands out. Once a luxury leisure activity that few rural Chinese could partake in, tourism has become an important part of people's lives over the past three decades, and many water conservancy projects—including dikes, reservoirs, and lakes—have become tourist destinations, another change that was unimaginable decades ago. Where dikes are adjacent to towns or run across cities, the local governments have built many riverside parks, from small ones in rural towns to gigantic parks in urban areas (especially in the Wuhan area), for the enjoyment of the local people. The PRC state has been so successful in maintaining the security of dikes over the past seven decades that the local people have taken the integrity of dikes for granted. This is particularly true of the youngest generation or the new millennial generation, which has grown up in an era of abundance and safety—for them, hunger and dike ruptures are alien or remote history.

There have been other changes as well in a wide variety of domains. The use of dikes as military works was particularly common in the Republic, but not much in the Qing. In the early PRC, the Great Jingjiang Dike was considered militarily important, but only from a strategic perspective. Cooperative management of dikes was a serious problem before 1949, but the early PRC government even transferred people from one region to build dikes in other regions under the slogan "all peasants under the heaven belong to one family." In the imperial era, the Jianghan people emigrated far and wide to escape water calamities at home, while in the PRC, many peasants have left the villages to work in cities all across China—as well as abroad in some instances—as dispatch laborers (that is, workers hired through labor dispatch companies). Before 1949, the local people were responsible for almost all high-water control materials; after 1949, the state begin to provide more and more materials, and since the early twentieth-first century, the state has entirely assumed this burden.

Changes have also appeared in popular culture. People in the Qing usually prayed to the dragon king for rain and to the water deities for deliverance from water calamities. In the Republic, although some people realized this was superstition, they nonetheless followed these practices as a part of tradition. In the PRC, however, people gradually abandoned what the Communists dismissed as "superstition." Changes also appeared in the role of the state. In the Qing and the Republic, the local residents frequently pled or asked for relief from the state, while in the PRC the state has taken full responsibility for providing relief to flood-stricken people by giving them food, clothing, and shelter, and also by hiring them to repair broken dikes through work-relief projects. The state itself has ordered cadres to send and distribute relief; there is no need for the peasants to ask for help. In the early PRC, most officials responsible for high-water control fulfilled their duty by patrolling dikes no matter what the weather, although a few neglected their responsibilities. But slackers were far less numerous than in the Republic. Unlike in the Qing and the Republic, when most high-water control laborers came from nearby villagers, in the PRC the government can call up laborers from distant villages and counties or even other provinces (e.g., PLA soldiers), while nearby villagers continue with life as normal. In short, the early PRC state deeply penetrated rural society, and at the same time, the peasants gradually came to rely on the state to take care of dike affairs.

In traditional times, local leaders regarded dike affairs as the first priority of governance, but in recent decades, though the dike system is still important, the first priority has shifted to economic development or GDP growth. Unlike GDP-driven strategies in the reform age, when cadres knew but ignored the environmental degradation that can arise when economic growth is valued over all else, in the early PRC the main goal of cadres, who applied subsistence-driven strategies and worked hard to feed the local population without considering the environmental consequences, was the development of agriculture—even the ban on the reclamation of river beaches and islets in the early PRC was aimed not at protecting the environment, but reducing water calamities, this for the purpose of protecting life and the stability of agriculture. Environmental protection only became a pressing issue in the reform era, mainly because of the increasingly negative consequences of industrial development and over-reclamation.

4 National Significance of the Jianghan Experience

The Jianghan experience, in all its multifaceted dimensions, calls for a reexamination of some general understandings of, and assumptions about, rural China in imperial times, under the rule of GMD, in the Mao era, and during the reform age.

Almost all students of Chinese history recognize the heavy population pressure China faced in its history, particularly in the Qing and afterward. Likewise, they know that China's "severe environmental pressure is exacted by merely feeding the people."[1] Population pressure manifests itself in various ways. Past scholarship once attributed many of late imperial China's socioeconomic problems to its huge population. The Jianghan experience shows that population pressure in the Jianghan Plain was also expressed through environmental challenges. From imperial bureaucrats in the Qing to Communist cadres in the PRC, all attributed the reclamation of private polders, at least partly, to population pressure. Population pressure was also one of the main reasons the ban on private polders failed in the Qing. The major raison d'être for the reclamation of river beaches and lakes on the Jianghan Plain, mostly in the 1960s–1970s, was population pressure as well. Thus, even an official gazetteer of Hubei admitted that the reclamation of lakes was an effective way of reducing population pressure in the collective era.[2]

The nationalization of wasteland, the deep penetration of state power into rural society, and the peasants' eagerness to bring more land under the plow in an agricultural society all contributed to widespread and extensive land reclamation in the first three decades of the PRC. This was also a time when the state built many large water conservancy projects on the Jianghan Plain. These projects not only improved the water control system, but also raised land productivity. Some of them, particularly the reclamation of lakes, had been tried by the locals before but mostly failed for various reasons. In other words, many of these projects were not forced on the public by a despotic state as some critics have claimed, but, on the contrary, were the state's response to what the people demanded.

1 Smil 2015: 9.

2 *Hubei shuili zhi* 2000: 734.

The reclamation of private polders was officially banned since 1748 because they blocked waterways and caused water calamities, but this ban never took hold. Before 1949, local officials usually did not enforce the ban partly because of population pressure, and even the new state after 1949 failed to ban private polders also partly for the same reason. The PRC state successfully banned many undesirable social customs (e.g., gambling, prostitution, and opium smoking) almost overnight in the early 1950s, but it failed to ban the reclamation of river beaches and lakeshore. Therefore, although the criticism of the over-reclamation of land in the early PRC has a basis in fact, it also is ahistorical and lacks context, since a person who does not have to worry about food and clothing certainly cannot understand a person whose stomach is empty.

With China's economic growth since the 1980s, however, millions of young people have left the villages in search of non-farming jobs and have lost any desire to reclaim land. At the same time, many lowlands have been returned to lakes for floodwater diversion and fishery (now more profitable than farming), and many villagers have been relocated from the river side to the land side of river dikes. Furthermore, both the state and society have gradually come to realize the importance of flood diversion and the protection of the "natural environment," particularly after the 1998 flood. In that flood, beach polders along the Yangzi River not only blocked waterways but also were costly to protect, leading the state to order the abandonment of many polders on the water side of river dikes (thereby creating space to discharge floodwater) and the return of many low-lying lands to lakes and rivers. As a result, lake acreage has been rising, and the widespread reclamation of land from lakesides and riversides—a thousand-year-old tradition that began in the Song dynasty on the Jianghan Plain—seems to have come to an end. The reclamation of private polders and how the state has handled this issue and its effects on the Jianghan Plain in different eras thus demonstrate the complex relationships between population pressure, environmental restrictions, and sociopolitical change in rural China.

Another major goal of these reclamation projects in the early PRC was the elimination of schistosomiasis, an epidemic disease that had plagued the Jianghan people for two millennia. The state vigorously pursued this goal by enacting policies to eradicate *Oncomelania* snails, the intermediate host of schistosome larvae. In other words, these were governmental efforts to feed more people and to improve their living conditions, the result of a combination of historical traditions and practical demands, not simply a "war against nature" as critics have claimed. Rather, if there was indeed such a war, it was a war against small worms (schistosome, or the blood fluke) and hunger. Furthermore, the notion that humans are not powerless before nature, expressed in a nearly ubiquitous slogan in the PRC under Mao—"Man can conquer nature" (*ren ding sheng tian*)—was a creation of neither Mao nor the CCP, but instead has a very long history.

Based on all these changes, particularly in the early PRC, a close relationship between the state and water conservancy in general and the dike systems in particular emerged. The PRC government at various levels frequently described water conservancy and dike work as "a political task," meaning they were of the highest priority. In practice, this translated into the requirement that such tasks had to be successfully completed at all costs. This phenomenon can be described as the politicization of water conservancy (where governance was focused on political considerations), a unique feature of the PRC, particularly in its early years. In recent decades, though the influence of politics has faded, the party-state system still prioritizes political considerations in decision-making. This is evident in both the Three Gorges project and the South-to-North water diversion project, which are widely regarded as politically inspired programs.

Many people believed that the large-scale water conservancy construction in 1957 presaged the Great Leap Forward. In fact, an emphasis on mutual-aid and cooperation had already emerged in the early 1950s, particularly regarding dike affairs since both annual repairs and high-water control required a great deal of cooperation: cooperation among the residents within the same polder, cooperation with neighboring polders, and even the cooperation of different administrative units. Before 1949, when land was privately owned, all these types of cooperation were very difficult to achieve. The CCP also assigned propaganda an important role in dike affairs, a decision based on experience that grew out of wartime. Therefore, even though the Jianghan people shouldered a heavy burden of working on numerous water conservancy projects (particularly diking) in the early 1950s, many of them still ardently supported those projects. And their support grew still firmer when it became clear that the early PRC successfully and significantly reduced the frequency of dike ruptures. The mass campaigns to eliminate illiteracy and to promote public health (the so-called patriotic health campaign) further incorporated diking—a purely economic activity—into a society-wide political and cultural movement. Through the building of water conservancy projects, centered on diking on the Jianghan Plain in the early PRC, the new state successfully convinced the general population that the CCP was better than the GMD, and the PRC was better than the Nationalist state. All of this laid the foundation for future mass campaigns and movements in the collective era.

In traditional China disaster relief was a responsibility of the government. The PRC inherited this tradition and made great efforts to prevent calamities. In the early years of the PRC, however, the main concern was reducing calamities and protecting the production of grain, not protecting the environment. The idea of environmental protection or sustainable development only became common in the developed world after the potentially disastrous side effects of modern industry were brought to the public's attention. Leading the way was Rachel Carson's celebrated *Silent Spring*. Carson's book was published in 1962: it would be unrealistic to complain that the people of China in the 1950s should have entertained ideas that only emerged later.

With the growth of environmental protectionism has come an increasing concern over the negative consequences of dams all over the world. It is probably a good idea to destroy dams in some places, but a dike is not a dam. It is simply impossible for the Jianghan people to abandon their river dikes. Those who criticized the local people for building too many private polders in the Qing were usually high-ranking bureaucrats who had to take the whole situation into account, or bystanders who did not even live in the region. Needless to say, their lives usually were not threatened by floods nor a shortage of land. For the local peasants, however, the first priority was and remains survival, and land is the basic resource essential to life. The Jianghan peasants who reclaimed lakeshore wasteland and river beaches usually did so to ensure their very survival, or at least improve their lives, even if only slightly. They were the ones who faced the threat of floods, who suffered from water calamities, and who searched for ways to ease the pressure of a land shortage. For such people, their sophisticated responses to their changing environment were the most rational options. There is no reason to doubt that they would certainly have changed their behavior if there were other options. This, in fact, was exactly what happened at the end of the twentieth century, when non-farming jobs lured young villagers away from the countryside and farming, ending polder enclosure and reclamation of lakeside wasteland. The river dikes, however, still must be there.

These Jianghan experiences also allow us to reevaluate Karl Wittfogel's notion of oriental despotism. Wittfogel claimed that the Chinese imperial government was built upon its full and despotic domination of water control, which, as we have seen, is totally untrue. He also claimed that the early PRC state was totalitarian, not different from any other hydraulic society, but he actually did not even discuss the construction of water conservancy projects—the core of an "oriental" or hydraulic society—in the early PRC. Although his book was first published in 1957, in 1955 he had already correctly observed that "the power of the ruling class is reaching an all-time high" and "the new monopoly bureaucracy seeks to establish full control everywhere."[3]

3 Wittfogel 1955: 478.

Thus, ironically, the Maoist state in the collective era is probably the closest to his notion of an "oriental despotism" (or the most "despotic" in Chinese history), as during this period, the Chinese state was able to deeply penetrate and exercise unprecedented control over rural society. At the same time, the state constructed and controlled a great many water conservancy projects. These projects, however, were not obstacles to the development of the Chinese socio-economy as Wittfogel would have us believe. Instead, they laid the foundation for China's agricultural development in the collective era and the rapid economic takeoff in the reform era that followed. The CCP's achievements in water conservancy also prompted the peasants to identify with the new state.

Yet even in this period, the state did not "fully" control rural society. On the Jianghan Plain, villagers continued to enclose private polders from lakeshore and river beaches and islets, in violation of state regulations. That is to say, the CCP's power was not, and still is not, omnipotent. Although it declared the reclamation of riverbanks and lakes without permission to be illegal—a step that could be easily misunderstood as an upper-level government policy that ignored peasants' resistance—local villagers ignored the official ban, and many times after reclamation the upper-level government yielded and accepted these "illegal" acts as *faits accomplis*.

With the withdrawal of state power from rural governance in the post-Mao era, particularly since 2005, when the agricultural tax was abolished, it is no longer possible for the state to mobilize hundreds and thousands of peasants to work on the construction of water conservancy projects as compulsory laborers. Today, many residents of rural China look back on the water conservancy projects of the collective era nostalgically. Also, contrary to Wittfogel's claim that traditional China needed a despotic central government to control water, it is because there was a powerful central government in the early PRC that a nationwide movement of water conservancy construction was possible.

Although since the 1979 reform the state has gradually withdrawn from rural governance, it is still deeply involved in water conservancy. At the same time, the state's guiding principle has gradually "turned from 'water conservancy is the lifeblood of agriculture' to 'water conservancy is the foundation of the national economy.'"[4] In 2011, the State Council issued the first Number 1 Central Document (the Number 1 Central Document is the first document issued by China's central government every year) on water conservancy, which declared that water conservancy was regarded not only crucial to modern agriculture, but also the foundation of socioeconomic development, and even of vital

4 *Xiaogan shi shuili zhi* 2011: 5.

importance to China's "national security."[5] In other words, the central government concluded that water conservancy was no longer simply an economic issue surrounding agricultural development, but a strategic issue involving national survival. The document therefore emphasized that the state would continue to play a leading role in the construction of water conservancy projects. That is to say, the market-oriented economic reform launched in 1979 did not change the nature of water conservancy projects as public goods. Thus, the state determined that water conservancy must not be subjected to market forces. It is indeed unthinkable that China's dike systems would be contracted to private companies. There is every reason to believe dikes will remain under the management of the state, continuing a millennia-long tradition.

Since the 1980s, and particularly after the 1998 flood, we have also witnessing many positive changes, such as the establishment of numerous nature reserves, the ban on logging on the upper reaches of the Yangzi River valley, the promotion of afforestation in the Northwest, the ending of the over-reclamation of lakes, the return of low-lying lands to water, the relocation of people from the water side of river dikes to the land side, an increasing reliance on clean and green energy (thus preserving trees and protecting the environment), the rapid disappearance of farm cattle, and more recently, a 10-year fishing ban in the Yangzi River (since 2020), and so on. These changes demonstrate the intention of the Chinese government to improve the environment, or switch to "eco-development,"[6] which heralds a better future for China's environment. Nationwide, "despite disparaging reports, the available figures show that China's growth is sustainable."[7] The future of China's environment is in the hands of the people of China, and it is reasonable to believe that they can and will find solutions for their environmental problems. China's environment should not and will not collapse. Therefore, from the perspective of sustainable development, we can be guardedly optimistic about the future of China's environment.

---

5 "Decision from the CPC Central Committee," 2010.

6 Harrell 2023.

7 Roumasset, Burnett, and Wang 2008: 281.

Glossary

Aiguo weisheng yundong 爱国卫生运动
badi 巴堤
bayuan 巴垸
Baiju *yuan* 白莒垸
bao 保
baozhang 保障（垸）
biaogong 标工
bian'e 匾额
Caidian 蔡店
cailan di 菜蓝堤
caopeng 草棚
Changjiang 长江
Changping cang 常平仓
Chenglingji 城陵矶
chi 尺
cun 寸
Dachaihu 大柴湖
dan 担
daotai 道台
Danjiangkou 丹江口
Dangyang 当阳
dibao 堤保
diding yin 地丁银
didong 堤董
difang zhi 堤防志
dijia 堤甲
dijiafa 堤甲法
dijie 堤街
dilin guanli yuan 堤林管理员
diyuan 堤垸
dizhang 堤长
Diaocha 汈汊（湖、垸）
Dongjinghe 东荆河
Dongting 洞庭
dongshi 董事
Dongxihu 东西湖

dou 斗
doufu di 豆腐堤
douzha di 豆渣堤
Dujiatai 杜家台
Dujiangyan 都江堰
E-Yu-Wan sansheng jiaofei zong silingbu 鄂豫皖三省剿匪总司令部
erpin 二品
fang 方
fu 府、夫
fudi 府堤
gandi 干堤
gong 弓、工
Gong'an 公安
Gongxiaoshe 供销社
gongzhen 工赈
guanyangdian 管养点
guanyangzu 管养组
guodi 国堤
guofu minming 国赋民命
Guomindang 国民党
Han/Hanjiang/Hanshui 汉/汉江/汉水
Hanchuan 汉川
Hanlu 寒露
Hankou 汉口
Hannan 汉南（区）
Hanyang 汉阳
Haoxue 郝穴
hebing 河兵
hequ zhi 河渠志
Honghu 洪湖
Honghu Gedi 洪湖隔堤
hu 湖
Huayuankou 花园口
Hudu 虎渡
Hulu *yuan* 葫芦垸
huyuan 湖垸
jia 甲
jianfeng chazhen 见缝插针
Jianli 监利

Jiang 江
Jianghan pingyuan 江汉平原
Jianghan gongchengju 江汉工程局
Jiangling 江陵
Jiangxia 江夏
jiaoduan 角端
jin 斤
Jingdi 荆堤
Jingjiang 荆江
Jingjiang dadi 荆江大堤
Jingjiang digongju 荆江堤工局
Jingmen 荆门
Jingshan 京山
Jingzhou 荆州（府、地区、市、区）
jiuxue shisankou 九穴十三口
juan 卷
jundi 军堤
kafang 卡房
Kaixiangong 开弦弓
li 里
lijin 厘金
liang 两
Longwang miao 龙王庙
Madian 马店
Mianyang 沔阳（州、县）
mieluo yuan 灭螺垸
mindi 民堤
minyuan 民垸
mu 亩
mulu 目录
muyou 幕友
Ouchi 藕池
pingyuan xinghong 平垸行洪
Qianjiang 潜江
ren ding sheng tian 人定胜天
Renwubu 人武部
qin 琴
quanzong 全宗
sagu 撒谷

sanxian 三线
Shahu Mianyang zhou, shinian jiu bushou 沙湖沔阳州，十年九不收
Shashi 沙市
Shayang 沙洋（镇、县）
shaopeng 哨棚
shaowu 哨屋
Shenbao 申报
sheng 升
shi 市、石
Shishou 石首
shuifu 水府
shuili tongzhi 水利同知
shuiqing 水情
si-yuan 私垸
Songzi 松滋
taipeng 抬棚
tanyuan 滩垸
Tanglin 唐林
tewu 特务
Tianjiakou 田家口
Tianmen 天门
tianren heyi 天人合一
Tiaoguan 调关
Tiaoxian 调弦
tufei 土费
tuanlian 团练
tuanyong 团勇
tuniu 土牛
Wagouzi 挖沟子
Wanchengdi 万城堤
wei 围、圩、卫
weidi 围堤
wen 文
wo 硪
Wuchang 武昌（府、县、区）
xian 县
xiandi 县堤
Xiantao 仙桃
xiang 乡

Xiaogan 孝感（地区、市、县）
Xiaonan 孝南（区）
xirang 息壤
xiufangchu 修防处
yanglian yin 养廉银
Yaodi 遥堤
yigong daizhen 以工代赈
yigong yinong 亦工亦农
yuan 垸、院
yuan 元、圆
yuanzhang 垸长
yuanzi 垸子
Yuekou 岳口
Yunmeng 云梦
zhang 丈
Zhanghe 漳河
zhen 镇
Zhicheng 枝城
zhihuibu 指挥部
Zhijiang 枝江
Zhongxiang 钟祥
zhou 州
zhoutong 州同
zhouyuan 洲垸
zhuangding 壮丁
zhubu 主簿
zidi 子堤
zong 总

Bibliography

Gazetteers

Traditional Gazetteers

Gong'an xian zhi [Gazetteer of Gong'an County]. 1874.

Hanchuan tuji zhengshi [Geography and products of Hanchuan County]. 1895.

Hanchuan xian zhi [Gazetteer of Hanchuan County]. 1873.

Hanyang fu zhi [Gazetteer of Hanyang Prefecture]. 1747.

Hanyang xian zhi [Gazetteer of Hanyang County]. 1868.

Hubei tongzhi [General gazetteer of Hubei] [Lan Wenjin, ed.]. 1921. Shanghai: Shangwu yinshuguan.

Jiangling xian zhi [Gazetteer of Jiangling County]. 1794.

Jiangling xian zhi [Gazetteer of Jiangling County]. 1876.

Jiangxia xian zhi [Gazetteer of Jiangxia County]. 1869.

Jianli xian zhi [Gazetteer of Jianli County]. 1872.

Jingjiang di zhi [Gazetteer of the Jingjiang Dike]. 1937. N.p.

Jingmen zhili zhou zhi [Gazetteer of Jingmen Independent Department]. 1868.

Jingshan xian zhi [Gazetteer of Jingshan County]. 1882.

Jingzhou fu zhi [Gazetteer of Jingzhou Prefecture]. 1685.

Jingzhou fu zhi [Gazetteer of Jingzhou Prefecture]. 1880.

Mianyang zhou zhi [Gazetteer of Mianyang Department]. 1894.

Qianjiang xian zhi [Gazetteer of Qianjiang County]. 1694.

Qianjiang xian zhi [Gazetteer of Qianjiang County]. 1879.

Shishou xian zhi [Gazetteer of Shishou County]. 1866.

Songzi xian zhi [Gazetteer of Songzi County]. 1869.

Songzi xian zhi [Gazetteer of Songzi County]. 1982 [1937]. N.p.

Xiaogan xian zhi [Gazetteer of Xiaogan County]. 1883.

Yunmeng xian zhi [Gazetteer of Yunmeng County]. 1840.

Yunmeng xian zhi [*Gazetteer of Yunmeng County*]. 1883.

Zhijiang xian zhi [Gazetteer of Zhijiang County]. 1866.

Zhongxiang xian zhi [Gazetteer of Zhongxiang County]. 1867.

Zhongxiang xian zhi [Gazetteer of Zhongxiang County]. 1937.

New Gazetteers

'98 Jiangling kanghong zhi [Gazetteer of high-water control in Jiangling in 1998]. 1999. N.p.

'98 Jingzhou kanghong zhi [Gazetteer of high-water control in Jingzhou in 1998]. 1999. Beijing: Zhongguo jingji chubanshe.

Caobuhu nongchang zhi [Gazetteer of Caobuhu Farm]. 1988. N.p.

Chaihu zhen zhi (1996–2010) [Gazetteer of Chaihu Township (1996–2010)]. 2013. N.p.

Changfeng xiang zhi [Gazetteer of Changfeng Township]. 2016. Wuhan: Wuhan chubanshe.

Changtangkou qu zhi [Gazetteer of Changtangkou Ward]. 1990. N.p.

Dangyang shi zhi (1979–2000) [Gazetteer of Dangyang City (1979–2000)]. 2008. Wuhan: Wuhan chubanshe.

Dangyang xian zhi [Gazetteer of Dangyang County]. 1992. Beijing: Zhongguo chengshi chubanshe.

Dongbao qu zhi (1985–2005) [Gazetteer of Dongbao District (1985–2005)]. 2013. Wuhan: Changjiang chubanshe.

Dongjinghe difang zhi [Gazetteer of the Dongjing River dikes]. 1994. Wuhan: Wuhan daxue chubanshe.

Dongxihu qu jianzhi [A brief gazetteer of Dongxihu District]. 2012. Wuhan: Wuhan chubanshe.

Duodao qu zhi (1949–2005) [Gazetteer of Duodao District (1949–2005)]. 2011. Wuhan: Hubei renmin chubanshe.

Fenshui zhen zhi [Gazetteer of Fenshui Town]. 1998. N.p.

Gong'an xian zhi [Gazetteer of Gong'an County]. 1990. Beijing: Hanyu dacidian chubanshe.

Gong'an xian zhi [Gazetteer of Gong'an County]. 2010. Beijing: Zhongguo huanjing kexue chubanshe.

Guanzhuanghu nongchang zhi [Gazetteer of Guanzhuanghu Farm]. 2000. N.p.

Hanchuan difang zhi [Gazetteer of dikes in Hanchuan]. 2007. N.p.

Hanchuan shi shuichan zhi (1442–2005) [Gazetteer of aquatic product of Hanchuan City (1442–2005)]. 2010. N.p.

Hanchuan shi zhi (1986–2005) [Gazetteer of Hanchuan City (1986–2005)]. 2010. Wuhan: Changjiang chubanshe.

Hanchuan shuili zhi (1950–2008) [Gazetteer of water conservancy in Hanchuan (1950–2008)]. 2013. Wuhan: Changjiang chubanshe.

Hanchuan xian zhi [Gazetteer of Hanchuan County]. 1992. Beijing: Zhongguo chengshi chubanshe.

Hanjiang difang zhi [Gazetteer of the Han River dikes]. 2015. Wuhan: Changjiang chubanshe.

Hanyang qu shuiwu zhi [Gazetteer of water affairs in Hanyang District]. 2015. Wuhan: Wuhan chubanshe.

Hanyang xian shuili zhi [Gazetteer of water conservancy in Hanyang County]. 1990. N.p.

Hanyang xian zhi [Gazetteer of Hanyang County]. 1989. Wuhan: Wuhan chubanshe.

Haoxue zhen zhi [Gazetteer of Haoxue Town]. 1986. N.p.

Honghu fenxuhong gongcheng zhi [Gazetteer of the Honghu flood diversion project]. 2016. Beijing: Zhongguo shuili shuidian chubanshe.

Honghu shi zhi (1987–2007) [Gazetteer of Honghu City (1987–2007)]. 2014. Wuhan: Hubei renmin chubanshe.

Honghu xian shuichan zhi [Gazetteer of the aquatic industry of Honghu County]. 1986. N.p.

Honghu xian zhi (1949–1959) [Gazetteer of Honghu County (1949–1959)]. 1963. N.p.

Honghu xian zhi [Gazetteer of Honghu County]. 1992. Wuhan: Wuhan daxue chubanshe.

Hongshan qu zhi [Gazetteer of Hongshan District]. 2009. Wuhan: Wuhan chubanshe.

Huanghu nongchang zhi (1957–1985) [Gazetteer of Huanghu Farm (1957–1985)]. 1988. N.p.

Huangjing qu zhi [Gazetteer of Huangjing Ward]. 1986. N.p.

Hubei gonglu zhi (1980–2005) [Gazetteer of highways in Hubei (1980–2005)]. 2013. Wuhan: Wuhan chubanshe.

Hubei sheng hupo zhi 1 [Gazetteer of lakes in Hubei 1]. 2014. Wuhan: Hubei kexue jishu chubanshe.

Hubei sheng Shishouxian diming zhi [Survey of the names of places in Shishou County, Hubei Province]. 1983. N.p.

Hubei sheng zhi: dili 1 [Gazetteer of Hubei Province: Geography, vol. 1.] 1997. Wuhan: Hubei renmin chubanshe.

Hubei sheng zhi: junshi [Gazetteer of Hubei Province: Military Affairs]. 1996. Wuhan: Hubei renmin chubanshe.

Hubei sheng zhi: minzheng [Gazetteer of Hubei Province: Civil Administration]. 1994. Wuhan: Hubei renmin chubanshe.

Hubei sheng zhi: nongye 2 [Gazetteer of Hubei Province: Agriculture, vol. 2]. 1999. Wuhan: Hubei renmin chubanshe.

Hubei sheng zhi: shuili [Gazetteer of Hubei Province: Water conservancy]. 1995. Wuhan: Hubei renmin chubanshe.

Hubei shuili zhi [Gazetteer of water conservancy in Hubei]. 2000. Beijing: Zhongguo shuili shuidian chubanshe.

Hubei shuiyun zhi [Gazetteer of water transportation of Hubei]. 1996. Wuhan: Wuhan chubanshe.

Huchang qu zhi [Gazetteer of Huchang Ward]. 1987. N.p.

Jiangdi xiang zhi [Gazetteer of Jiangdi Township]. 2006. Wuhan: Wuhan chubanshe.

Jianghan qu zhi [Gazetteer of Jianghan District]. 2007. Wuhan: Wuhan chubanshe.

Jianghu nongchang zhi [Gazetteer of Jianghu Farm]. 1988. N.p.

Jiangling difang zhi [Gazetteer of the dikes in Jiangling]. 1984. N.p.

Jiangling xian shuili zhi [Gazetteer of water conservancy in Jiangling County]. 1984. N.p.

Jiangling xian zhi [Gazetteer of Jiangling County]. 1990. Wuhan: Hubei renmin chubanshe.

Jiangling xian zhi (1994–2007) [Gazetteer of Jiangling County (1994–2007)]. 2015. Wuhan: Chongwen shuju.

Jiangxia qu zhi (1980–2004) [Gazetteer of Jiangxia District (1980–2004)]. 2007. Wuhan: Wuhan chubanshe.

Jianli difang zhi [Gazetteer of the dikes in Jianli]. 1991. Wuhan: Hubei renmin chubanshe.

Jianli shuili zhi [Gazetteer of water conservancy in Jianli]. 2005. Beijing: Zhongguo shuili shuidian chubanshe.

Jianli xian nongye zhi [Gazetteer of agriculture of Jianli County]. 2011. Wuhan: Chongwen shuju.

Jianli xian zhi [Gazetteer of Jianli County]. 1959. N.p.

Jianli xian zhi [Gazetteer of Jianli County]. 1994. Wuhan: Hubei renmin chubanshe.

Jianli xian zhi (1979–2006) [Gazetteer of Jianli County (1979–2006)]. 2013. Wuhan: Hubei renmin chubanshe.

Jingjiang dadi zhi [Gazetteer of the Great Jingjiang Dike]. 1989. Nanjing: Hehai daxue chubanshe.

Jingjiang difang zhi [Gazetteer of the Jing River dikes]. 2012. Beijing: Zhongguo shuili shuidian chubanshe.

Jingjiang fenhong gongcheng zhi [Gazetteer of the Jing River flood diversion project]. 2000. Beijing: Zhongguo shuili shuidian chubanshe.

Jingmen shi minzheng zhi (1869–2007) [Gazetteer of civil administration of Jingmen City (1869–2007)]. 2016. Wuhan: Hubei renmin chubanshe.

Jingmen shi shuili zhi [Gazetteer of water conservancy in Jingmen City]. 1989. Wuhan: Hubei jiaoyu chubanshe.

Jingmen shi shuili zhi (1986–2008) [Gazetteer of water conservancy in Jingmen City (1986–2008)]. 2012. Wuhan: Changjiang chubanshe.

Jingmen weisheng zhi [Gazetteer of public health in Jingmen]. 1990. Beijing: Zhongguo wenshi chubanshe.

Jingshan xian zhi [Gazetteer of Jingshan County]. 1990. Wuhan: Hubei renmin chubanshe.

Jingzhou diqu shuiyun zhi [Gazetteer of water transportation in Jingzhou Prefecture]. 1989. Nanning: Guangxi renmin chubanshe.

Jingzhou diqu zhi [Gazetteer of Jingzhou Prefecture]. 1996. Beijing: Hongqi chubanshe.

Jingzhou qu zhi [Gazetteer of Jingzhou District]. 2013. Wuhan: Wuhan chubanshe.

Jingzhou shi nongken zhi (1994–2005) [Gazetteer of land reclamation of Jingzhou City (1994–2005)]. 2010. N.p.

Jingzhou shi zhi (1994–2005) [Gazetteer of Jingzhou City (1994–2005)]. 2015. Beijing: Zhongguo wenshi chubanshe.

Jingzhou shuili zhi 1 [Gazetteer of water conservancy of Jingzhou, vol. 1]. 2016. Beijing: Zhongguo shuili shuidian chubanshe.

Jingzhou weisheng zhi [Gazetteer of public health in Jingzhou]. 1990. N.p.

Lengshui zhen zhi (1986–2009) [Gazetteer of Lengshui Town (1986–2009)]. 2012. N.p.

Lingjiaohu nongchang zhi [Gazetteer of Lingjiaohu Farm]. 1991. Beijing: Zhongguo wenshi chubanshe.

Longdian qu zhi [Gazetteer of Longdian Ward]. 1989. N.p.

Madian cun zhi [Gazetteer of Madian Village]. 2004. N.p.

Maliang zhen zhi [Gazetteer of Maliang Town]. 2013. N.p.

Maozui zhen zhi [Gazetteer of Maozui Town]. 1990. N.p.

Miancheng zhi [Gazetteer of Miancheng]. 2000. Wuhan: Hubei kexue jishu chubanshe.

Mianyang Chenchang qu xiangtu zhi [Gazetteer of Chenchang Ward, Mianyang]. 1987. N.p.

Mianyang jiaotong zhi [Gazetteer of transportation in Mianyang]. 1985. N.p.

Mianyang xian zhi [Gazetteer of Mianyang County]. 1989. Wuhan: Huazhong shifan daxue chubanshe.

Miaotou zhen zhi (1949–2008) [Gazetteer of Miaotou Town (1949–2008)]. 2010. N.p.

Ouchi zhen zhi [Gazetteer of Ouchi Town]. 1985. N.p.

Pengchang zhen zhi [Gazetteer of Pengchang Town]. 1994. N.p.

Pengxing xiang zhi (1949–2009) [Gazetteer of Pengxing Township (1949–2009)]. 2011. N.p.

Qianjiang shi zhi [Gazetteer of Qianjiang City]. 2011. Wuhan: Changjiang chubanshe.

Qianjiang shuili zhi [Gazetteer of water conservancy in Qianjiang]. 1997. Beijing: Zhongguo shuili shuidian chubanshe.

Qianjiang xian zhi [Gazetteer of Qianjiang County]. 1990. Beijing: Zhongguo wenshi chubanshe.

Qingchuan jie zhi [Gazetteer of Qingchuan Subdistrict]. 2005. Wuhan: Wuhan chubanshe.

Qingshan qu zhi 1 [Gazetteer of Qingshan District 1]. 2006. Wuhan: Wuhan chubanshe.

Qujialing guanliqu zhi (1985–2007) [Gazetteer of Qujialing Administrative District (1985–2007)]. 2014. Wuhan: Hubei renmin chubanshe.

Qujiawan zhi [Gazetteer of Qujiawan]. 1986. N.p.

Renmin Dayuan nongchang zhi (1957–1987) [Gazetteer of People's Dayuan Farm (1957–1987)]. 1988. N.p.

Shadaoguan zhen zhi [Gazetteer of Shadaoguan Town]. 2009. N.p.

Shashi qu zhi [Gazetteer of Shashi District]. 2007. Wuhan: Hubei renmin chubanshe.

Shashi shi zhi 1 [Gazetteer of Shashi City, vol. 1]. 1992. Beijing: Zhongguo jingji chubanshe.

Shashi shuili difang zhi [Gazetteer of water conservancy and dikes in Shashi]. 1994. Taiyuan: Shanxi gaoxiao lianhe chubanshe.

Shayang xian minzheng zhi [Gazetteer of civil administration of Shayang County]. 2012. N.p.

Shayang xian zhi [Gazetteer of Shayang County]. 2013. Wuhan: Changjiang chubanshe.

Shishou shi zhi (1986–2005) [Gazetteer of Shishou City (1986–2005)]. 2014. Wuhan: Hubei renmin chubanshe.

Shishou xian zhi [Gazetteer of Shishou County]. 1990. Beijing: Hongqi chubanshe.

Songzi shi zhi (1986–2005) [Gazetteer of Songzi City (1986–2005)]. 2012. Wuhan: Wuhan chubanshe.

Songzi shuili zhi [Gazetteer of water conservancy in Songzi]. 2008. Beijing: Zhongguo huanjing kexue chubanshe.

Songzi xian minzheng zhi [Gazetteer of civil administration of Songzi County]. 1985. N.p.

Songzi xian weisheng zhi (1911–1985) [Gazetteer of public health, Songzi County (1911–1985)]. 1986. N.p.

Songzi xian zhi [Gazetteer of Songzi County]. 1986. N.p.

Songzi xumu shouyi zhi (1949–2011) [Gazetteer of animal husbandry and veterinary science in Songzi (1949–2011)]. 2013. Wuhan: Hubei renmin chubanshe.

Taihugang nongchang zhi [Gazetteer of Taihugang Farm]. 1988. N.p.

Tanglin cun zhi (1950–2005) [Gazetteer of Tanglin Village (1950–2005)]. 2008. N.p.

Tianmen shi zhi (1979–2003) [Gazetteer of Tianmen City (1979–2003)]. 2013. Wuhan: Changjiang chubanshe.

Tianmen shuili zhi [Gazetteer of water conservancy in Tianmen]. 1999. Beijing: Zhonghua shuju.

Tianmen xian weisheng zhi [Gazetteer of public health in Tianmen County]. 1984. N.p.

Tianmen xian zhi [Gazetteer of Tianmen County]. 1989. Wuhan: Hubei renmin chubanshe.

Tiaoguan zhen zhi [Gazetteer of Tiaoguan Town]. 2001. N.p.

Tuoshi zhen zhi [Gazetteer of Tuoshi Town]. 2008. N.p.

Wagouzi zhi [Gazetteer of Wagouzi (village)]. 2013. Wuhan: Hubei renmin chubanshe.

Weishui liuyu zhi [Gazetteer of the Wei River Valley]. 2010. Beijing: Zhongguo yanhuang wenhua chubanshe.

Weishui shuiku zhi [Gazetteer of the Wei River Reservoir]. 1996. Beijing: Zhongguo shuili shuidian chubanshe.

Wuchang qu zhi [Gazetteer of Wuchang District]. 2008. Wuhan: Wuhan chubanshe.

Wuchang xian zhi [Gazetteer of Wuchang County]. 1989. Wuhan: Wuhan daxue chubanshe.

Wuhan difang zhi [Gazetteer of the dikes in Wuhan]. 1986. N.p.

Wuhan shi zhi (1980–2000) 1 [Gazetteer of Wuhan Municipality (1980–2000) 1]. 2006. Wuhan: Wuhan chubanshe.

Wuhan shi zhi: nongye zhi [Gazetteer of Wuhan Municipality: Gazetteer of Agriculture]. 1989. Wuhan: Wuhan daxue chubanshe.

Wusan nongchang zhi (1953–1985) [Gazetteer of Wusan Farm]. 1987. N.p.

Wuyin Gong'an kanghong zhi [Gazetteer of high-water control in Gong'an in 1998]. 1999. Beijing: Zhonghua shuju.

Xiantao shi zhi (1986–2002) [Gazetteer of Xiantao City (1986–2002)]. 2011. Wuhan: Hubei renmin chubanshe.

Xiantao shuili zhi [Gazetteer of water conservancy in Xiantao]. 2008. Wuhan: Changjiang chubanshe.

Xiaochang xian zhi (1993–2012) [Gazetteer of Xiaochang County (1993–2012)]. 2013. Wuhan: Hubei renmin chubanshe.

Xiaogan diqu shuili zhi [Gazetteer of water conservancy in Xiaogan Prefecture]. 1996. Wuhan: Wuhan daxue chubanshe.

Xiaogan shi nongye zhi (1949–2009) [Gazetteer of agriculture in Xiaogan City (1949–2009)]. 2011. Wuhan: Chongwen shuju.

Xiaogan shi shuili zhi (1990–2007) [Gazetteer of water conservancy in Xiaogan City (1990–2007)]. 2011. Wuhan: Hubei renmin chubanshe.

Xiaogan shi zhi [Gazetteer of Xiaogan City]. 1992. Beijing: Xinhua chubanshe.

Xiaogan shi zhi (1949–2005) 1 [Gazetteer of Xiaogan City (1949–2005) 1]. 2013. Wuhan: Hubei renmin chubanshe.

Xiaogan xian jianzhi [A brief gazetteer of Xiaogan County]. 1959. Wuhan: Hubei renmin chubanshe.

Xiaonan qu zhi (1990–2008) [Gazetteer of Xiaonan District (1990–2008)]. 2015. Wuhan: Hubei renmin chubanshe.

Xiaxindian zhen zhi [Gazetteer of Xiaxindian Town]. 2012. N.p.

Xiliuhe qu zhi [Gazetteer of Xiliuhe Ward]. 1986. N.p.

Yanglinwei qu zhi (1840–1985) [Gazetteer of Yanglinwei Ward (1840–1985)]. 1989. N.p.

Yingcheng xian zhi [Gazetteer of Yingcheng County]. 1992. Beijing: Zhongguo chengshi chubanshe.

Yongjiu cun zhi (1949–2011) [Gazetteer of Yongjiu Village (1949–2011)]. 2013. Wuhan: Hubei renmin chubanshe.

Yunmeng xian zhi [Gazetteer of Yunmeng County]. 1994. Beijing: Sanlian shudian.

Zaoshi zhen zhi [Gazetteer of Zaoshi Town]. 2009. N.p.

Zhanghe shuiku yimin zhi [Gazetteer of Zhanghe Reservoir Relocatees]. 1996. Beijing: Zhongguo shuili shuidian chubanshe.

Zhanghe shuiku zhi [Gazetteer of Zhanghe Reservoir]. 1991. Beijing: Xinhua chubanshe.

Zhengchang qu zhi [Gazetteer of Zhengchang Ward]. 1985. N.p.

Zhijiang minzheng zhi [Gazetteer of civil administration of Zhijiang]. 1990. N.p.

Zhijiang shi zhi (1979–2002) [Gazetteer of Zhijiang City (1979–2002)]. 2012. Beijing: Fangzhi chubanshe.

Zhijiang xian zhi [Gazetteer of Zhijiang County]. 1990. Beijing: Zhongguo chengshi jingji shehui chubanshe.

Zhongxiang shi zhi (1979–2005) [Gazetteer of Zhongxiang City (1979–2005)]. 2013. Wuhan: Changjiang chubanshe.

Zhongxiang shuili zhi [Gazetteer of water conservancy in Zhongxiang]. 1998. Beijing: Zhongguo shuili shuidian chubanshe.

Zhongxiang xian zhi [Gazetteer of Zhongxiang County]. 1990. Wuhan: Hubei renmin chubanshe.

Zhongzhouyuan nongchang zhi (1957–1985) [Gazetteer of Zhongzhouyuan Farm (1957–1985)]. 1989. N.p.

Zhouji nongchang zhi [Gazetteer of Zhouji Farm]. 1991. N.p.

Zongkou nongchang zhi (1955–1985) [Gazetteer of Zongkou Farm (1955–1985)]. 1986. N.p.

Archival Sources

Every archival file has a serial number, which usually includes three parts: the first part is a category (*quanzong*) number, the second is a subcategory (*mulu*) number, and the third is a file (*juan*) number. Some digital files also include the page number or file number of the digitalized file (not the page number of the original file)—that is the fourth part (for archives of Shishou, both the third and the fourth parts are the file number). Following the serial number is the title of a document in that file (one file may include many documents), except for archives of Songzi, where they are file titles. Since some documents have no title or the title is extraordinarily long, I have shortened or created a title based on the contents of the document.

Caidian District Archives [Formerly Hanyang County Archives]

30-1-1 *Guanyu fangxun gongzuo jiangcheng banfa de guiding* [On rules of rewards and punishments related to high-water control].

30-1-4a *Hanyang xian digong baogao* [Report on dike work in Hanyang County].

30-1-4b *Hanyang xian guanyu 1955 niandu suixiu jihua caoan* [Draft plan of the 1955 annual repairs in Hanyang County].

30-1-5 *Hanyang xian fangxun zhihuibu chubu gongzuo qingkuang zonghe baogao* [A comprehensive report on the initial work of the high-water control headquarters of Hanyang County].

30-1-14 *Guanyu gongbu xundi guize de guiding* [Regulations on the announcement on rules of dike patrolling].

30-1-16 *Wu Nengguang shuji zai mingong daibiao dahui shang de zongjie (caoan)* [Party secretary Wu Nengguang's summary of the general meeting of labor representatives (draft)].

30-1-24a *Guanyu 1960 niandu gongzuo chubu zongjie* [A preliminary summary of work in 1960].

30-1-24b *Fangxun yuangong jiangcheng tiaoli* [Regulations on rewards and punishments for high-water control personnel].

30-1-24c *Guanyu Guanlianhu weiken gongcheng de baogao* [Report of the enclosure of Lake Guanlian].

30-1-37 *Guanyu ganweidi zhongdian xiujian zhuanmu jiegou fangxun shaopeng de tongzhi* [Notice on prioritizing the construction of brick-and-wood shaopeng for high-water control on the main and embanked dikes].

Gong'an County Archives

38-1-4a *Gong'an xian fangxun zhihuibu binjiang gandi Zuolouwan louzha qiangxian baogao* [Report on the emergency repair of the sluice gate in Zuolouwan on the major Yangzi River dike, high-water control headquarters of Gong'an County].

38-1-4b *Gong'an xian Mahaokou qu fangxun zhihuibu chengbao* [Report from the high-water control headquarters of Mahaokou Ward, Gong'an County].

38-1-7a *Gong'an xian gandi gongcheng weiyuan huiyi suixiu zongjie baogao* [Summary report on annual repairs of the engineering committee of the major river dikes in Gong'an County].

38-1-7b *Guanyu zai dishang jinxing xuanchuan jiaoyu gongzuo de zhishi* [Instructions regarding conducting propaganda and education on dikes].

Hanchuan City Archives [Formerly Hanchuan County Archives]

50-1-1a *Mi Liao liangyuan xieshui jiufen* [Dispute between Mi polder and Liao polder over the discharge of water].

50-1-1b *Huayansi kuikou qianhou* [The whole story of the rupture in Huayansi].

50-1-15a *Wei fangxun fudan xu zhengshou fangchanshui baoqing heshi you* [A request to collect house property tax for high-water control].

50-1-15b *Guanyu 1954 nian fangxun gongzuo de chubu yijian* [Preliminary advice on high-water control in 1954].

50-1-16a *Hanchuan xian 1954 nian fangxun zhongde zhengzhi sixiang jiaoyu gongzuo zongjie* [Summary of political and ideological education in the 1954 high-water control in Hanchuan County].

50-1-16b *Hanchuan xian 1954 nian fangxun zongjie* [Summary of the 1954 high-water control in Hanchuan County].

50-1-18 *Hanjiang fangxun gongzuo de jidian jiben jingyan* [Some basic experience in high-water control of the Han River].

50-1-22 *Guanyu dukou fudi zhong cunzai jige zhuyao wenti de tongbao* [Report on some major problems in blockage and dike restoration].

50-1-24 *Hubei sheng fangxun zong zhihuibu guanyu fangxun caiwu ruogan wenti de dafu* [Answers to several issues on the financial affairs of high-water control, high-water control general headquarters of Hubei Province].

50-1-28a *Hanchuan xian dierci fangxun huiyi zongjie baogao* [Summary report on Hanchuan Country's second high-water control meeting].

50-1-28b *Hanchuan xian 1955 nian fangxun huiyi zongjie baogao* [Summary report on Hanchuan County's 1955 high-water control meeting].

50-1-35a (Diyiqu) *1955 nian suixiu gongzuo zongjie* [Summary of the annual repairs in 1955 (First Ward)].

50-1-35b *Hanchuan xian disan qu shuili gongcheng zhihuibu 1955 nian gandi suixiu zongjie* [Summary of the annual repairs of the major dikes in 1955, Third Ward Water Conservancy Engineering Headquarters, Hanchuan County].

50-1-42 *Hanchuan xian Baohu yuan bingyuan hexiu zongjie baogao* [Summary report of the consolidation and co-repair of Baohu polder in Hanchuan County].

50-1-45 *Hanjiang xiayou fenhong gongcheng Hanchuan tugong zhihuibu zongjie baogao* [Summary report of Hanchuan's civil engineering headquarters of the flood diversion project of the lower Han River valley].

50-1-50a *Xinjie xiang guanli yanghu gongzuo baogao* [Report on management and maintenance in Xinjie Township].

50-1-50b *Hanchuan xian Sanlong yuan hedi bingyuan de zongjie baogao* [Summary report on the consolidation of the dikes of Sanlong polder in Hanchuan County].

50-1-52a *Guanyu Ma'an xiang Huangguahu shenqing weidi de chuli jiancha baogao* [Investigative report on the application to enclose Lake Huanggua of Ma'an Township].

50-1-52b *Guanyu Lianghe xiang Xinsheng gongshe shenqing weiyuan de chuli jiancha baogao* [Investigative report on the application to enclose a polder by Xinsheng Commune, Lianghe Township].

50-1-65 *Hongqi chazai fangxun zhanchang, shengli shuyu laodong renmin* [Red flags are planted on the battlefield of high-water control, and victory belongs to working people].

50-1-67 *Xuanchuan tigang* [Publicity outline].

50-1-71 *Xiong xianzhang zai xiang Diaochahu jinjun shishi dahui shangde zongjie baogao* [County head Xiong's summary report on the mobilization rally for the march to Lake Diaocha].

50-1-87 *Guanyu shuili gongcheng guanli yanghu gongzuo de baogao* [Report on the management and maintenance of water conservancy projects].

Hanyang District Archives

27-2-2 *Guanyu Wuhan shiqu fangxun jingfei fudan wenti ji kaizhi biaozhun de guiding* [Regulations on the burden of high-water control funds and standards of expenditures in downtown Wuhan].

27-2-4 *Guanyu jindong mingchun zhishu zaolin de baogao* [Report on afforestation this winter and the following spring].

27-2-8 *Wuhan shi shiqu hedao difang guanli shishi banfa* [Implementation measures to manage the river course and dikes in downtown Wuhan].

Hubei Provincial Archives

LS1-2-0484-007-001 *Xiuzheng gexian mindi xiufangchu zuzhi guicheng tiaowen duizhao biao* [A table comparing revised articles in the organizational regulations of the annual repairs and high-water control agencies of people's dikes in each county].

LS1-3-0214-007 *Guanyu yigong daizhen de wanglai gongwen* [Current official documents on work relief].

LS1-3-0234-001 *Hanchuan shuizai qizhen* [Begging for relief from the flood in Hanchuan].

LS1-3-0234-002 *Hanchuan xianzhang qiuzhen* [A request for relief from the Hanchuan County head].

LS1-3-0236-005 *Tianxingzhou yuan zaimin qing jizhen* [A request for emergency relief from victims in Tianxingzhou polder].

LS1-3-0237-001 *Mianyang xian 15 yuan qingzhen* [A request for relief by 15 polders in Mianyang County].

LS1-3-0297-004 *Qing daibu qiantao zhi xiufang zhuren Tang Guifang* [A request to arrest absconder Tang Guifang, the director of annual repairs and high-water control].

LS1-3-566 *Zhu Bingkun tunmo dikuan qiangzhu buli* [Zhu Bingkun embezzled dike fees and was derelict in emergency building].

LS1-4-452 *Guanyu qian Hubei shuiliju juzhang Chen Keming deng bei fu chengjie an yijue shu* [Resolution on the punishment of Chen Keming, the former director of the Hubei Bureau of Water Conservancy, and others].

LS1-5-598 *Difang zaolin ji xianzhi qingxiedi kenzhi banfa* [Afforestation measures on dikes and the restriction on reclamation on slopes].

LS1-5-606 *Tongyi shuili xingzheng shiye banfa* [Measures to unify water conservancy administration].

LS1-5-0607-0018 *Songzi xian zhengfu guanyu ge diyuan sheli xiufangchu bing duling renzhen xiufang de chengwen* [Report on the establishment of an annual repairs and high-water control agency in each polder and the order to carefully maintain dikes, Songzi County government].

LS1-5-608 *Youdi gexian zunzhang fangxun* [Counties with dikes are following regulations to control high water].

LS1-5-614 *Jianghan gandi fangxun banfa* [Methods of high-water control on the major dikes along the Yangzi and Han Rivers].

LS1-5-639 *Hubei shenghui fangxunchu zuzhi guicheng* [Rules of organization of the high-water control office in the capital of Hubei Province].

LS1-5-640 *Cheng wei zunling chengbao banli fangxun qingxing* [Following orders to submit a report on the handling of high-water control].

LS1-5-644 *Hubei sheng shuizai ji zhenji diaocha baogao* [Investigative report on the floods and relief in Hubei Province].

LS1-5-658 *Fangxun qinei tongxun banfa* [Methods of communication during high-water control].

LS1-5-0764-001 *Hanyang Xihu yuan yuanmin jiufen* [Disputes among residents of Xihu polder in Hanyang].

LS1-5-784 *Wei chengbao jiangshui baozhang qingpai fangxun zhuanyuan qiangjiu* [Requesting that the high-water control commissioner be sent as floodwater rose suddenly].

LS1-5-785 *Cheng wei tunzhen feidi wangshang yamin qiken lianqing chaban buji yizhong zhenwu er xu zaili jianmian kuilei shi* [Report on embezzling relief, abandoning dikes, deceiving superiors, and oppressing people, and a plea to investigate and replenish the (means of) relief to focus on relief and to relieve victims and to avoid repeated losses].

LS1-5-0790-001 *Wei chengbao shuxian zuzhi xiufangchu ji huafen dongbao jingguo qingxing huiju tubiao jiqing jianhe you* [Report on the process of organizing the annual repairs and high-water control agency and the divisions of dike management in our county, with a chart and request for approval].

LS1-5-849 *Wei Jingjiang digong fei guibing Caizhengju fuzhengshi ainan xing you* [It is really difficult to arrange for the Collection Department of the Bureau of Finance to take over the Jingjiang dike fees].

LS1-5-851 *Qing bian Jingjiang dizhi you* [Reasons for the request to compile a gazetteer of the Jing River dikes].

LS1-5-852 *Zhun han yi Jingdi huagui Jianghanju tongchou banli* [The approval letter on the transfer of the Jing River dikes to the jurisdiction of the Bureau of Engineering Work on the Yangzi and the Han Rivers].

LS1-5-0854-001 *Xiufang zhuren Zhao Qinglong guyi bizha chengzai* [Zhao Qinglong, the director of annual repairs and high-water control, deliberately closed sluice gates and caused a disaster].

LS1-5-855A *Guanyu mufei yu qianfei de chuli banfa ji didu* [On the treatment of dike fees and arrears, as well as dike vermin].

LS1-5-855B *Zhijiang qingjiang mindi huagui gandi* [Zhijiang implored to list people's dikes as major dikes].

LS7-2-2707[1]-001 *Wei nieci wugao zisi haigong qingqiu kaiting bianlun panding quanli yiwu yiming zhenxiang er jing diaowan shi* [Because of false accusations and selfishness, a court debate to determine rights and obligations to warn the cunning is requested].

LS7-3-00003675-069 *Shishou xian Xingxue yuan di xiufangchu guanyu wei quzhang Huang Huazhong jianggong zhecuo de daidian, zhengmingshu* [Official document and certificate for puppet ward chief Huang Huazhong to atone for crimes by meritorious service, annual repairs and high-water control agency of Xingxue polder, Shishou].

LS7-6-0006269-003 *Qing xieji anfan Jiang Zhenzhi* [Please help arrest criminal Jiang Zhenzhi].

LS19-2-2295 [1] *Jianghan gongchengju dianqing jiang digong zhuankuan shifen zhiyi yongyu mindi* [The Bureau of Engineering Work on the Yangzi and Han Rivers is asked by telegraph to use ten percent of the special dike fund on people's dikes].

LS19-4-6552 *Shishou xian Luocheng yuan tudi suoyouquan jiufen* [Disputes over land ownership in Luocheng polder, Shishou County].

LS20-11-00001132-001 *Li Wanqing shoutianqi* [Li Wanqing's land sales deed].

LS20-11-00001162-001 *Hu Santai deng shoutianqi* [The land sales deed of Hu Santai et al].

LS23-3-0866-001 *Jianghan gongchengju diqi gongwusuo 1946 niandu Shishou dufu gongcheng* [The 1946 projects of blockage and restoration in Shishou by the Seventh Work Office of the Bureau of Engineering Work on the Yangzi and the Han Rivers].

LS24-3-00002802-009 *Hanyang xian Moliu yuan xiufang zhuren qingzhen* [The director of the annual repairs and high-water control agency of Moliu polder in Hanyang County requests relief].

LS30-1-0200-001 *Wang Qingquan simai zhenliang qintun gongkuan* [Wang Qingquan secretly sold relief grain and embezzled project funds].

LS30-1-0203-003 *Jiangling xian Wancheng xiufang banshichu zhuren Wang Qingquan simai zhenliang qinshi gongkuan* [Wang Qingquan, director of the Bureau of Annual Repairs and High-Water Control of Wancheng, Jiangling County, secretly sold relief grain and embezzled project funds].

LS30-1-0207-017 *Liu Wenkun tunshi zhenmian an* [The case of Liu Wenkun's embezzlement of relief flour].

LS30-1-0889-004 *Jianghan gandi gongzhen gongcheng gaijin yijian shu* [Suggestions for improving work-relief projects on the main dikes along the Yangzi and Han Rivers].

LS30-1-1445-002 *Hanyang xian Xihu dayuan xiufang zhuren qingzhen* [The director of the annual repairs and high-water control agency of Xihu polder in Hanyang County requests relief].

LS31-4-58 *Hubei dikuan jielüe* [A brief history of the collection of dike fees in Hubei].

LS31-4-60A *Hanchuan xianzhang bian Wangjiayuan gandi xiuzhu an* [1946] [The Hanchuan County head defends his role in the building and maintenance of the Wangjia polder dike].

LS31-4-60B *Shishou gongmin Chen Zuwu deng chengkong Zhao Nan'gai an* [Chen Zuwu and others, citizens of Shishou, accuse Zhao Nan'gai (of corruption)].

LS31-4-63 *Huang Kepei zheng chanquan an* [The case of Huang Kepei, who fought for his property rights].

LS31-4-221 *Zhengli Hanshui xiayou hedi jihua caoan* [Draft plan for the realignment of dikes along the lower Han River valley].

LS31-4-347 *Wei chengji shuxian ge xiufangchu jingfei yilanbiao* [To submit the table of expenditures of each polder in the county].

LS31-4-371 *Shishou xian geyuan xiufangchu zhengfu zhuren xuanju banfa* [Procedures for the election of the director and deputy director of the annual repairs and high-water control agency of each polder in Shishou County].

LS31-4-1170 *Hubei Jianghan gandi suixiu fanghu zhangcheng* [Regulations on annual repairs and protection of the major dikes of the Yangzi and Han Rivers in Hubei].

LS31-4-1403 *Hanyang Daxing yuan diwu jiufen* [Disputes over dike affairs in Daxing polder, Hanyang County].

LS31-4-1404 *Hanyang Yaodi xiang yuanmin jiufen* [Disputes among polder residents of Yaodi Township, Hanyang County].

LS31-4-1405 *Hanyang Changfeng yuan diwu jiufen* [Disputes over dike affairs in Changfeng polder, Hanyang].

LS31-4-1406 *Hanyang Xinhe yuan yemin daibiao kong xiufang zhuren* [Resident representatives of Xinhe polder, Hanyang, accuse its director of the annual repairs and high-water control agency].

LS31-4-1407 *Hanyang Xihu yuan diwu jiufen* [Disputes over dike affairs in Xihu polder, Hanyang].

LS31-4-1408 *Mianyang Yaoyuan xiufang zhuren yurou xiangmin* [The director of the annual repairs and high-water control agency in Yao polder, Mianyang, bullies the local people].

LS31-4-1531 *Hanchuan minyuan hezuo jiufen* [Disputes over cooperation in people's polders in Hanchuan].

LS31-4-1724 *Shishou Yonghu dayuan gongmin kong xiufang zhuren an* [Citizens of Yonghu polder in Shishou accuse the director of the annual repairs and high-water control agency].

LS35-3-007-002 *Guanyu chaban Shishou xian Zhao Nan'gai wubi an* [On the investigation of Zhao Nangai's embezzlement in Shishou County].

LS36-1-69 *Hanchuan Hanyang Tongxing yuan diwu jiufen* [Disputes over dike affairs in Tongxing polder, Hanchuan and Hanyang].

LS36-1-75 *Wei chengbao Jiangyongdi Junshandi dishen bei zhujun juewa gouzhu zhanhao qingxing* [Report on the excavation of the dike body and the construction of trenches on the Jiangyong dike and the Junshan dike by garrison troops].

LS36-1-122 *Pai yuanjing daiqiang ducu minfu qiangxian* [Sending armed police officers to urge laborers to conduct emergency repairs].

LS36-2-15 *Hubei sheng gexian minyuan xiufangchu zuzhi tongze* [General rules on the organization of the annual repairs and high-water control agencies of people's polders in the counties of Hubei].

LS36-2-18 *Wei binjiang minyuan dinian bengtan xiufu weijian zaidian chengqing* [Asking again to repair riverside polder dikes that have collapsed over the years and were difficult to repair].

LS36-2-066-003 *Peixing yuan qian xiufang zhuren Yang Xibo tanwu an* [The corruption case of Yang Xibo, the former director of the annual repairs and high-water control agency of Peixing polder].

LS36-2-081-002 *Guanyu Hanchuan Yongsheng yuan xiufang zhuren shangai dixian kenyu paiyuan gengzheng de cheng, pi* [Petition and comments on the request of sending people to investigate the director of the annual repairs and high-water control agency of Yongsheng polder in Hanchuan, who changed the dike line without authorization].

LS63-2-005-006 *Peixing yuan xiufang zhuren Yuan Kexian qinshi gongkuan* [Yuan Kexian, the director of the annual repairs and high-water control agency of Peixing polder, embezzled public funds].

SZ1-02-0216-007 *Honghu kuikou xiangqing baogao* [Report on the details of the rupture in Honghu].

SZ1-2-155 *Digong fudan wenti de chuli fang'an (caoan)* [Solution to the problem of the dike work burden (draft)].

SZ1-5-59A *Guanyu jinnian xiudi gongzuo de zhishi (caoan)* [Instructions on the building of dikes this year (draft)].

SZ1-5-59B *Diwei Xuanchuanbu guanyu xiudi xuanchuan gongzuo zhishi* [Instructions on publicity on dike building, Department of Publicity, Prefectural Committee].

SZ1-5-65 *Mianyang diwei yuanyuefen zonghe baogao* [January's comprehensive report of the Mianyang Prefectural Committee].

SZ18-2-92 *Fangxun jianbao (dishiqi)* [Brief reports on high-water control (Issue 10)].

SZ34-02-0112-028 *Changjiang yan'an hudi guiyue* [Regulations on the protection of dikes along the Yangzi River].

SZ34-2-116 *Yaodi Xiaojianghu qu xuhong ji Hanjiang xiayou fenxie shuidao chakan baogao* [A survey report of the flood diversion in the Xiaojianghu area, the Distance Dike section, and flood diversion channels in the Lower Han River Valley].

SZ34-2-117 *Hubei Shishou xian Jiaoziyuan xiaohe dusai jingguo he chuli banfa* [The process for blocking the Jiaoziyuan riverlet in Shishou County, Hubei, and the method of handling the issue].

SZ34-2-220 *Xi xun chengli geji fangxun zhihuibu* [Hope to establish high-water control headquarters at all levels soon].

SZ34-2-223 *Wei Gong'an Nanwuzhou yuandi jueding paohui guanyu jumin qianyi ruhe jiejue qing heshi you* [Request on how to deal with the people of Nanwu Islet after its polder dike was destroyed].

SZ34-2-225 *Wei fengling benxian "xuhong kenzhiqu" hou jiang benxian juti qingkuang yu qunzhong yaoqiu chengqing heshi you* [Upon being requested to open land for discharging floodwater, report on the situation and people's opinions].

SZ34-2-404 *Youguan shuili zhengce wenti de jidian zhuyi shixiang* [Some notes regarding the policy of water conservancy].

SZ34-2-406 *Guanyu 1953 nian difang dongxiu gongzuo de zhishi* [Instructions on the repair of dikes in the winter of 1953].

SZ34-2-412A *Hanyang xian digong fudan diaocha baogao* [A survey report on the burden of dike works in Hanyang County].

SZ34-2-412B *Qianjiang xian jianchazu digong fudan diaocha zongjie* [Survey summary of the burden of dike work, Qianjiang County survey team].

SZ34-2-412C *Guanyin xiang digong wendi diaocha ziliao* [Survey data on the problems of dike work in Guanyin Township].

SZ34-2-412D *Yongxing xiang 12 zu dianxinghu 52 nian jiben qingkuang diaocha biao* [Survey table of the basic situation of typical households in the twelfth team, Yongxing Township, 1952].

SZ34-2-414 *Hubei sheng san nian lai shuili jianshe zongjie baogao* [Summary report on the construction of water conservancy in Hubei Province in the past three years].

SZ34-5-210 *Buzhun ken Tianxingzhou dian* [Telegram on the ban on reclamation of Tianxing Islet].

SZ34-5-708 *Guanyu Jingjiang dadi zhanbei jiagu wenti de baogao* [Report on the reinforcement of the Great Jingjiang Dike for war preparedness].

SZ37-1-9 *Ti'an* [Proposal].

SZ039-001-0012-0002 *1954 nian Hubei sheng zaiqu renkou siwang diaocha tongjibiao* [Table of a survey of statistics on death in disaster areas in Hubei in 1954].

SZ67-01-0965-004 *Xichuan yimin ban Jingzhou zongjie baogao* [Summary report on the relocation of people from Xichuan to Jingzhou].

SZ67-04-0515-007 *Danjiangkou shuiku yimin jiben qingkuang* [The basic situation of relocatees of the Danjiangkou Reservoir].

SZ68-2-102 *Guanyu qiangxian jiuzai wuzi ying suidao suiyun* [Materials for emergency repairs and disaster relief should be transported immediately after they arrive].

SZ68-2-226 *Hubei sheng renmin zhengfu shuiliju suoshu gejuchu 1950 nian xiudi gongzuo gaikuang* [An outline of dike construction of all bureaus of the Department of Water Conservancy, People's Government of Hubei Province].

SZ68-2-227 *Hubei sheng shuiliju 1949 nian gongzuo zongjie* [Work summary of 1949, Bureau of Water Conservancy of Hubei Province].

SZ73-05-0215-014 *Guanyu Jingjiang dadi difang suixiu gongcheng shiliao yunshu wenti de diaocha baogao* [Investigative report on problems in the transportation of stone in annual repairs of the Great Jingjiang Dike].

SZ113-1-268 *Guanyu shuiku yimin anzhi gongzuo yijian de baogao* [Report of advice on the placement of reservoir relocatees].

SZ113-2-11 *Hubei sheng 1950 nian yinian lai de difang gongzuo zongjie baogao* [Summary report on dike works in Hubei Province in the year 1950].

SZ113-2-022-003 *1952 nian dong zhi 1953 nian chun yigong daizhen xingxiu shuili de baogao* [A report on water conservancy construction via work relief from the winter of 1952 to the spring of 1953].

SZ113-2-35A *Hanyang xian 1952 nian Jianghan gandi fangxun zongjie baogao* [Summary report on high-water control on the major dikes of the Yangzi and Han Rivers in Hanyang County in 1952].

SZ113-2-35B *Hanjiang fangxun zongjie baogao* [Summary report on high-water control on the Han River].

SZ113-2-60-001 *Hubei sheng Jianghan ganzhidi kuikou chushu ji gaikuang biao* [Table of breaches of the dikes along the Yangzi and Han Rivers and their major branches in Hubei Province].

SZ113-2-077-001 *Hanjiang liang'an jin bainian lai kuikou qingkuang diaochabiao* [A survey table of ruptures of the Han River [dikes] on both banks in the past two hundred years].

SZ113-2-144-009 *Dongjinghe di 10 nian xiufang gongzuo chengjiu zongjie* [Summary of a decade of achievement in annual repairs and high-water control of the Dongjing River dikes].

SZ113-2-195A *Guanyu difang jiancha gongzuo de baogao* [Report on a dike inspection].

SZ113-2-195B *Guanyu difang huihuai wenti de disanci jiancha baogao* [The third inspection report on issues of dike destruction].

SZ113-2-195C *Suixiu donggong he fudi xiaojie baogao* [A summary report on the winter's annual repairs and dike restoration].

SZ113-2-195-004 *Guanyu Hanjiang difang jiancha qingkuang de baogao* [Report on the investigation of the Han River dikes].

SZ113-2-195-005 *Guanyu Hanjiang gandi bei pohuai hou huifu qingkuang de zongjie baogao* [A summary report on the restoration of the major dikes of the Han River].

SZ113-2-274-003 *Hubei sheng Jingjiang dadi gongcheng guanli qingkuang jieshao* [Introduction to the management of the Great Jingjiang Dike project of Hubei province].

SZ113-2-274-009 *Hanjiang gandi kaizhan fudi fulin gongzuo de zongjie* [Summary of the development of the restoration of the major Han River dikes and their forests].

SZ113-2-311-006 *Mianyang xian Hanjiang gandi ji hongdao liang'an difang guanyang gongzuo zhuanti baogao* [Special report on the maintenance of the major Han River dikes and dikes along the flood-relief channel in Mianyang County].

SZ113-2-342-001 *Guanyu Tianjiakou kuikou qingkuang de diaocha baogao* [An investigative report on the rupture in Tianjiakou].

SZ113-2-342-003 *Guanyu baosong Changjiang gandi Tianjiakou kuikouhou de diaocha baogao* [In regard to sending the investigative report on the rupture of the Yangzi River dike in Tianjiakou].

SZ113-3-231-008 *Hanjiang gandi jinian lai difang guanli yanghu jingyan zongjie* [Summary of the experiences of the maintenance of the major Han River dikes in the past several years].

SZ113-3-285-006 *Guanyu zhuyao jianghe zhoutan weiwei qingkuang he chuli yijian de baogao* [Reports on the situation of the enclosure of river beaches of the major rivers and suggestions how to deal with them].

SZ113-3-393-001 *Guanyu shuiku yimin gongzuo ruogan wenti de guiding* [Regulations on several issues concerning reservoir relocatees].

SZ113-3-393-002 *Shuili dianli bu guanyu bufen sheng (qu) jiancha qingli shuili jingfei shiyong he guanli qingkuang de baogao* [Report on the investigation and clearance of the use and management of water conservancy funds in some provinces, Ministry of Water Conservancy and Electricity].

SZ113-3-542-009 *Hanjiang difang guanli gongzuo jianjie* [A brief introduction to the management of the Han River dikes].

SZ113-6-150 *Guanyu dui Changban "Jingbei fangyu gongcheng sheji" yijian de baogao* [Report on the suggestion to release silt to northern Jingzhou by the Office of the Yangzi River Committee].

SZ139-1-000007-018 *Guanyu Danjiangkou shuiku Henan Xichuan xian yimin qian'an wenti de baogao* [Report on the issues of relocation and placement of relocatees of Xichuan County, Henan Province, due to the Danjiangkou Reservoir].

SZ139-1-000360-012 *Guanyu Danjiangkou shuiku yimin banqian gongzuo jixiang juti wenti de buchong guiding* [Supplement regulations on several issues about the resettlement of the Danjiangkou Reservoir relocatees].

SZ139-6-0028-002 *Guanyu jiaqiang difang guanli, baozhang difang anquan de tongzhi* [Notice on strengthening dike management to ensure dike safety].

Jiangxia District Archives [*Formerly Wuchang County Archives*]

49-1-2a *1955 nian difang suixiu zongjie he fangxun renwu de chubu yijian* [Summary of annual repairs and preliminary suggestions on the task of high-water control in 1955].

49-1-2b *Hubei sheng 1955 nian fangxun gongzuo zongjie* [Summary of high-water control in Hubei Province in 1955].

49-1-4a *Zhang Tixue tongzhi zai quansheng shuili gongzuo huiyi shangde baogao* (*jilugao*) [Comrade Zhang Tixue's report on the provincial water conservancy conference (notes)].

49-1-4b *Wei jishi chuli dukou fudi zhong tanwu, nuoyong, cuokuan, daoqie, langfei deng weibei caijing jilü anjian de tongbao* [Report on the timely treatment of cases of the violation of financial discipline such as corruption, embezzlement, wrong payment, theft, and waste in blockage and dike restoration].

49-1-5 *Fangxun xundi guize* [Rules on patrolling dikes during high-water control].

49-1-6 *Wuchang xian shuili gongcheng zhihuibu difang suixiu weisheng gongzuo zongjie baogao* [Summary report on health work in annual repairs of dikes, Water Conservancy Project Headquarters of Wuchang County].

49-1-61 *Wuchang xian shiyi chu zhongdian weihu gongcheng diaocha ziliao* [Survey data of eleven key lake enclosure projects in Wuchang County].

49-1-65 *Quanxian weihu gongcheng peitao qingkuang* [The situation with accessories needed for lake-reclamation projects in the whole county].

49-2-1 *Yanfang feiteba pohuai fangxun gongzuo ji shaohui fangxun cailiao* [Take strict precautions against the subversion of high-water control and burning of high-water control materials by bandits, spies, and tyrants].

49-2-8 *Xiaogan xiufangchu wusan nian suixiu gongzuo zongjie* [Summary of the 1953 annual repairs, Bureau of Annual Repairs and High-Water Control of Xiaogan].

49-2-14 *Guanyu dui nongmin shuili jishuyuan shiyong de jidian yijian de han* [Letter with advice on the use of peasant hydraulic technicians].

49-2-18 *Shenghuo zhuadehao ganjin riri gao—guanyu zuohao mingong shenghuo de jidian zuofa* [Working enthusiasm is rising day after day due to good living services: On several methods to better serve laborers].

49-2-21 *Guanyu ganmindi jige juti wenti de qingshi baogao* [A request report on several concrete issues regarding major dikes and people's dikes].

49-2-42 *Guanyu Wuchang xian fangxun gongzuo de jinji baogao* [An urgent report on high-water control in Wuchang County].

49-2-44a *Guanyu chuli jiangdi yanxian fangkongdong de jiancha baogao* [An inspection report on the treatment of air raid shelters along river dikes].

49-2-44b *Jiangdi yanxian fangkongdong zheng zai jiji jinxing chuli* [Air raid shelters along the river dikes are being dealt with now].

49-2-78 *Wuchang xian Changjiang gandi qingkuang* [The situation of the major Yangzi River dike in Wuchang County].

Jianli County Archives

1-1-010-004 *Guanyu dali xiudi zhishui de zhishi* [Instructions regarding devoting major efforts to building dikes and controlling water].

1-1-029-022 *Xiudi zongjie baogao* [Summary report on dike building].

1-1-033-026 *Ge qu xiang gan daibiaohui guanyu guanche xiudi qingkuang* [The meeting of representative cadres of each ward and township on carrying out dike building].

1-1-092-008 *Xianwei guanyu jiaqiang xiudi gongzuo lingdao zuohao mingong zhengzhi sixiang gongzuo xunsu kefu gezhong hunluan xianxiang de zhishi* [Instruction of the county party committee on reinforcing the leadership of dike building, making a success of laborers' ideological political work, and quickly overcoming various disorderly situations].

1-1-092-009 *Xianwei guanyu jiaqiang xiudi gongzuo lingdao de zaici zhishi* [The second instruction of the county party committee to reinforce the leadership of dike building].

1-1-271-012 *Dui dangqian fangxun gongzuo de zaici zhishi* [Second instruction on current high-water control].

1-1-271-013 *Guanyu 1959 nian fangxun gongzuo ganbu mingong jiangcheng tiaoli* [Regulations on rewards and punishments of cadres and laborers in the 1959 high-water control].

1-1-502-009 *Min Likun tongzhi zai xian jinji fangxun huiyi shangde baogao* [Comrade Min Likun's report on the county's emergency meeting on high-water control].

22-1-006-008 *Wei fengling yanjin wajue dipo er fang shuihuan xi qieshi zunzhao bing bugao zhouzhi you* [In order to strictly prohibit the excavation of dikes and prevent floods, we hope to follow up by announcing the rationale].

22-1-006-010 *Fangxun jiangcheng zanxing banfa* [Temporary methods regarding rewards and punishments in high-water control].

22-1-006-012 *Hubei sheng gexian zhidi mindi hudi guiyue* [Regulations on the protection of branch dikes and people's dikes in the counties of Hubei Province].

22-1-006-016 *Guanyu jindong mingchun xiudi gongzuo zhishi* [Instructions on dike building this winter and the coming spring].

22-1-013-010 *Xiudi huibao tigang* [Outline of reporting dike building].

22-1-021-012 *Wei tongbao Jianli xian xiudi qingkuang* [Reporting the situation of dike building in Jianli County].

22-1-085-016 *Peng Daqi tongzhi zai xian zhihuibu mingong zhengzhi gongzuo huiyi shangde zongjie* [Comrade Peng Daqi's summary on the laborers' political work conference of the county headquarters].

22-1-266-027 *Hubei sheng Jingzhou zhuanqu shuili gongcheng guanyang zanxing banfa* (caoan) [Temporary maintenance methods of water conservancy projects in Jingzhou Prefecture, Hubei Province (draft)].

22-1-329-005 *Guanyu difang guanyang qingkuang de baogao* [Report on the dike maintenance situation].

Jingshan County Archives

00046-001-0007 *Hubei sheng xunlian nongmin shuili jishuyuan jihua fang'an* [The plan to train peasant hydraulic technicians in Hubei Province].

00046-001-0017 *Guanyu nongmin shuili jishuyuan xunlianban de zongjie baogao* [Summary report on the training class of peasant hydraulic technicians].

00046-001-0032 *Guanyu Yaodi suixiu tufang gongcheng jungong de zongjie baogao* [Summary report on the completion of the annual repairs earthwork of the Distance Dike].

00046-001-0046 *1957 niandu difang suixiu qicai gongzuo zongjie* [A summary of materials on dike affairs, 1957].

00046-001-0088 *Guanyu pizhuan benbu gongzuozu jiancha Shishou dangqian xingxiu shuili de qingkuang* [Transferring our headquarters' work team on the examination of the current situation of water conservancy construction in Shishou].

00046-001-0117 *Tixue tongzhi zai quansheng fangxun gongzuo huiyi shang dui gedi fuze tongzhi de jianghua (jilugao, weijing benren shenyue)* [Comrade Tixue's speech to regional leaders on the provincial conference on high-water control (notes, not reviewed by Tixue)].

00046-001-0124 *Hubei sheng Jianghan gandi hudilin caiwu shouzhi guanli guiding* [Regulations on the financial management of income and expenditures regarding the protection forests along the major dikes of the Yangzi and Han Rivers in Hubei Province].

00046-001-0144 *Dangqian jianghe difang fangxun cunzai wenti he yijian* [Current problems of high-water control of river dikes and suggestions].

Jingzhou District Archives [Formerly Jiangling County Archives]

48-1-5a *Jiangling xian 1952 nian shuili gongcheng shishi banfa jihua caoan* [A draft plan for the methods of implementing water conservancy projects in Jiangling County in 1952].

48-1-5b *Kaizhan Jingjiang dadi qunzhong xing de hudi yundong wenti* [Issues of the implementation of a mass campaign to protect the Great Jingjiang Dike].

48-1-8 *Jiangling xian 1951 nian fangxun gongzuo baogao* [A work report of the 1951 high-water control in Jiangling County].

48-1-13 *Digong shouce* [Manual of dike works].

48-1-21 *Jingjiang fenhong Jingzhou qu Jiangling xian zhihuibu gongzuo baogao* [A work report of the headquarters of Jingling County, Jingzhou Prefecture, on the Jingjiang flood diversion project].

48-1-32 *Jiangling xian shuili gongcheng zhihuibu suixiu gongcheng zongjie* [Summary of annual repairs projects, Water Conservancy Project Headquarter of Jiangling County].

48-1-54a *Jiangling xian shuili jishuyuan xunlianban gongzuo zongjie* [Summary of the training class of hydraulic technicians in Jiangling County].

48-1-54b *Xunlian nongmin shuili jishuyuan de baogao* [Report on the training of peasant hydraulic technicians].

48-1-95 *Rao Mintai tongzhi er yue ershier ri zai quanqu shuili dianhua huiyi shangde baogao* [Comrade Rao Mintai's report on the prefectural telephone conference on water conservancy, February 22].

48-1-183 *Gaohao difang guanli, fuwu nongye shengchan—Jiangling xian difang guanli gongzuo de tihui* [Handle dike management well, serve agricultural production: Experience with dike management in Jiangling County].

48-1-342 *Guanyu qingli shouhui ganbu zhigong jiezhi nuoyong gongkuan de yijian* [Suggestions on the clearance and recall of public funds borrowed and embezzled by cadres and workers].

Jingzhou Municipal Archives [Formerly Jingzhou Prefectural Archives]

D35-1-4a *Jingzhou zhuanqu 1951 niandu fangxun zongjie baogao* [Summary report on the 1951 high-water control in Jingzhou Prefecture].

D35-1-4b *Hubei sheng shuiliju Jingzhou zhuanqu shuili fenju shuili gongzuo zonghe baogao* [The comprehensive report on water conservancy of the Jingzhou Branch Bureau of Water Conservancy of the Bureau of Water Conservancy of Hubei Province].

D35-1-10 *Mianyang xian 52 niandu shuili gongcheng donggong zongjie ji chungong buzhi baogao* [Summary report on the winter work and the assignment of spring work of water conservancy projects in 1952 in Mianyang County].

D35-1-12 *Gong'an xian shengchan daibiao dahui fangxun buzhi zongjie* [Summary of the assignments of high-water control in the conference of production representatives of Gong'an County].

D35-1-21 *Jingjiang fenhong Huduhe xiqu paishui gongcheng zhihuibu gexiang gongzi biaozhun* [Standards of various wages of the headquarters of the drainage project, western district of the Hudu River, Jingjiang flood diversion (project)].

D35-1-37 *Songzi xian shuiliju 1953 niandu shuili gongzuo jihua* [The 1953 annual plan of water conservancy, the Bureau of Water Conservancy, Songzi County].

D35-1-38a *Mianyang xian donggong zongjie he chungong buzhi baogao* [Report on winter work and report on the assignment of spring work in Mianyang County].

D35-1-38b *Guanyu fangxun qiangxian zhong de jidian jiben jingyan* [Some basic experiences in high-water control and emergency repairs].

D35-1-41a *Jingjiang fenhongqu nanxiandi bakou xiehong zhihuibu zongjie baogao* [Summary report of the flood diversion headquarters, the southern line, Jingjiang flood diversion zone].

D35-1-41b *Jingjiang fenhong gongcheng 1954 nian chungong ge xian zhihuibuzhang kezhang huiyi taolun shixiang huibian* [A collection of items discussed in the meeting of directors of the headquarters of different counties on the 1954 spring work of the Jingjiang flood diversion project].

D35-1-41c *Jingjiang fangxun fenhong gongzuo zongjie* [Summary of high-water control and flood diversion on the Jing River].

D35-1-50a *Jingjiang dadi Qianjiang xian gongcheng zhihuibu guanyu 1954 nian suixiu gongcheng zongjie baogao* [Summary report on the 1954 annual repairs, project headquarters of Qianjiang County, Great Jingjiang Dike].

D35-1-50b *1954 niandu suixiu gongcheng zongjie* [Summary of the 1954 annual repairs].

D35-1-50c *Songzi xian shuili gongcheng zhihuibu 1954 nian suixiu gongcheng zongjie* [Summary of the 1954 annual repair projects, the headquarters of water conservancy projects, Songzi County].

D35-1-89 *Changshou xiang nongtian shuili guanche pinggong jifen heli fudan de jieshao* [Introduction to the implementation of work-points and a reasonable burden on farmland water conservancy in Changshou Township].

D35-1-127a *1957 nian suixiu gongcheng zongjie* [Summary of the 1957 annual repairs].

D35-1-127b *Guanyu 1957 niandu suixiu tufang gongcheng caiwu gongzuo de zongjie baogao* [Summary report on the earthwork of the 1957 annual repairs and finances].

D35-1-148a *Hubei sheng shuiliting Jingzhou qu Hanjiang xiufangchu guanyang zhuitan huiyi zongjie* [A summary of the drill exploration conference of the Bureau of Annual Repairs and High-Water Control of the Han River of Jingzhou, Department of Water Conservancy, Hubei].

D35-1-148b *Difang guanyang gongzuo liangnian guihua—58 zhi 59* [A two-year plan for the upkeep of dikes, 1958 to 1959].

D35-1-170 *Guanyu caishi jixiang caijing kaizhi wenti de baogao* [A report on financial issues related to quarries].

D35-1-175 *Honghu xian shinian shuili jianshe chengjiu tongji 1959* [Statistics on the achievements of water conservancy construction in Honghu County in the past decade, 1959].

D35-1-207a *Rao Mintai zhihuizhang sanyue shisiri wanshang zai tuan yishang shuji dianhua hui shangde baogao (jilugao)* [Commander Rao Mintai's report on the telephone meeting of secretaries at the regiment level and above, evening of March 14 (notes)].

D35-1-207b *Rao Mintai tongzhi zai dangwei (kuoda) huiyi shangde zongjie baogao (jilugao)* [Comrade Rao Mintai's summary report on the party committee's (enlarged) meeting (notes)].

D35-1-255 *Guanyu xiada jindong mingchun difang suixiu gongcheng tushifang renwu ji jingfei zhibiao de tongzhi* [A notice regarding the workload of earth and stone and funding of the annual repairs in this winter and next spring].

D35-1-285 *Guanyu 1963 niandu difang suixiu gongcheng zongjie (chugao)* [Summary of the 1963 annual repairs project (first draft)].

D35-1-319 *Jingzhou qu shuiku yimin gongzuo qingkuang de baogao* [Report on the work situation of reservoir immigration in Jingzhou District].

D35-1-412a *Guanyu Shishou zhoutan minyuan paohui de chakan baogao* [An investigative report on the destruction of beach polders in Shishou].

D35-1-412b *Guanyu Qianjiang xian Sanjiang qu, Banghu qu weijing qingshi zai waitan weiyuan qingkuang de huibao* [A report on the unauthorized reclamation of beachlands in Sanjiang Ward and Banghu Ward, Qianjiang County].

D35-1-412c *Qing ji zhuanzhi youguan danwei pairen ducu chedi paohui waitan weiken minyuan de baogao* [Report on informing the relevant unit to send people to supervise the destruction of beach polders].

Qianjiang City Archives [Formerly Qianjiang County Archives]

40-1-10a *Qianjiang xian fangxun huiyi zongjie* [Summary report on the Qianjiang County high-water control meeting].

40-1-10b *Li xianzhang zai quxiangzhang lianxi huiyi shang guanyu fangxun gongzuo de buzhi* [The assignment of high-water control by county head Li at the high-water control meeting of sub-county leaders].

40-1-13 *Hubei sheng shuiliju Dongjinghe di xiufangchu wuyi nian suixiu gongcheng zongjie* [Summary of the 1951 annual repairs, Bureau of Annual Repairs and High-Water Control of the Dongjing River Dikes, Department of Water Conservancy, Hubei Province].

40-1-14 *Jingzhou zhuanqu shuili fenju wuyi nian chungong zongjie fangxun fanghan huiyi zongjie* [Summary of the spring 1951 work, summary of the flood and drought prevention meeting, Bureau of Water Conservancy, Jingzhou Prefecture].

40-1-24-1 *Guanyu quebao fangxun dianhua changtong de tongzhi* [Notice to ensure that the high-water control phone line remains unblocked].

40-1-35a *Qianjiang xian xiufang zhihuibu 1952 nian donggong zongjie* [Summary of work in the winter of 1952, annual repairs and high-water control headquarters of Qianjiang County].

40-1-35b *Qianjiang xian xiufang zhihuibu wuer niandu suixiu zongjie ji fangxun buzhi huiyi zongjie* [Summary of the 1952 annual repairs and summary of the assignments at the high-water control meeting, annual repairs and high-water control headquarters of Qianjiang County].

40-1-35c *Wang xianzhang zai shuili daibiao dahui shang guanyu wuer nian shuili gongcheng jige wenti de baogao* [County head Wang's report to the water conservancy representatives conference about several problems regarding water conservancy projects in 1952].

40-1-35d *Qianjiang xian 1952 nian fangxun weisheng gongzuo ji 1953 nian suixiu buzhi weisheng gongzuo huiyi zongjie* [Summary of the conference on public health in the 1952 high-water control and the assignment of public health in the 1953 annual repairs].

40-1-52 *Qianjiang xian diqiqu fangxun zhihuibu diyici fangxun xiaojie baogao* [Summary report of the first high-water control, high-water control headquarters of the Seventh Ward, Qianjiang County].

40-1-55a *Hubei sheng 1954 nian shuili gongzuo huiyi zongjie* (*caoan*) [Summary of the 1954 water conservancy conference of Hubei Province (draft)].

40-1-55b *Rao Mintai fu zhihuizhang shi yi yue er shi ri zai quanqu shuili gongcheng zhengzhi gongzuo huiyi shangde zongjie baogao* [Deputy commander Rao Mintai's summary report on the political working conference on prefecture-wide water conservancy projects, November 20].

40-1-55c *Jingzhou zhuanqu shuili gongcheng mingong ganbu jiangcheng tiaoli* [Regulations on rewards and punishments of laborers and cadres in water conservancy projects, Jingzhou Prefecture].

40-1-57a *Baogao qian yi jieduan fangxun zhunbei gongzuo you* [Reporting on preparations for high-water control in the previous stage].

40-1-57b *Cong chuzhang zai Hanjiang fangxun zongjie huiyi shangde fayan* [Director Cong's speech at the summary meeting on high-water control of the Han River].

40-1-57c *Xie juzhang zai Hanjiang bayue fangxun zongjie huiyi shangde zongjie baogao* [Director Xie's summary report on the summary meeting on high-water control of the Han River in August].

40-1-73 *1954 niandu chunxiu gongcheng zongjie baogao* [Summary report on the spring dike-building project of 1954].

40-1-93 *Qianjiang xian diqiqu fangxun zhihuibu zongjie baogao* [A summary report of the high-water control headquarters of the Seventh Ward, Qianjiang County].

40-1-111 *Qianjiang xian Banghu qu shuili gongcheng zhihuibu zongjie baogao* [A summary report of the water conservancy projects headquarters, Banghu Ward, Qianjiang County].

40-1-113 *Guanyu 1956 nian fangxun fudan banfa caoan* [Draft plan of the burden of the 1956 high-water control].

40-1-125 *Guanyu 1956 nian 10 yue fen zhengxian jiagu gongzuo baogao* [Work report on the rebuilding and reinforcement in October 1956].

40-1-129 *1957 nian suixiu gongcheng zongjie* [Summary of the annual repairs project of 1957].

40-1-148a *Qianjiang xian fangxun zhihuibu 1957 nian fangxun fudan banfa yijian* [Advice of the high-water control headquarters of Qianjiang County on the burden of the 1957 high-water control].

40-1-148b *Fangxun zuozhan jihua de yijian* [Advice on the high-water control combat plan].

40-1-156 *Sheng renmin weiyuanhui zhaokai Jingzhou, Xiaogan, Huanggang, Xiangyang sige zhuanqu fangxun jinji huiyi de zongjie* [Summary of the urgent meeting on high-water control of four prefectures (Jingzhou, Xiaogan, Huanggang, and Xiangyang), summoned by the Provincial People's Committee].

40-1-180 *Hanjiang gandi shinian xiufang gongzuo chengjiu zongjie (1950–1959)* [Summary of the achievements in annual repairs and high-water control of the major Han River dikes in the past ten years (1950–1959)].

40-1-195 *Qianjiang xian difang yan'ge* [The evolution of the dikes in Qianjiang].

40-1-234 *Guanyu dui difang guanyang gongzuo jixu jiejue de qingshi baogao* [Request regarding issues of dike maintenance that are in need of immediate solution].

40-1-305 *Banghu qu liusi nian disici fangxun gongzuo zongjie* [Summary of the fourth high-water control in 1964 in Banghu Ward].

40-1-413 *Dongjinghe pan pi Lin Kong, fangxun kangzai zhan hongfeng—Zongkou nongchang fangxun zhihuibu huibao* [Criticizing Lin Biao and Confucius on the banks of the Dongjing River, battling the peak of the flood in high-water control and disaster resistance: Report of the high-water control headquarters of Zongkou Farm].

40-1-532a *Jianghe difang duxun fangmian de jige wenti* [Some issues regarding high-water control on the river dikes].

40-1-532b *Guanyu Hanjiang fangxun zhunbei gongzuo zhong cunzai de jige wenti he yijian de baogao* [Report on some issues and suggestions regarding the preparation of high-water control on the Han River].

Shishou City Archives [Formerly Shishou County Archives]

1-1-543-46 *Guanyu difang jinjiao de diaocha* [Survey of dike safety zones].

1-1-556-67 *Guanyu Daloukou bakou kuijue chengzai shijian dui youguan shizhi renyuan chufen yijian de baogao* [Report on the suggestions on the punishment of those who were derelict in their duty in the case of the break at Daloukou].

20-1-053-005 *Jianli xian geming weiyuanhui guanyu fangxun qiangxian de jinji dongyuan ling* [Emergency mobilization order for high-water control and emergency repairs, Jianli County Revolutionary Committee].

20-1-394-24 *Shishou xian shuili gongcheng guanli yanghu banfa (caoan)* [Methods of maintaining water conservancy projects in Shishou County (draft)].

39-1-6-2 *Hubei sheng Shishou xian shuiliju wuyi niandu gongzuo zongjie* [Work summary of 1951, Bureau of Water Conservancy of Shishou County, Hubei Province].

39-1-15-3 *Guanyu wusi nian fangxun gongzuo chubu qingkuang zongjie baogao* [A summary report on the initial situation of the 1954 high-water control].

39-1-16-4 *Shishou xian diliuqu fangxun zhihuibu wusi nian fangxun xiaojie baogao* [A summary report on the 1954 high-water control, high-water control headquarters of the Sixth Ward of Shishou County].

39-1-19-3 *1955 nian fangxun gongzuo zongjie baogao* [Summary report on high-water control in 1955].

39-1-36-1 *Hubei sheng Jingzhou zhuanqu shuili gongcheng zhihuibu guanyu weiken zhoutan shi ying zunzhao de jidian* [On some regulations to follow when reclaiming river beaches, the water conservancy projects headquarters of Jingzhou Prefecture, Hubei Province].

39-1-38-3 *Guanyu dicheng guanli yanghu de qingkuang baogao* [Report on the dike management and maintenance situation].

39-1-68-12 *Guanyu shuili fangmian de qingkuang* [The situation regarding water conservancy].

39-1-95-2 *Guanyu jianshe zhanbei difang guihua de baogao* [Report on the plan to build dikes for war preparedness].

39-1-122-1 *Guanyu Fuxingzhou weiyuan ying liji paohui de han* [Letter regarding the need to immediately destroy the polder dike in Fuxingzhou].

40-1-10-1 *Mao Zedong sixiang de weida shengli—Jingjiang dadi zhanbei zhengxian jiagu gongcheng de zongjie* [The great victory of Mao Zedong Thought: A summary of the reinforcement project of the Great Jingjiang Dike for war preparedness].

40-1-20-3 *Guanyu fanglanglin wenti de diaocha baogao* [An investigative report on the issues of the breakwater forests].

40-1-39-1 *Guanyu zenyang zai difang shang gao duozhong jingying wenti de chubu diaocha* [A preliminary investigation of the operation of a diversified business on dikes].

Songzi City Archives [Formerly Songzi County Archives]

33-1-2 *Songzi xian renmin zhengfu shuiliju zongjie suixiu ji zhi fangxun gongzuo baogao* [A summary of annual repairs and a working report on high-water control arrangements, Songzi County Bureau of Water Conservancy].

33-1-5 (1)a *Guanyu fangxun fanghan gongzuo de zhishi* [Instructions on high-water control and drought prevention].

33-1-5 (1)b *Songzi xian liangnian lai shuili gongzuo zongjie* [A summary of the past two years of water conservancy in Songzi County].

33-1-5 (1)c *Songzi xian 1951 niandu suixiu gongzuo chubu zongjie* [A preliminary summary of the 1951 annual repairs in Songzi County].

33-1-6 *1951 nian suixiu gongcheng donggong zongjie* [Summary of annual repairs in the winter of 1951].

33-1-8 *1952 nian suixiu zongjie yu fangxun buzhi* [Summary of annual repairs and high-water control arrangements in 1952].

33-1-10 *Songzi xian diyiqu digong weiyuanhui 1952 nian fangxun kanghan zongjie* [Summary of high-water control and drought relief in the winter of 1952, the Dike Committee of the First Ward of Songzi County].

33-1-16 *Wei chexiao diyuan guanliduan jiaqiang qu digong weiyuanhui baoqing heshi you* [A quest to abolish the maintenance department of the dike-polder administration].

33-1-63 *Zheng Dingzhi fu xianzhang zai quan xian fangxun huiyi shang de baogao* [Deputy county head Zheng Dingzhi's report on the county meeting on high-water control].

33-1-80 *Guanyu yan jianghe zhoutan weiken xiji paohui de tongzhi* [Notice on the immediate destruction of enclosures on beaches along rivers].
100-1-97 *Yonghe yuan* [Yonghe polder].
100-1-99 *Desheng yuan* [Desheng polder].
100-1-182 *Datong yuan* [Datong polder].
100-2-47 *Quanmei yuan* [Quanmei polder].
100-2-52 *Wei Songzi xian zhengfu caigou shaochai junshi zhenggu fuma deng* [Purchasing firewood, the military rounding up laborers and horses, etc., by the Songzi County phony government].
100-2-198 *Qing ling baojia zhang xun diao minfu gansu xiuzhu bing ken pai jing xiezhu* [Please order baojia heads to quickly deploy laborers to build dikes and plea for policemen to help].
100-2-242 *Xingzheng yihui juan* [Dossier of administration, County Council].
100-2-256 *Peixing yuan* [Peixing polder].
100-2-364 *Canyihui diyijie disanci dahui caizheng lei yi'an* [Financial proposals, the third meeting of the first session of the County Council].
100-2-371 *Jianshe juan* [Dossier of construction].
100-2-466 *Shuangxing yuan* [Shuangxing polder].
100-2-469 *Datong yuan* [Datong polder].
100-4-1651 *Zhongcheng yuan bao Du Qifa kangwan dizheng juan* [Dossier of Zhongcheng polder's reports on Du Qifa's resistance to dike administration].
100-5-280 *Jiuji weiyuanhui juan* [Dossier of the Relief Commission].
100-6-1060 *Tianliang juan* [Dossier of land taxation].
100-8-2 *Sanhe yuan* [Sanhe polder].
100-8-30 *Caixue yuan* [Caixue polder].
100-8-32 *Xian canyihui anjuan* [Files of the County Council].
100-8-38 *Sanxing yuan* [Sanxing polder].
100-8-42 *Futai yuan* [Futai polder].
100-8-43 *Shangxing yuan* [Shangxing polder].
100-8-49 *Jiang-Song Taibao yuan* [Taibao polder of Jiangling and Songzi counties].
100-8-69 *Song-Zhi-Yi Shaqipeng Tongxin yuan* [Tongxin polder in Shaqipeng, [shared by] Songzi, Zhijiang, and Yidu].
100-8-87 *Sanhe yuan (diyiben)* [Sanhe polder (book 1)].
100-8-93 *Xida yuan xiufangchu wubi dizheng juan* [Dossier of malpractices and dike administration of the annual repairs and high-water control agency of Xida polder].
100-8-95 *Diwei zhengzhi dang'an anjuan* [The enemy's political archives].
100-8-97 *Changshou yuan* [Changshou polder].
100-8-99 *Songzi xian binjiang gandi xiufangchu* [The Annual Repairs and High-Water Control Agency of major river dikes in Songzi County].
100-8-103 *Quanmei yuan* [Quanmei polder].

100-8-104 *Xida yuan* [Xida polder].
100-8-106 *Shanhe yuan* [Shanhe polder].
100-8-108 *Peixing yuan digong* [Dike work in Peixing polder].
100-8-120 *Yongfeng yuan* [Yongfeng polder].
100-8-125 *Sanhe yuan (dierben)* [Sanhe polder (book II)].
100-8-149 *Zhi-Song Zhongcheng yuan* [Zhongcheng polder of Zhijiang and Songzi].
100-8-154 *Yongfu yuan* [Yongfu polder].
100-8-155 *Datong yuan* [Datong polder].
100-8-160 *Xida yuan* [Xida polder].
100-8-170 *Xida yuan (disanben)* [Xida polder (book III)].
100-8-172 *Changshou yuan (dierben)* [Changshou polder (book II)].
100-10-186 *Wei xian gandi jianyan biaoce deng* [The phony county government's inspection forms on major dikes, etc.].
100-10-204 *Songzi didang zi 1944 nian, caijing yusuan chengwen* [The enemy's archives, Songzi, financial submissions, since 1944].
100-10-307 *Wei jianshe fangmian wenjian deng* [The phony government's documents on construction, etc.].
100-10-393 *Wei jianshe yewu wanglai wenjian ji jianshe fangmian de daidian, zhiling* [The phony government's construction business documents and construction-related telegrams and instructions].
100-10-421 *Wei minzheng sifa jingcha wenjian ji yiban xingmin anjian* [The phony government's civil, judicial, and police documents, and general criminal and civil cases].
100-10-423 *Wei tongji hanjian anjian ji xianzhi ge jiguan yewu gongzuo baogao* [The phony government's wanted traitor cases and business work reports of various county agencies.]
100-10-482 *Wei si zhuanshu gongwen deng* [Official documents from the phony Office of the Fourth Administrative Inspectorate, etc.]
100-30-15 *Bennian mindi fangxun (1947 nian)* [This year's high-water control of people's dikes (1947)].

Tianmen City Archives [Formerly Tianmen County Archives]

42-1-149 *Guanyu quanxian nongtian jiben jianshe liangyou buzhu de tongzhi* [Notice on the grain and cooking oil subsidy for the county's farmland capital construction].
42-1-155 *Guanyu pizhun "Hanjiang gandi fangxun qingkuang de jiancha baogao"* [Forwarding "Inspection report on the high-water control situation of the major Han River dikes"].
42-1-158a *Jianchi dangde jiben luxian dada zhishui renmin zhanzheng—Honghu fanghong pailao gongcheng Tianmen gongduan jungong zongjie* [Insisting on the basic

line of the party, wage a water control people's war: A summary of the completion of the Tianmen section of the Honghu flood control and drainage project].

42-1-158b *1974 nian fangxun zongjie* [Summary of high-water control in 1974].

42-1-169 *Guanyu yanjin zai weiji fanghong anquan de hedao nei shezhi zushuiwu de tongzhi* [Notice of the ban on placing objects in the river course that block the water and endangered safety].

42-1-172 *Guanyu 1975 nian shi yue fen fangxun gongzuo zongjie baogao* [Summary report on the high-water control in October 1975].

42-1-183 *Guanyu xiada 1976 niandu fangxun wuzi de tongzhi* [Notice on the release of materials for the 1976 high-water control].

42-1-188 *Luo Heshan: jianchi shijian diyi de guandian, nuli zhangwo fanglanglin bingchong fangzhi guilü* [Luo Heshan: Adhere to the standpoint of practice first, and strive to master the rules of controlling insects in breakwater forests].

42-1-216 *Guanyu xiada 1979 nian fangxun jingfei de tongzhi* [Notice on the release of the 1979 high-water control funds].

Wuhan Municipal Archives

XX000065-WS01-0156-0001 *Wuhan shi fanghong gongzuo zongjie* [Summary of flood control in Wuhan Municipality].

XX000065-WS01-1138-0034 *Guanyu zhizhi zai difang he shizhen zhuyao gandao liangbian zhongzhi nongzuowu de jiancha qingkuang ji chuli yijian de baogao* [Report on the investigation on stopping the raising of crops on dikes and along major city streets and advice on handling the issue].

XX000065-WS01-1272-0016 *Guanyu difang xuncha qiangxian duiwu zhengzhi shencha de tongzhi* [Notice on the political investigation of the dike patrol and emergency repair team members].

XX000065-WS01-1273-0004 *Guanyu Dongxihu difang jige wenti de baogao* [Report on several issues regarding dikes in Dongxihu].

XX000065-WS01-1273-0010 *Guanyu gongcheng guanli he guan'gai guanli de baogao* [Report on engineering management and irrigation management].

XX000065-WS01-1281-0020 *Guanyu zhengdun difang jianfeng chazhen de huiyi jilu* [Minutes of a meeting on rectifying "jianfeng chazhen" (make use of every bit of space) on dikes].

XX000065-WS01-1281-0030 *Guanyu chuli difang jinjiao fanwei nei suo zhong nongzuowu de qingshi baogao* [A report requesting instructions on the treatment of crops grown within dike safety zones].

XX000065-WS01-1281-0036 *Pifu Wuchang qu renwei guanyu chuli difang jinjiao xian yinei de nongzuowu* [Reply to (the request to) handle crops planted within the dike safety zone in Wuchang District].

xx000065-ws01-1281-0062 *Wei weihai difang anquan, lüjiaobugai de weifa fenzi Lin Liyun, qing yifa yancheng de baogao* [Report on the request to severely punish offender Lin Liyun who harmed the security of dikes and refused to mend his ways despite repeated admonitions].

xx000065-ws01-1385-0002 *Guanyu zuzhi xundidui he qiangxiandui de tongzhi* [Notice to organize the dike patrol team and the emergency repair team].

xx000065-ws01-1386-0011 *Guanyu difang jinjiao xiannei zhongzhi shuitian qingkuang de baogao* [Report on the cultivation of paddies within dike safety zones].

xx000065-ws01-1386-0028 *Qing yu linxian lianxi, zhizhi nongmin toukan Dongxihu fanglanglin de baogao* [Report on the request to contact neighboring counties to stop peasants from stealthily chopping down the breakwater forests of Dongxihu].

xx000065-ws01-1391-0019 *Guanyu fangxun qiangxian shiyong liangshi zhuanyong madai wenti de tongzhi* [Notice on the use of special grain sacks in high-water control and emergency repairs].

xx000065-ws01-1439-0007 *Guanyu Zhanggongdi fangxun qicai de baogao (jianju xin)* [Report on high-water control materials on the Revered Mr. Zhang dike (accusation letter)].

xx000065-ws01-1488-0005 *Qiaokou qu guanyu zhengdun difang neiwai pojiao 8 gongchi yinei suo zhongzhi de nongzuowu gongzuo zongjie* [A work summary of the rectification of crops planted within 8 meters of the foot of the dike in Qiaokou District].

xx000065-ws01-1489-0008 *Quanyu youguan pohuai difang de baogao* [A report on the destruction of dikes].

xx000065-ws01-1589-0001 *Jingjiang dadi fangzhi tuqi baiyi xianchang huiyi kaimuci* [The opening speech at the on-the-spot meeting on termite control on the Great Jingjiang Dike].

xx000065-ws01-1691-0021 *Guanyu yanhe nongmin he jumin zai difang jinjiao xiannei wa fangkongdong qingkuang fanying de baogao* [Report on the excavation of air raid shelters within dike safety zones by peasants and residents along the river].

xx000065-ws01-1829-0008 *Yikao dangde lingdao, dada difang jianshe de renmin zhanzheng* [Relying on the party's leadership, launching a dike-building people's war].

xx000065-ws01-1946-0014 *Guanyu yigong daizhen gongcheng jinzhan qingkuang de baogao* [Report on the progress of work-relief projects].

xx000065-ws01-1954-0016 *Guanyu zhizhi sheyuan zai difang shang chancao de qingshi baogao* [A request report on stopping commune members from rooting up grass on dikes].

xx000084-ws01-0089-0004 *Wuhan difang xiangong ziliao* [Data on dangerous sections of Wuhan's dikes].

xx000084-ws03-0083-0030 *Guanyu yaoqiu jiejue xinhe dongdi huidi zhongzhi de qingqiu baogao* [A request report on the demand to solve the destruction of the eastern section of the new river dike to grow crops].

XX000084-WS03-0137-0004 *Wuhan shijiao sixian yiqu difang jianshe guihua yijian shu* [Suggestions on the dike-building plan in the four counties and one district of suburban Wuhan].

XX000129-WS01-0005-0003 *Fangxun qianfang gongnong guanxi qingkuang de baogao* [Report on the situation of worker-peasant relations on the front lines of high-water control].

XX000129-WS01-0043-0015 *Mouchang qingniantuan zongzhishu Liang Wanrong zai qiangxian zhong linzhen tuotao shoudao jilü chufen* [Liang Wanrong, the general secretary of the Communist Youth League of a certain factory, fled during emergency repairs and was punished].

XX000129-WS01-0060-0006 *Wuhan shi fangxun zong zhihuibu fangxun yuangong jiangcheng tiaoli* [Regulations on rewards and punishments of high-water control employees, General Headquarters of High-Water Control of Wuhan Municipality].

XX000129-WS01-0086-0001 *Bairi fangxun douzheng zongjie* [Summary of one hundred days of high-water control].

XX000129-WS01-0100-0001 *Wuchang fangxun gongzuo jiben zongjie* [A basic summary of high-water control in Wuchang].

Xiaonan District Archives [Formerly Xiaogan County Archives]

81-1-3a *Xiaogan xian xiudi fudan qingkuang baogao* [Report on the burden of dike building in Xiaogan County].

81-1-3b *Xiaogan xian xiudi watang gongzuo zongjie* [Summary of dike building and pond excavation in Xiaogan County].

81-1-31 *Guanyu huitong Huangpi xian diaocha Tongjiahu weiken gongcheng de baogao* [Report on the co-survey with Huangpi County regarding the enclosure of Lake Tongjia].

81-1-45 *Xiaogan xian 1956 nian difang suixiu gongzuo zongjie baogao* [Summary report on annual repairs of the dikes of Xiaogan County in 1956].

81-1-53 *Guanyu zhengqiu chubu niding de "diyuan guanli jigou mingcheng, guanxia fanwei ji zucheng renyuan yijian" de han* [Letter to solicit draft "opinions on the institution's name, jurisdiction and dike management staff"].

Yingcheng City Archives [Formerly Yingcheng County Archives]

49-1-1 *Yingcheng xian mindi diaocha biao* [A survey table of people's dikes in Yingcheng County].

49-1-2a *Digong fudan banfa caoan* [Draft methods on (dealing with) the dike work burden].

49-1-2b *Guanyu xiudi gongzuo zhong jige juti wenti de zhishi* [Instructions on several concrete issues in dike building].

49-1-3 *Minguo 11 nian zhi 25 nian Daxing yuan suoyou shouzhi qingce* [All accounts of income and expenditures of Daxing polder, 1922 to 1936].

49-1-4a *Fangxun jiangcheng zanxing banfa* [Temporary methods of rewards and punishments in high-water control].

49-1-4b *Hubei Jianghan gandi suixiu fanghu zanxing guize* [Temporary regulations on annual repairs and high-water control of the major dikes along the Yangzi and Han Rivers in Hubei].

49-1-9 *Yingcheng xian wuyi nian diyuan suixiu gongcheng zongjie* [Summary of the annual repairs project of the dikes in Yingcheng County in 1951].

49-1-26 *Yingcheng xian shengchan jiuzai gongzuo zongjie baogao* [Summary report on the work of production disaster relief in Yingcheng County].

49-1-45 *Yingcheng xian xunlian nongmin shuili yu shuitu baochi jishuyuan de zongjie baogao* [Summary report on the training of peasant technicians of water conservancy and soil and water conservation in Yingcheng County].

49-1-51a *Yingcheng xian 1957 nian difang guanyang gongzuo baogao* [Report on the maintenance of dikes in Yingcheng County in 1957].

49-1-51b *Yingcheng xian zhimindi fangxun zhunbei gongzuo qingkuang de baogao* [Report on the preparations for high-water control of the branch and people's dikes in Yingcheng County].

49-1-61 *Yingcheng xian Xiaojiafan weiken gongcheng shigong qianhou fadong qunzhong de jiben jingyan* [Basic experiences in motivating people involved in the process of the project to enclose Xiaojiafan of Yingcheng County].

49-1-75 *Xianhe mindi Yujiadaokou kuikou qianhou* [Before and after the rupture of the people's dike along the Xian River on Yujiadaokou].

49-1-90a *Zhonggong Yingcheng xianwei guanyu weiken Tian'edang de jueding* [The decision of the Yingcheng County party committee to enclose Swan Lake].

49-1-90b *Tian'edang de laili* [The origins of Swan Lake].

Yunmeng County Archives

46-1-13 *1972 nian fangxun wenti yijian de baogao* [Report of advice on issues of high-water control in 1972].

46-1-15a *Women shi zenyang fazhan fanglang hudilin de* [How we developed the breakwater forests].

46-1-15b *Xue Dazhai wei geming guandi zaolin* [Learn from Dazhai to manage dikes and plant forests for the revolution].

46-1-15c *Hanchuan xian dazao fanglang hudilin de qingkuang he zuofa* [The situation and practice of the growth of a large number of breakwater forests in Hanchuan County].

Zhijiang City Archives [Formerly Zhijiang County Archives]

45-1-1 *Zhijiang xian Wen'an qu Gonghe yuan ji Wayaohu yuan gongcheng zongjie* [Summary of the project in Gonghe polder and Wayaohu polder of Wen'an Ward, Zhijiang County].

45-1-2 *Zhijiang xian 1952 nian fangxun gongzuo zongjie* [Summary of high-water control in Zhijiang County in 1952].

45-1-3 *Zhijiang xian 1953 nian fangxun gongzuo zongjie* [Summary of high-water control in Zhijiang County in 1953].

45-1-6 *Zhijiang xian Shangbailizhou yuan xiufangchu 1950 nian san zhi shi'er yuefen jingchang, jishi, fangxian gefei zhichu yusuan shu* [Expenditure budget of the bureau of annual repairs and high-water control of Shangbailizhou polder in Zhijiang County on recurring items, stone, and emergency repairs, from March to December in 1950].

45-1-8 *Zhijiang xian 1954 nian fangxun gongzuo jiben zongjie* [Basic summary of high-water control in 1954 in Zhijiang County].

45-1-11a *Zhijiang xian renmin zhengfu dijiu qugongsuo dier difang guanliduan suixiu gongcheng zongjie* [Summary of annual repairs of the Second Dike Upkeep Section, Ninth Ward Office, People's Government of Zhijiang County].

45-1-11b *Zhijiang xian diyi qu difang guanliduan Jiangnan chengguan fuxun gongzuo zongjie* [Summary of the summer high-water control, county seat on the southern side of the Yangzi River, Dike Upkeep Section of the First Ward, Zhijiang County].

45-1-11c *Zhijiang xian diwuqu difang guanliduan 1955 nian suixiu zongjie* [Summary report of the 1955 annual repairs of the dike management section, Fifth Ward, Zhijiang County].

45-1-15 *Zhijiang xian shuiliju zhuanfa zhongyang shuili dianli bu guanyu shuili guanli danwei kaizhan "wuhao" pingbi zanxing banfa de tongzhi* [Notice on temporary methods for selecting five-goodness units of water conservancy management, Ministry of Water Conservancy and Electricity, forwarded by the Bureau of Water Conservancy of Zhijiang County].

45-1-61 *Guanyu difang jiezhizha she guanli jigou de baogao* [Report on the establishment of a management institute to [manage] dikes and check sluice gates].

45-1-104 *Sifang qingkuang fanying, 1972/6/2* [Situation reflection on the four preventions, June 2, 1972].

45-1-116 *Sifang qingkuang fanying, 1973/7/5* [Situation reflection on the four preventions, July 5, 1973].

45-1-216 *Guanyu zhengshou guoying, jiti qishiye danwei ji chengzhen jumin difang jingfei de tongzhi* [Notice on the collection of dike fees from state-owned and collective enterprises and institutions and urban residents].

45-1-233 *Guanyu zhengshou difangfei biaozhun de baogao* [Report on the collection standards for dike fees].

45-1-310a *Fangxun kanghong jianbao (dishisiqi), 1981/7/19* [Bulletin on high-water control, no. 14, July 19, 1981].

45-1-310b *Fangxun kanghong mingling (diyihao), 1981/7/15* [Order on high-water control, no. 1, July 15, 1981].

Zhongxiang City Archives [Formerly Zhongxiang County Archives]

54-1-3a *Zhongxiang xian shuiliju 1951 nian chungong zongjie baogao* [Summary report on the spring 1951 work, Bureau of Water Conservancy of Zhongxiang County].

54-1-3b *Zhongxiang xian 1951 nian fangxun gongzuo zongjie baogao* [Summary report on the 1951 high-water control in Zhongxiang County].

54-1-3c *Zhongxiang xian 1951 nian shang bannian shuili gongzuo zongjie baogao* [Summary report on water conservancy in Zhongxiang County in the first half of 1951].

54-1-5a *Zhongxiang xian shuiliju 1952 niandu zongjie baogao* [The 1955 annual summary report of the Bureau of Water Conservancy of Zhongxiang County].

54-1-5b *1952 nian shuili gongzuo baogao* [The 1952 work report on water conservancy].

54-1-5c *Zhongxiang xian Hanjiang gandi gongcheng weiyuanhui 1952 nian suixiu zongjie* [Report on the 1952 annual repairs of the Engineering Committee of the Han River dikes of Zhongxiang County].

54-1-15 *Bai xianzhang zai 1955 nian fangxun kanghan fanglao huiyi shangde zongjie baogao* [County head Bai's summary report on the 1955 meeting on high-water control, drought-resistance, and prevention of waterlogging].

54-1-27 *Guanyu chen shuiluo shiji jiaqiang quanmian zhunbei yubei zaizhan de zhishi* [Instructions on strengthening comprehensive preparations when water is withdrawing and preparing for another battle].

54-1-35a *Zhongxiang xian 1957 niandu mindi suixiu zongjie* [Summary of the 1957 annual repairs of people's dikes in Zhongxiang County].

54-1-35b *Guanyu xinjian minyuan de qingshi* [A request regarding the building of new people's polders].

54-1-58 *Wang Xianqing tongzhi zai shuilihui shangde baogao* [Comrade Wang Xianqing's report at the water conservancy conference].

54-1-108 *1971 niandu gongzuo zongjie* [Work summary of 1971].

54-1-129 *Yikao qunzhong guanli difang* [Rely on the masses to manage the dikes].

54-1-232 *Jingzhou diqu xingzheng gongshu biaozhang 1983 nian fangxun kangzai xianjin jiti, xianjin geren jiangpin qingdan* [List of awards to advanced groups and individuals in high-water control and disaster prevention, recognized by the Administrative Office of Jingzhou Prefecture, 1983].

Other Sources

Alley, Rewi. 1956. *Man Against Flood: A Story of the 1954 Flood on the Yangtse and of the Reconstruction That Followed It*. Peking: New World Press.

Anonymous. 1991. "Hubei sheng Jiangling xian fu xianzhang Li Dahan yong shenti yingyong duzhu Jingjiang dadi loudong" [Li Dahan, deputy county head of Jiangling, Hubei, heroically blocked a leak in the Great Jingjiang Dike with his body]. *Jiangling wenshi ziliao*, 6: 77 (The report originally appeared in the *China Youth Daily*, August 1, 1954).

Ao Wenwei and Li Xiang. 2003. "Minguo shiqi de wushu jiuzai" [Disaster relief by sorcery in the Republican era]. *Hubei wenshi*, 1: 165–69.

Bello, David A. 2016. *Across Forest, Steppe, and Mountain: Environment, Identity, and Empire in Qing China's Borderlands*. New York: Cambridge University Press.

Benedict, Carol. 1996. *Bubonic Plague in Nineteenth-Century China*. Stanford, CA: Stanford University Press.

Cai Dongzhou. 1999. "Guan Yu xianxiang tanyuan" [Exploring the origins of the Guan Yu phenomenon]. *Zhonghua wenhua luntan*, 1: 53–62.

Cao Le'an. 1990. "Wode shuili shengya" [My water conservancy career]. *Hubei wenshi ziliao*, 3: 17–28.

Cao Shuji and Li Yushang. 2006. *Shuyi: Zhanzheng yu heping—Zhongguo de huanjing yu shehui bianqian (1230–1960 nian)* [Plague: War and peace—China's environment and social change, 1230–1960]. Ji'nan: Shandong huabao chubanshe.

Carson, Rachel. 1962. *Silent Spring*. Boston: Houghton Mifflin.

Changjiang shuili weiyuanhui Changjiang kance guihua sheji yanjiuyuan, jianshe yu guanliju, ed. 2005. *Changjiang zhi—shuiku yimin yu kuqu jianshe* [Gazetteer of the Yangzi River: Reservoir relocatees and construction]. Beijing: Zhongguo dabaike quanshu chubanshe.

Chang, Michael G. 2003. "Fathoming Qianlong: Imperial Activism, the Southern Tours, and the Politics of Water Control, 1736–1765." *Late Imperial China*, 2: 51–108.

Chen Guoxiao. 2000. "Cong Tangkeng dao Liugang—Zhongguo renmin jiefangjun Guangzhou junqu di 31 yubei yiyuan jiti zhuanye renmin dayuan nongchang ji" [From Tangkeng to Liugang: On the transfer of the PLA's No. 31 Hospital, Guangzhou Military Region, to People's Dayuan Farm]. *Jingzhou wenshi zhiliao*, 3: 143–50.

Chen Tao. 2007. "Hongshui yu cunluo kongjian jiegou de yanbian: Jianghan pingyuan Yangma cun de Taizi de shehuixue kaocha" [Floods and the evolution of the spatial structure of villages: A sociological survey of Taizi in Yangma Village on the Jianghan Plain]. *Gujin nongye*, 2: 9–14.

Chen Zhangguan. 1982. "Difang gonglu wangshi shiling" [Tidbits on dikes and highways of the past]. *Hubei wenshi ziliao*, 5: 155–60.

Chen Zhenhan, Xiong Zhengwen, Li Chen, and Yin Hanzhang, eds. 1989. *Qingshilu jingjishi ziliao (Shunzhi-Jiaqing chao), nongye bian, 1–3* [Data on economic history from the Qing records, agriculture, 1644–1820. Vol. 1–3]. Beijing: Beijing daxue chubanshe.

Cheng Gongneng. 2004. "Jingjiang Shishou duan jinxi" [The past and present of the Shishou section of the Jing River]. *Jingzhou wenshi zhiliao*, 6: 15–19.

Cheng Qijing and Xu Yiping. 1987. "Huxiang pingzhang" [A defense of lake areas]. Pp. 11–15 in Hubei renmin guangbo diantai, ed., *Huxiang xing* [Travels to lake areas]. Wuhan: Changjiang wenyi chubanshe.

Chi, Ch'ao-ting. 1963 [1936]. *Key Economic Areas in Chinese History: As Revealed in the Development of Public Works for Water-Control*. New York: Paragon Book Reprint Corp.

Courtney, Chris. 2018a. *The Nature of Disaster in China: The 1931 Yangzi River Flood*. Cambridge: Cambridge University Press.

Courtney, Chris. 2018b. "At War with Water: The Maoist State and the 1954 Yangzi Floods." *Modern Asian Studies*, 52 (6): 1807–36.

Dai Qing. 1998. *The River Dragon Has Come!: The Three Gorges Dam and the Fate of China's Yangtze River and Its People*. Translated by Yi Ming. New York: M.E. Sharpe.

Davies, Jeremy. 2016. *The Birth of the Anthropocene*. Oakland: University of California Press.

"Decision from the CPC Central Committee and the State Council on Accelerating Water Conservancy Reform and Development." 2010. *China Daily*, June 16, 2017. http://www.chinadaily.com.cn/china/19thcpcnationalcongress/2017-06/16/content_29775385.htm. (Accessed, March 9, 2024).

DeMare, Brian. 2012. "Local Actors and National Politics: Rural Amateur Drama Troupes and Mass Campaigns in Hubei Province, 1949–1953." *Modern Chinese Literature and Culture*, 24 (2): 129–78.

Ding, Xiangli. 2024. *Hydropower Nation: Dams, Energy, and Political Changes in Twentieth-Century China*. Cambridge: Cambridge University Press.

Dodgen, Randall. 1999. "Hydraulic Religion: 'Great King' Cults in the Ming and Qing." *Modern Asian Studies*, 33 (4): 815–33.

Dodgen, Randall A. 2001. *Controlling the Dragon: Confucian Engineers and the Yellow River in Late Imperial China*. Honolulu: University of Hawai'i Press.

Dong Longkai. 2001. "Huanghe zaihai yu jindai Shandong de heshen xinyang, shehui shenghuo xisu." Pp. 488–515 in Fudan daxue lishi dili yanjiu zhongxin, ed., *Ziran zaihai yu Zhongguo shehui lishi jiegou* [Natural calamities and the historical structure of Chinese society]. Shanghai: Fudan daxue chubanshe.

Duara, Prasenjit. 1988. *Culture, Power, and the State: Rural North China, 1900–1942*. Stanford, CA: Stanford University Press.

Edgerton-Tarpley, Kathryn. 2008. *Tears from Iron: Cultural Responses to Famine in Nineteenth-Century China*. Berkeley and Los Angeles: University of California Press.

Edgerton-Tarpley, Kathryn. 2016. "Between War and Water: Farmer, City, and State in China's Yellow River Flood of 1938–1947." *Agricultural History*, 1: 94–116.

Elvin, Mark. 1993. "Three Thousand Years of Unsustainable Development: China's Environment from Archaic Times to the Present." *East Asian History*, 6: 7–46.

Elvin, Mark. 1998. "Who Was Responsible for the Weather? Moral Meteorology in Late Imperial China." *Osiris*, 13: 213–37.

Elvin, Mark. 2004. *The Retreat of the Elephants: An Environmental History of China*. New Haven, CT: Yale University Press.

Emerson, Bo. 2007. "Putting Faith in Call for Rain." *Atlanta Journal-Constitution*, November 13.

Fang Bin. 2012. "Wuhan shi Caidian qu lian'ou chanye jidi de xingcheng" [The formation of a lotus roots base in Caidian District, Wuhan]. *Wuhan wenshi ziliao*, 9: 49–53.

Fang, Jingyun, Sheng Rao, and Shuqing Zhao. 2005. "Human-Induced Long-Term Changes in the Lakes of the Jianghan Plain, Central Yangtze." *Frontiers in Ecology and the Environment*, 3 (4): 186–92.

Fang Ming and Dong Yumei. 2003. *Zhuang huai jiyue gu Jingzhou* [Splendid history of ancient Jingzhou]. Beijing: Tuanjie chubanshe.

Fang Mu. 1995. "Shuishen chongbai de moliu—tan Ming-Qing heyuan de mixin" [The last school of water deity worship: On the superstition of officials in charge of river affairs in the Ming-Qing era]. *Sichuan shuili*, 2: 49–50.

Fei Hsiao-tung. 1962 [1939]. *Peasant Life in China: A Field Study of Country Life in the Yangtze Valley*. London: Routledge and Kegan Paul.

Friedman, Edward, Paul G. Pickowicz, and Mark Selden, with Kay Ann Johnson. 1991. *Chinese Village, Socialist State*. New Haven, CT: Yale University Press.

Fu Zhongjin. 2006. "Danjiangkou shuiku gongdi shangde Mianyang dibashi" [The Eighth Division of Mianyang on the Danjiangkou Reservoir worksite]. *Xiantao wenshi*, 12: 50–57.

Gailiang digong yijian shu [Suggestions on reforming dike works]. N.d. [Republic]. Pp. 1–5 in E-Yu-Wan sansheng jiaofei zong silingbu and Hubei shuili digong shiwu qingli weiyuanhui, ed., *Qingli baogaoshu* [Report on checking up (on dike affairs)].

Gan Xinghong. 2013. "Gongheguo diyige 'Dujiangyan' gongcheng de yunniang yu juece" [Discussion and decision of the PRC's first "Dujiangyan" project]. *Hubei wenshi*, 2: 155–62.

Gao, Mobo C.F. 2018. *Gao Village Revisited: The Life of Rural People in Contemporary China*. Hong Kong: Chinese University Press.

Gao, Yan. 2012. "Transformation of the Water Regime: State, Society and Ecology of the Jianghan Plain in Late Imperial and Modern China." PhD diss., Carnegie Mellon University. ProQuest, 3520084: 69–97.

Gao, Yan. 2022. *Yangzi Waters: Transforming the Water Regime of the Jianghan Plain in Late Imperial China*. Leiden: Brill.

Gross, Miriam. 2016. *Farewell to the God of Plague: Chairman Mao's Campaign to Deworm China*. Oakland: University of California Press.

Gu Hao, ed. 2006. *Zhongguo zhishui shijian* [The history of water conservancy in China], 2nd ed. Beijing: Zhongguo shuili shuidian chubanshe.

Guan Qingtao. 1983. "Jianghan pingyuan jianghu lishi bianqian yu shuili shiye de fazhan" [Historical changes in rivers and lakes and the development of water conservancy on the Jianghan Plain]. *Hubei shuili zhi tongxun*, 5–6: 30–36.

Guo Yuneng. 1989. "Yihai shuizai shiqing lu" [Records of the actual situation of the 1935 flood]. *Zhijiang wenshi ziliao*, 4: 119–23.

Hanchuan nianjian (di yi juan) 1997–2006 [Hanchuan yearbook (volume 1) 1997–2006]. 2008. N.p.

Harrell, Stevan. 2023. *An Ecological History of Modern China*. Seattle: University of Washington Press.

He Bingdi [Ping-ti Ho]. 1988. *Zhongguo gujin tudi shuzi de kaoshi he pingjia* [An examination and evaluation of land data in China in the past and present]. Beijing: Zhongguo shehui kexue chubanshe.

Hershatter, Gail. 2011. *The Gender of Memory: Rural Women and China's Collective Past*. Berkeley: University of California Press.

Hong Qingyu. 1999. "Guanyu Sanxia gongcheng lunzheng de lishi huiyi" [Historical memories of the debate over the Three Gorges Project]. Pp. 452–462 in Zhengxie Hubei sheng weiyuanhui wenshi ziliao weiyuanhui, ed., *Hubei wenshi jicui* [Collection of literature and history of Hubei]. Wuhan: Hubei renmin chubanshe.

Hsiao Kung-ch'uan. 1960. *Rural China: Imperial Control in the Nineteenth Century*. Seattle: University of Washington Press.

Hu, Ch'ang-tu. 1955. "The Yellow River Administration in the Ch'ing Dynasty." *Far Eastern Quarterly*, 14 (4): 505–13.

Hu Qiwei. 2008. "Caoyun xingfei yu shuishen chongbai de shengshuai—yi Ming-Qing shiqi Xuzhou wei zhongxin de kaocha" [The rise and decline of tribute grain transportation and the vicissitudes of water deities worship: An investigation based on the Xuzhou area in the Ming-Qing era]. *Zhongguo kuangye daxue xue bao (shehui kexue ban)*, 2: 108–12.

Hu Yingze. 2006. "Shuijing yu beifang xiangcun shehui—jiyu Shanxi, Shaanxi, Henan sheng bufen diqu xiangcun shuijing de tianye kaocha" [Water wells and Northern China's rural society: Based on field research on rural water wells in some areas of the provinces of Shanxi, Shaanxi, and Henan]. *Jindaishi yanjiu*, 1: 55–78.

Hu Zuhe, ed. 1999 [1838]. *Jingchu xiushu zhiyao* [Key to the management of the dike systems in Hubei]. Wuhan: Hubei renmin chubanshe.

Huang, Philip C.C. 1990. *The Peasant Family and Rural Development in the Yangzi Delta, 1350–1988*. Stanford, CA: Stanford University Press.

Huang, Philip C.C. 2008. "Centralized Minimalism: Semiformal Governance by Quasi Officials and Dispute Resolution in China." *Modern China*, 31 (1): 9–35.

Huang, Philip C.C., and Yuan Gao. 2015. "Should Social Science and Jurisprudence Imitate Natural Science?" *Modern China*, 41 (2): 131–67.

Huang Yonggui. 1993. "Gezhouba gongcheng de yimin dongqian he anzhi" [The relocation and placement of immigrants of the Gezhou Dam project]. *Hubei wenshi ziliao*, 1: 232–40.

Huang Zhigang. 1973 [1934]. *Zhongguo de shuishen* [Chinese water deities]. Taipei: Dongfang wenhua shuju.

Hubei An Xiang Yun dao shuili ji'an [A collection of papers on water conservancy in the circuits of Anlu, Xiangyang, and Yunyang]. N.d. [Qing] N.p.

Hubei hangyun shi [History of shipping in Hubei]. 1995. Beijing: Renmin jiaotong chubanshe.

Hubei nongye dili [Agricultural geography of Hubei]. 1980. Wuhan: Hubei renmin chubanshe.

Hubei sheng ditu yuan, comp. 1998. *Hubei sheng ditu ce* [Atlas of Hubei Province]. Ji'nan: Shandong sheng ditu chubanshe.

Hubei sheng shuiliting, ed. 2003. *Fengbei—Hubei Changjiang difang jianshe da xiezhen* [Monumental work: Photo album of diking along the Yangzi River in Hubei]. Wuhan: Hubei meishu chubanshe.

Hubei sheng shuiliting and Hubei sheng dang'anju (guan), eds. 1998. *Hubei sheng 1954 nian fangxun kanghong dang'an xuanbian* [Selection of archives on the 1954 high-water control in Hubei Province]. N.p.

Hubei sheng shuiliting and Hubei sheng dang'anju (guan), eds. 1999. *Hubei sheng 1931 nian shuizai dang'an xuanbian* [Selection of archives on the 1931 flood in Hubei Province]. N.p.

Hubei sheng shuiliting and Hubei sheng fangxun kanghan zhihuibu bangongshi, eds. 2000. *Hubei Changjiang fangxun* [High-water control of the Yangzi River in Hubei]. Wuhan: Hubei renmin chubanshe.

Hubei sheng zhengfu mishuchu tongjishi, comp. 1938. *Hubei nongcun diaocha baogao 8: Jianli xian* [Report of a rural survey of Hubei, vol. 8: Jianli County]. Wuchang: Department of Statistics, Office of the Secretariat, Government of Hubei.

Hubei xianzheng gaikuang [A survey of county administration in Hubei]. 1934. Compiled and published by the Department of Civil Affairs, Government of Hubei.

Hunan sheng binhu zhoutu shichatuan baogaoshu [Report of the investigation mission of lakeshore land in Hunan]. 1947. N.p.

Jiang Tanyu, ed. *Zhongguo xin nongcun diaocha* [Surveys of China's new countryside]. 2007. Beijing: Renmin chubanshe.

Jiang Yongwu and Zhang Jiawei, eds. 2002. *Jingzhou bai jing* [One hundred scenes of Jingzhou]. Shanghai: Shanghai shudian chubanshe.

Jianghe xiufang jiyao [Summary of annual repairs and high-water control of rivers]. 1971 [1936]. Taipei: Zhuanji wenxue chubanshe [reprint].

Jin Feng. 1996. *Jiangxia shi zhi* [A history of Jiangxia]. N.p.

Jing, Jun. 2000. "Environmental Protests in Rural China." Pp. 143–60 in Elizabeth J. Perry and Mark Selden, eds., *Chinese Society: Change, Conflicts and Resistance*. London: Routledge.

Jingzhou diqu 40 nian 1949–1989 [Forty years of Jingzhou Prefecture 1949–1989]. 1989. Wuhan: Hubei renmin chubanshe.

"Jingzhou Folk Custom: Temple Fair at Guan Yu Temple." 2011. *China Daily*, December 29. https://www.chinadaily.com.cn/m/hubei/travel/2011-12/29/content_14351548.htm. (Accessed, December 17, 2024).

Jingzhou fu Jiangling xian tuanlian qingce [Detailed list of the militia of Jiangling County, Jingzhou Prefecture]. N.d. [Qing] N.p.

Jingzhou hangyun shi [The history of shipping in Jingzhou]. 1996. Beijing: Renmin jiaotong chubanshe.

Jingzhou shi shuiliju, Jingzhou shi fangxun zhihuibu bangongshi, Hubei sheng Zhanghe gongcheng guanliju, eds. 2008. *Rao Mintai shiliao zhuanji* [Special collection of historical data on Rao Mintai]. N.p.

Johnson, Kay Ann. 1983. *Women, the Family and Peasant Revolution in China*. Chicago: University of Chicago Press.

Kinzley, Judd C. 2018. *Natural Resources and the New Frontier: Constructing Modern China's Borderlands*. Chicago: University of Chicago Press.

LaManna, Eva Melody. 2010. "Three's a Crowd: Examining Georgia's Options in the Tri-state Water Wars under Principles of International Law." *Georgia Journal of International and Comparative Law*, 39 (1): 215–39.

Lavelle, Peter B. 2020. *The Profits of Nature: Colonial Development and the Quest for Resources in Nineteenth-Century China*. New York: Columbia University Press.

Leonard, Jane Kate. 1996. *Controlling from Afar: The Daoguang Emperor's Management of the Grand Canal Crisis, 1824–1826*. Ann Arbor: Center for Chinese Studies, University of Michigan.

Li Guozhang. 1986. "Weihu Jingjiang dadi qianhou" [The whole story of the protection of the Great Jingjiang Dike]. *Shishou wenshi ziliao*, 1: 72–79.

Li, Huaiyin. 2005. *Village Governance in North China, 1875–1936*. Stanford, CA: Stanford University Press.

Li, Huaiyin. 2009. *Village China under Socialism and Reform: A Micro-history, 1948–2008*. Stanford, CA: Stanford University Press.

Li Jialin. 2014. "Nanshui beidiao yimin Shayang anzhi shimo" [The whole story of the resettlement of the relocatees of the South-to-North water diversion project to

Shayang]. Pp. 115–20 in Liu Xihai and Zhu Junqi, eds., *Chao yong Dachaihu*. Beijing: Zhongguo wenhua chubanshe.

Li Jiansheng, ed. 1999. *Zhongguo jianghe fanghong congshu—zonglun juan* [Collection of high-water control of rivers in China—general discussion volume]. Beijing: Zhongguo shuili shuidian chubanshe.

Li, Lillian M. 2007. *Fighting Famine in North China: State, Market, and Environmental Decline, 1690s–1990s*. Stanford, CA: Stanford University Press.

Li Rui. 1996. "'Da yuejin' qinli ji" [Records of personal experience of the Great Leap Forward]. Shanghai: Shanghai yuandong chubanshe.

Li Shouhe. 2001. "Jingjiang fenhong gongcheng shouci qiyong ji" [The first use of the Jingjiang flood diversion project]. *Hubei wenshi ziliao*, 1: 90–102.

Li Xiangjun. 1995. *Qingdai huangzheng yanjiu* [Studies on the famine relief policies of the Qing dynasty]. Beijing: Zhongguo nongye chubanshe.

Li Yousong. 2007. "Xuexichong bing zhi Cao Cao bingbai Chibi" [Snail fever caused the defeat of Cao Cao at the Red Cliff]. *Lishi xuejia chazuo*, 7: 151–60.

Li Zinan. 2001. "1935 nian hongshui weikun Jingzhou cheng jishi" [On-the-spot record of the inundation of Jingzhou city in 1935]. *Hubei wenshi ziliao*, 2: 110–15.

Liang Jiamian, ed. 1989. *Zhongguo nongye kexue jishu shigao* [A draft history of agricultural science and technology in China]. Beijing: Nongye chubanshe.

Liang Yin'guo. 1995. "Fuhe gaidao gongcheng kuzhan ji" [Memories of the hard battle of the Fuhe (river) diversion project]. *Yingcheng wenshi ziliao*, 9: 94–116.

Lin Yishan. 2011. *Lin Yishan zhishui wenji II* [Collection of Lin Yishan's works on water conservancy, II]. Wuhan: Changjiang chubanshe.

Lin Yuan. n.d. (Shishou) *Chongkai guxue ji* [Notes on reopening the old outlets (in Shishou)]. In *Shishou xian zhi* 1866, vol. 7: 30a.

Lin Zexu. 1935 [1885]. *Lin wen zhong gong zhengshu* [Memorials of Lin Zexu]. Shanghai: Shangwu yinshuguan.

Liu Keyi. 2000. "Guanyu Jingjiang fangxun kanghong ruogan weiti de sikao" [Thoughts on issues related to high-water control of the Jing River]. *Wuhan shuili dianli daxue xuebao (shehui kexue ban)*, 1: 17–21.

Liu Qiansheng, ed. N.d. *Hongshui taotao, Jiangdi weiwei—1998 Majiazhai xiang zhansheng teda hongshui shilu* [Torrential flooding, lofty river dike: A record of the conquest of the great flood in Majiazhai Township in 1998]. N.p.

Liu, Ts'ui-jung. 1970. "Dike Construction in Ching-chou, A Study Based on the 'T'i-fang chih' Section of the *Ching-chou fu-chih*." *Papers on China*, 23: 1–27. Cambridge, MA: East Asian Research Center, Harvard University.

Liu Wei. 1987. "Renmin dayuan xingwan qianhou" [The whole story of the enclosure of the People's Dayuan polder]. *Shishou wenshi ziliao*, 3: 176–87.

Liu Yun. 1982. "Hubei Jianghe difang hudilin fazhan qingkuang" [The developmental situation of the dike protection forest along the river dikes in Hubei]. *Hubei shuili zhi tongxun*, 2: 19–20, 23.

Liu Zhenggang [oral], Cao Jinliang [transcript]. 2008. "Dongjinghe xiayou gaidao gongcheng er san shi" [Something about the diversion project of the lower reaches of the Dongjing River]. *Zhishui zuji—Xiantao wenshi*, 13: 103–10.

Lu Xiqi. 2011. "Ming-Qing shiqi Jianghan pingyuan de weiyuan: cong 'shuili gongcheng' dao 'shuili gongtongti'" [Polders of the Jianghan Plain in the Ming-Qing era: From "water control projects" to "water control communities"]. Pp. 348–439 in Zhang Jianmin and Lu Xiqi, eds., *Lishi shiqi Changjiang zhongyou diqu renlei huodong yu huanjing bianqian zhuanti yanjiu* [Monographic studies on human activities and environmental change in the history of the middle of the Yangzi River valley]. Wuhan: Wuhan daxue chubanshe.

Lü Xingbang. 2011. "Jianghan pingyuan de diyuan shuili yu jiceng shehui (1942–1949)—yi Hubei sheng Songzi xian Sanhe yuan wei zhongxin" [Water control of dike-polders and grassroots society in the Jianghan Plain (1942–1949), centered on Sanhe polder, Songzi County, Hubei]. *Gujin nongye*, 1: 105–17.

Lü Xingbang. 2015. "Yi 'kangzhan' weiming: Minguo Hubei Gong-Song Changgeng di'an shulun" [In the name of the "War of Resistance Against Japan": Discussion of the case of the Changgeng dike between Gong'an and Songzi, Hubei, in the Republic]. *Gujin nongye*, 2: 90–97.

Lü Xingbang. 2019a. "Cong husheng dao gongxing: Minguo wanqi Jianghan pingyuan de yuan yu shizhen—yi Hubei sheng Songzi xian wei kaocha zhongxin" [English title in the original: "From Alternation to Common Prosperity: 'Yuan' and Town in Jianghan Plain in Late Republic of China—Survey of Songzi County in Hubei Province"]. *Gujin nongye*, 1: 96–107.

Lü Xingbang. 2019b. "Yuan de shengcheng—yi qing zhi minguo shiqi de Hubei sheng Songzi xian weili" [English title in the original: "Formation of 'Yuan'—Taking Songzi County of Hubei Province in the Republic of China as an Example"]. *Xihua shifan daxue xuebao (zhexue shehui kexue ban)*, 4: 38–44.

Lü Xingbang. 2020. "Diehe yihuo bingxing: Diyuan shuili xingzheng yu xiangbao xingzheng zhi guanxi—yi minguo wanqi Hubei sheng Songzi xian wei kaocha zhongxin" [English title in the original: "Congruence or Parallel: on the relationship between embankment yuan water conservancy administration and township constable administration—Taking Songzi County of Hubei in the Late Republic of China as the Reserarch Center"]. *Xihua shifan daxue xuebao (zhexue shehui kexue ban)*, 4: 101–108.

Luo Jianjin. 2001. "Dui 'ren ding sheng tian' de lishi fansi" [Some reflections on "Through his efforts man can conquer nature"]. *Ziran bianzhengfa tongxun*, 5: 68–73.

Luo Senlin. 1970. "Jiangling huajiu 4" [Talking about old times in Jiangling 4]. *Hubei wenxian*, 14: 53–58.

Luo Tinggui. 2004. "Xu" [Preface]. *Jiangling wenshi ziliao*, 1: 1–3.

Luo Xianqun, ed. 2013. "Mao Zedong liangguo Jingjiang" [Mao Zedong passed along the Jing River twice]. *Hubei wenshi*, 1: 71–75.

Ma Daofu. 2009. "Yimin gongzuo de suantian kula" [The sweet and the bitter in relocation work]. Pp. 187–89 in *Zhengfa wangshi: Xiantao wenshi ziliao xuanji*. Beijing: Zhongguo wenshi chubanshe.

MacKinnon, Stephen R. 2008. *Wuhan, 1938: War, Refugees, and the Making of Modern China*. Berkeley: University of California Press.

Marks, Robert B. 1998. *Tigers, Rice, Silk, and Silt: Environment and Economy in Late Imperial South China*. New York: Cambridge University Press.

Marks, Robert B. 2017. *China: An Environmental History*. Lanham, MD: Rowman & Littlefield.

McCully, Patrick. 1996. *Silenced Rivers: The Ecology and Politics of Large Dams*. London: Zed Books.

McNeill, J.R. 2000. *Something New under the Sun: An Environmental History of the Twentieth-Century World*. New York: W.W. Norton.

Mei Li, Zhang Guoxiong, and Yan Changgui. 1995. *Lianghu pingyuan kaifa tanyuan* [Exploring the origins of the development of the Hunan-Hubei Plain]. Nanchang: Jiangxi jiaoyu chubanshe.

Miao Qiyu. 1982. "Taihu diqu tangpu weitian de xingcheng he fazhan" [Formation and development of Lake Tai's canals and embanked fields]. *Zhongguo nongshi*, 1: 12–32.

Miller, Ian M. 2020. *Fir and Empire: the Transformation of Forests in Early Modern China*. Seattle: University of Washington Press.

Morita Akira. 1974. *Shindai suiri shi kenkyū* [Studies in the history of water control during the Qing dynasty]. Tokyo: Aki shobō.

Mostern, Ruth. 2021. *The Yellow River: A Natural and Unnatural History*. New Haven, CT: Yale University Press.

Mulingshu [Handbook of the magistrate]. 1990 [1848]. Yangzhou: Jiangsu Guangling guji keyinshe.

Murphey, Rhoads. 1967. "Man and Nature in China." *Modern Asian Studies*, 1 (4): 313–33.

Murphey, Rhoads. 1983. "Deforestation in Modern China." Pp. 111–28 in Richard P. Tucker and J.F. Richards, eds., *Global Deforestation and the 19th Century World Economy*. Durham, NC: Duke University Press.

Muscolino, Micah S. 2015. *The Ecology of War in China: Henan Province, the Yellow River, and Beyond, 1938–1950*. Cambridge: Cambridge University Press.

National Flood Relief Commission. 1933. *Report of the National Flood Relief Commission, 1931–1932*. Shanghai: Commercial Press.

Nedostup, Rebecca. 2009. *Superstitious Regimes: Religion and the Politics of Chinese Modernity*. Cambridge, MA: Harvard University Asian Center.

Ni Wenwei, ed. 2002 [1885]. *Jingzhou wanchengdi zhi* [Gazetteer of the Wancheng Dike in Jingzhou]. Wuhan: Hubei renmin chubanshe.

Oakes, Tim, and Donald S. Sutton, eds. 2010. *Faiths on Display: Religion, Tourism, and the Chinese State*. Lanham, MD: Rowman & Littlefield.

Paltemaa, Lauri. 2011. "The Maoist Urban State and Crisis: Comparing Disaster Management in the Great Tianjin Flood in 1963 and the Great Leap Forward Famine." *The China Journal*, 66: 25–51.

Peng Yuxin. 1990. *Qingdai tudi kaiken shi* [The history of land reclamation in the Qing dynasty]. Beijing: Nongye chubanshe.

Peng Yuxin and Zhang Jianmin. 1993. *Ming-Qing Changjiang liuyu nongye shuili yanjiu* [Studies on agricultural water conservancy in the Yangzi River valley in the Ming-Qing period]. Wuhan: Wuhan daxue chubanshe.

Perdue, Peter. 1987. *Exhausting the Earth: State and Peasant in Hunan, 1500–1850*. Cambridge, MA: Council on East Asia Studies, Harvard University.

Pietz, David A. 2002. *Engineering the State: The Huai River and Reconstruction in Nationalist China, 1927–1937*. New York: Routledge.

Pietz, David A. 2015. *The Yellow River: The Problem of Water in Modern China*. Cambridge, MA: Harvard University Press.

Pomeranz, Kenneth. 1991. "Water to Iron, Widows to Warlords: The Handan Rain Shrine in Modern Chinese History." *Late Imperial China*, 12 (1): 62–99.

Qi Lafa. 2009. "Daxing shuili zaofu Puyangren" [Constructing many water conservancy projects, bringing benefits to the people of Yingcheng]. *Yingcheng liushinian—Yingcheng wenshi*, 29: 225–35.

Qin Zhongxiang. 1998. "Xu Guorui yu Jingjiang dadi de kuijue" [Xu Guorui and the rupture of the Great Jingjiang Dike]. *Hubei wenshi ziliao*, 4: 18–21.

Qing huidian shili [Statutes and cases of the Qing dynasty]. 1991 [1899]. Beijing: Zhonghua shuju.

Quan Runhua (oral account), Tang Yongping and Wang Guangliang (transcription). 2013. "Ban'ge shiji yimin lu" [A half century road of immigration]. Pp. 63–66 in Liu Xihai and Zhu Junqi, eds., *Danshui liuchang—Jingmen Danjiang yimin zhuanji* [The long flow of the Danshui: Special collection on population relocation from the Dan River in Jingmen]. Beijing: Zhongguo wenhua chubanshe.

Rogaski, Ruth. 2004. *Hygienic Modernity: Meanings of Health and Disease in Treaty-Port China*. Berkeley: University of California Press.

Roumasset, James, Kimberly Burnett, and Hua Wang. 2008. "Environmental Resources and Economic Growth." Pp. 250–85 in Loren Brandt and Thomas G. Rawski, eds., *China's Great Economic Transformation*. Cambridge: Cambridge University Press.

Rowe, William T. 2001. *Saving the World: Chen Hongmou and Elite Consciousness in Eighteenth-Century China*. Stanford, CA: Stanford University Press.

Schmalzer, Sigrid. 2016. *Red Revolution, Green Revolution: Scientific Farming in Socialist China*. Chicago: University of Chicago Press.

Schlesinger, Jonathan. 2017. *A World Trimmed with Fur: Wild Things, Pristine Places, and the Natural Fringes of Qing Rule*. Stanford, CA: Stanford University Press.

Scott, James C. 1987. *Weapons of the Weak: Everyday Forms of Peasant Resistance*. New Haven, CT: Yale University Press.

Shahar, Meir, and Robert P. Weller, eds. 1996. *Unruly Gods: Divinity and Society in China*. Honolulu: University of Hawai'i Press.

Shapiro, Judith. 2001. *Mao's War against Nature: Politics and the Environment in Revolutionary China*. Cambridge: Cambridge University Press.

Shapiro, Judith. 2012. *China's Environmental Challenges*. Cambridge: Polity Press.

Shu Hui, ed. 2002 [1896]. *Jingzhou wancheng di xuzhi* [The sequel to the gazetteer of the Wancheng Dike in Jingzhou]. Wuhan: Hubei renmin chubanshe.

Shuili dianlibu shuiguan si keji si, Shuili shuidian kexue yanjiuyuan, eds. 1991. *Qingdai Changjiang liuyu xi'nan guoji heliu honglao dang'an shiliao* [Archival materials on flooding and water-logging of the Yangzi River and the international rivers in southwestern China in the Qing dynasty]. Beijing: Zhonghua shuju.

Smil, Vaclav. 1984. *The Bad Earth: Environmental Degradation in China*. Armonk, NY: M.E. Sharpe.

Smil, Vaclav. 2015. *China's Environmental Crisis: An Inquiry into the Limits of National Development*. London: Routledge. (First published by M.E. Sharpe in 1993.)

Snyder-Reinke, Jeffrey. 2009. *Dry Spells: State Rainmaking and Local Governance in Late Imperial China*. Cambridge, MA: Harvard University Asia Center.

Song Xishang, ed. 1954. *Lidai zhishui wenxian* [Historical literature on water control]. Taipei: Zhonghua wenhua chuban shiye weiyuanhui.

Su Xiangrong. 2007. *Sanxia juece lunbian: zhengce lunbian de jiazhi tanxun* [Debates on the Three Gorges project decision: Search for the value of policy debate]. Beijing: Zhongyang bianyi chubanshe.

Sun Fushi. 1939. *Yangzijiang zhi shuili* [Water conservancy along the Yangzi River]. Shanghai: Shangwu yinshuguan.

Sun Fushi. 1987. "Lun Changjiang zhongyou fanghong de lishi guocheng jiqi dangqian renwu" [On the historical process of flood control in the middle of the Yangzi River valley and the current task]. *Zhongguo shuili*, 2: 22–23, 24.

Tan Xuming. 2009. "Gudai quyu shuishen chongbai jiqi shehui xue jiazhi—yi dujiangyan shuili qu weili" [On regional water deity worship and its sociological value: A case study of the Dujiang weir area]. *Hehai daxue xuebao (zhexue shehui kexue ban)*, 1: 9–15.

Tang Zheng. 2007. "Song wenshen—'xue fang': xin Zhongguo diyichang yikong zhanzheng" [The control of snail fever: New China's first war to control epidemic disease]. *Beijing Daily*, August 28, 2007, cited from *Xinhua wenzhai*, 21: 98–103.

Tilt, Bryan. 2015. *Dams and Development in China: The Moral Economy of Water and Power*. New York: Columbia University Press.

Tu Deshen and Yang Zhichao. 2002. "Wuhan Longwang miao de bianqian" [The vicissitudes of the Dragon King Temple in Wuhan]. *Hubei wenshi ziliao*, 3: 1–8.

Tu Mingxin. 2014. "60 nianqian, huizhi buqu de fangxun jiyi" [Lingering memories of high-water control sixty years ago]. *Hubei wenshi*, 1: 82–87.

Wan Zhaolong. 1991. "Canjia xiuzhu Caobuhu difang suoji" [Fragmentary recollections of my participation in the construction of the Caobuhu dike]. *Dangyang wenshi*, 9: 62–66.

Wang Baoxin. 2002 [1933]. *Xu Hankou congtan* [Sequel to the collected talks about Hankou]. Wuhan: Hubei jiaoyu chubanshe.

Wang Fengsheng. 1832. *Chubei Jianghan xuanfang beilan* [Guidelines for the diversion and protection of the Yangzi and the Han Rivers in Hubei]. N.p.

Wang Minfu, ed. 1989. *Hubei sheng Jingzhou diqu zonghe nongye quhua* [The comprehensive agricultural divisions of Jingzhou Prefecture, Hubei]. Wuhan: Hubei kexue jishu chubanshe.

Wang Peijun. 2012. "Zhenshui shou yu Zhongguo chuantong zhenshui xisu" [Containing water beasts and the traditional Chinese custom of containing water]. *Hehai daxue xuebao (Zhexue shehui kexue ban)*, 2: 53–57.

Wang Pingsheng, ed. 2004. *Dongxihu shihua* [History of Dongxihu]. Wuhan: Wuhan chubanshe.

Wang Qinzeng. 1982. "Diaochahu de bianqian" [The vicissitudes of Lake Diaocha]. *Hubei shuli zhi tongxun*, 3–4: 70–71, 72.

Wang Shengfu. 1985. "Hubei sheng pingyuan diqu fanghong wenti de tantao" [Discussion of high-water control on the plains of Hubei Province]. *Hubei shuili*, 1: 4–7.

Wang Wenli, Wang Hongzhi, Liu Hu, and Wang Chengcheng. 2012. "'Tuitian huanhu' shishi qianhou Jianghan huqun bianhua bijiao yanjiu" [Comparative studies on the changes of Jianghan lakes before and after the implantation of the "return land to lakes" policy]. *Anhui nongye daxue xuebao*, 39 (3): 456–60.

Wang Yingming. 1986. "Jinri Shahu chang xin'ge" [Today Shahu sings a new song]. Pp. 196–200 in Hubei renmin guangbo diantai Hanjiang xing baodao zu, *Hanjiang xing* [Touring the Han River]. Wuhan: Hubei renmin chubanshe.

Wang Yingrui. 2006. "1954 nian Yanglin fangxun sanri" [Three days of high-water control in Yanglin in 1954]. *Shahe fanying—Yunmeng wenshi ziliao*, 24: 249–57.

Wang Yingzheng. 2010. *Zhongguo shuili shuidian gongcheng yimin wenti yanjiu* [Research on the problems of relocatees from water conservancy and hydropower projects in China]. Beijing: Zhongguo shuili shuidian chubanshe.

Wang Zhaohu, Ding Shoushi, and Yu Huiyun, comp. 1925. *Hubei difang jiyao* [Summary of Hubei's dikes]. N.p.

Wang Zhengben. 1999. "1931 nian Wuhan dashui suoji" [Miscellaneous records of the 1931 flood in Wuhan]. Pp. 511–17 in Zhengxie Hubei sheng weiyuanhui wenshi ziliao weiyuanhui, ed., *Hubei wenshi jicui* [Collection of literature and history of Hubei]. Wuhan: Hubei renmin chubanshe.

Wang Zhihua. 1986a. "Lüse changcheng" [The green great wall]. Pp. 161–64 in Hubei renmin guangbo diantai Hanjiang xing baodao zu, *Hanjiang xing* [Touring the Han River]. Wuhan: Hubei renmin chubanshe.

Wang Zhihua. 1986b. "Yaodi manbu" [Roaming on the Distance Dike]. Pp. 156–60 in Hubei renmin guangbo diantai Hanjiang xing baodao zu, *Hanjiang xing* [Touring the Han River]. Wuhan: Hubei renmin chubanshe.

Wei Ziren. 2002. "Zhongguo gudai de shuishen chongbai" [Water deity worship in ancient China]. *Huaxia wenhua*, 2: 31–33.

Will, Pierre-Etienne. 1985. "State Intervention in the Administration of a Hydraulic Infrastructure: The Example of Hubei Province in Late Imperial Times." Pp. 295–347 in Stuart Schram, ed., *The Scope of State Power in China*. London: School of Oriental and African Studies.

Will, Pierre-Etienne. 1990. *Bureaucracy and Famine in Eighteenth-Century China*. Translated from the French by Elborg Forster. Stanford, CA: Stanford University Press.

Will, Pierre-Etienne. 2006. "Zhonghua diguo wanqi guojia dui shuili de guanli" [The management of water conservancy in late imperial China]. Pp. 796–810 in Chen Feng, ed., *Ming Qing yilai Changjiang liuyu shehui fazhan shilun* [On the social development of the Yangzi River valley since the Ming-Qing era]. Wuhan: Wuhan daxue chubanshe.

Will, Pierre-Etienne, and R. Bin Wong. 1991. *Nourish the People: The State Civilian Granary System in China, 1650–1850*. Ann Arbor: Center for Chinese Studies, University of Michigan.

Wittfogel, Karl A. 1955. "Oriental Society in Transition with Special Reference to Pre-Communist and Communist China." *Far Eastern Quarterly*, 14 (4), Special Number on Chinese History and Society: 469–78.

Wittfogel, Karl August. 1981 [1957]. *Oriental Despotism: A Comparative Study of Total Power*. New York: Vintage Books.

Wu Shisi. 2001. "Diaochahu de lishi bianqian" (The historical changes of Lake Diaocha). Pp. 17–23 in Wu Shisi, ed., *Hanchuan lansheng* (Roving through Hanchuan), N.p.

Wu Xiaofeng. 2000. "Changjiang paoshi hudi gongcheng wubi xingwei chumu jingxin" [The shocking malpractices in the throw-stones-to-protect-dike project of the Yangzi River]. *Nanfang zhoumo*, Dec. 14.

Xi Yuan et al., eds. 1990 [1879]. *Jingzhou zhufang baqi zhi* [Gazetteer of the Eight Banners of the Jingzhou garrison]. Shenyang: Liaoning daxue chubanshe.

Xiang Xin. 1998. "60 niandai fangxun qiangxian de liangge xiao gushi" [Two minor stories about high-water control and emergency repairs in the 1960s]. *Hubei wenshi ziliao*, 4: 67–69.

Xiangdi cheng'an [Records on the Han River dikes]. 1969 [Qing]. Taipei: Wenhai chubanshe [reprint].

Xiao Zijun. 1991. "Zhao Nanjie" [Zhao Nanjie]. *Shishou wenshi ziliao*, 4: 114–18.

Xie Qianmao, ed. 1931. *1931 Hankou dashui ji* [A record of the 1931 Hankou flood]. Hankou: Jianghan yinshuguan.

Xin Zhiying. 2005. "Wo yu shuili de bujie zhi yuan" [My indissoluble relationship with water conservancy]. *Hubei wenshi*, 2: 206–20.

Xiong Chuanxian. 2006. "Yangling gongshe shuili jianshe zhong 'zuo' de jiaoxun" ["Left" lessons from the water conservancy construction in Yangling Commune]. *Yingcheng wenshi*, 26: 60–71.

Xiong Daorui, ed. N.d. [Republic] *Hubei tianfu gaiyao* [Outline of the land tax in Hubei]. Hankou: Xinchang yinshuguan.

Xiong Fanlin. 2008. "Lianhe yuan podi xuxie hongshui jilüe" [Brief chronicle on the opening of Lianhe polder's dike to discharge floodwater]. *Zhishui zuji—Xiantao wenshi*, 13: 94–98.

Xu Linguan. 2006. "1954 nian hongzai zhong de sanqu gongxiaoshe" [The supply and marketing cooperatives of the Third Ward in the 1954 flood]. *Shahe fanying—Yunmeng wenshi ziliao*, 24: 417–22.

Xu Qianqing. 1992. *Fanghong* [High-water control]. Beijing: Shuili dianli chubanshe.

Xu Xinzhou. 2006. "Xianhe kuikou he jiuzai de huiyi" [Recollections of the ruptures along the county's rivers and disaster relief]. *Shahe fanying—Yunmeng wenshi ziliao*, 24: 271–75.

Yang Hua. 2007. "Chudi shuishen yanjiu" [Studies on water deities of the Chu region]. *Jianghan luntan*, 8: 98–104.

Yang Guo and Chen Xi. 2008. *Jingji kaifa yu huanjing bianqian yanjiu: Song Yuan Ming Qing shiqi de Jianghan pingyuan* [Studies on economic development and environmental change: The Jianghan Plain in the Song-Yuan-Ming-Qing era]. Wuhan: Wuhan daxue chubanshe.

Yang Jingqu. 1982. "Shayang zhudi" [Dike building in Shayang] (posthumous manuscript). *Hubei wenshi ziliao*, 6: 168–70.

Yang Liming and Jian Xing'an. 2000. "Sanxia baqu yimin jishi" [Records of relocatees in the dam area of the Three Gorges project]. *Sanxia wenshi zongheng*, 2: 65–72.

Yang Tong. 1977. *Shahu Mianyang zhou de bianqian* [Changes in Shahu and Mianyang]. Wuhan: Hubei renmin chubanshe.

Yang Yongkui. 1999. "Minguo shiqi de wanjuan dang'an shi zenyang baocun xialai de" [How ten thousand volumes of Republican archives were preserved]. *Jingzhou wenshi ziliao*, 2: 345–48.

Yao Hanyuan. 1986. "Bei Song Jiangnan yutian ji Zhexi weitian" [Yu land in Jiangnan and wei land in western Zhejiang in the Northern Song]. Pp. 135–45 in *Kexue yanjiu lunwen ji* [Collections of essays of scientific research]. Beijing: Shuili dianli chubanshe.

Ye Hengchang. 2004. "Honghu shi jingnei hangdao mantan" [A discussion of channels in Honghu City]. *Jingzhou wenshi ziliao*, 6: 40–43.

Yin Lingling. 2019. "Jiangling 'xirang' yu Gun-Yu zhi shui" [The Jiangling xirang and the Gun-Yu flood control myth]. *Lishi yanjiu*, 4: 42–60.

Ying Xing. 2001. *Dahe yimin shangfang de gushi—cong "taoge shuofa" dao "baiping lishun"* [A story of petitioning by people displaced by the Dahe dam: From "asking for an explanation" to "settling the issue"]. Beijing: Sanlian shudian.

Yu Changlie. 1999 [1840]. *Chubei shuili difang jiyao* [Important notes on water conservancy and dike systems in Hubei]. Wuhan: Hubei renmin chubanshe.

Yu Tao. 2005. "20 shiji 30 niandai Hubei de shuili jianshe" [Water conservancy construction in Hubei in the 1930s]. *Hubei wenshi*, 2: 57–63.

Yuan Li. 2014. *Diyuan yu yibing: Jingjiang liuyu shuili de shengtai renleixue yanjiu* [Dikes and diseases: An eco-anthropological study of water conservancy in the Jing River valley]. Beijing: Zhongguo shehui kexue chubanshe.

Yue Xinfa. 2006. "1954 nian Zhangjinhe yiliao jiuzai pianduan huiyi" [Fragmentary memories of medical relief in Zhangjinhe in 1954]. *Jingzhou wenshi*, 8: 365–78.

Zaixu xingshui jinjian [The second sequel to the golden guidance of inland navigation]. Vol. 2, Zhongguo shuili shuidian kexue yanjiuyuan, shuili shi yanjiushi (ed.) 2004. Wuhan: Hubei renmin chubanshe.

Zeng Fanrong, ed. 1997. *Hubei tianda de shi* [The extremely important matter in Hubei]. Wuhan: Hubei kexue jishu chubanshe.

Zhang Fang, ed. 2007. *Er shi wu shi shuili ziliao zonghui* [Comprehensive collection of water conservancy data from the Twenty-five Histories]. Beijing: Zhongguo sanxia chubanshe.

Zhang Guoxiong. 1995. *Ming-Qing shiqi de lianghu yimin* [Immigration into Hunan and Hubei in Ming and Qing times]. Xi'an: Shaanxi renmin jiaoyu chubanshe.

Zhang Jianmin. 1999. *Hubei tongshi: Ming-Qing juan* [A comprehensive history of Hubei: The Ming and Qing]. Wuhan: Huazhong shifan daxue chubanshe.

Zhang Jianmin and Tang Gangmao, eds. 2014. *Hubei Tianmen Xiongshi qiyue wenshu* [Deeds and documents of the Xiong clan of Tianmen, Hubei]. Wuhan: Hubei renmin chubanshe.

Zhang Jiayan. 1997. "Shinian lai lianghu diqu ji Jianghan pingyuan Ming-Qing jingjishi yanjiu zongshu" [A review of research in the past decade on the economic history of Hunan-Hubei and the Jianghan Plain in the Ming-Qing era]. *Zhongguo shi yanjiu dongtai*, 1: 2–11.

Zhang Jiayan. 2001. "Qingdai Jianghan pingyuan yuantian nongye jingji texing fenxi" [Analyzing the features of the agrarian economy of polders in the Jianghan Plain in the Qing]. *Zhongguo shi yanjiu*, 1: 133–42.

Zhang Jiayan. 2011. "Yimin yundong, huanjing bianqian yu wuzi jiaoliu—Qingdai ji minguo shiqi Jianghan pingyuan yu waidi de guanxi" [Immigration, environmental change and material exchange: The relationship between the Jianghan Plain and other parts of the country in the Qing and the Republic]. *Zhongguo jingjishi yanjiu*, 1: 57–66.

Zhang, Jiayan. 2013a. "Environmental Adaptation and Peasant Subsistence in the Twentieth-Century Jianghan Plain: The Great Leap Famine Revisited." *American Review of China Studies*, 14 (1): 99–123.

Zhang, Jiayan. 2013b. "Organized Violent Conflicts over Water Control in Rural China: The Jianghan Plain, 1839–1979." *Rural China*, 10 (1): 129–58.

Zhang, Jiayan. 2014. *Coping with Calamity: Environmental Change and Peasant Response in Central China, 1736–1949*. Vancouver: University of British Columbia Press.

Zhang, Ling. 2016. *The River, the Plain, and the State: An Environmental Drama in Northern Song China, 1048–1128*. Cambridge: Cambridge University Press.

Zhang, Meng. 2021. *Timber and Forestry in Qing China: Sustaining the Market*. Seattle: University of Washington Press.

Zhang Xiantang. 2004. "Haoxue tieniuji ji" [Notes of Tieniuji in Haoxue]. *Jiangling wenshi ziliao*, 1: 194–201.

Zhang Youyu and Gao Chao, eds. 1991. *Zhonghua lüling jicheng—Qing* [A collection of Chinese laws and regulations—the Qing dynasty]. Jilin: Jilin renmin chubanshe.

Zhang Yunqian. 2004. "Jingchu mingzhu Mabuguai" [Mabuguai, the pearl of Jingzhou]. *Jiangling wenshi ziliao*, 1: 175–78.

Zheng Longchang. 2004. "Baili zhou fenhong qinli ji" [A personal account of flood diversion on Baili Islet]. *Sanxia wenshi zongheng*, 3: 157–61.

Zheng Zhaojing, ed. 1987. *Taihu shuili jishu shi* [The history of water conservancy technology on Lake Tai]. Beijing: Nongye chubanshe.

Zhongguo kexueyuan dili yanjiusuo, Shuilibu Changjiang shuili weiyuanhui Hanjiang gongzuodui. 1957. *Hanjiang liuyu dili diaocha baogao* [An investigative report on the geography of the Han River valley]. Beijing: Kexue chubanshe.

Zhou Hongbo and Gong Tianzhao, eds. 1999. *Jingjiang fanghong shouce* [Manual of high-water control on the Jing River]. Wuhan: Wuhan cehui keji daxue chubanshe.

Zhou Kuiyi. 2002. *Zhongguo kexue jishu shi: Shuili juan* [History of China's science and technology: Water conservancy]. Beijing: Kexue chubanshe.

Zhou Kuiyi et al. 1990. *Er shi wu shi hequzhi zhushi* [Annotation of the gazetteers of rivers in the Twenty-five Histories]. Beijing: Zhongguo shudian.

Zhou Shangfan. 1984. "Jingjiang dadi de yichang baodi douzheng" [A fight to protect the Great Jingjiang Dike]. *Hubei wenshi ziliao*, 9: 181–84.

Zhou Xiancai. 2008. "1954 nian Yuwanggong fenhong ji dukou fudi gongcheng" [The 1954 flood diversion at Yuwanggong and the restoration project]. *Zhishui zuji—Xiantao wenshi*, 13: 45–49.

Zhou Xiangsheng. 2008. "1983 nian Dujiataizha fenhong de qianqian houhou" [The whole story of the use of the Dujiatai sluice to discharge floodwater in 1983]. *Zhishui zuji—Xiantao wenshi*, 13: 1–17.

Zhou, Xun. 2020. *The People's Health: Health Intervention and Delivery in Mao's China, 1949–1983*. Montreal: McGill-Queen's University Press.

Zou Dongting and Zheng Zehua. 1998. "Jingjiang fenhong zhunbei daojishi qianhou" [The moment the Jingjiang flood diversion zone was ready to be used]. *Hubei wenshi ziliao*, 4: 146–52.

Index

www.ingramcontent.com/pod-product-compliance
Lightning Source LLC
LaVergne TN
LVHW010658300926
849161LV00005B/1161
* 9 7 8 9 0 0 4 7 2 7 2 3 6 *